POCKET

MANAGING EDITOR
Mike Froud

CONTRIBUTORS
Michael Fridjhon,
Tony Mossop CWM

WINE TASTING PANELS
Air France
John Platter
Michelangelo
South African Airways
Veritas Awards
Diners Club
Juliet Cullinan
Pinotage Association
SA Trophy Wine Show
WINE magazine

PUBLISHED BY WINE MAGAZINE (SA)
of Ramsay Son & Parker

CHAIRMAN
Alan Ramsay
MANAGING DIRECTOR
Harold Eedes
DIRECTORS
Brian Burnett
Rob Cowan
Stirling Kotze
BOOK MARKETING
Stuart Lowe
Liane Posthumus
Jo Bestic
Anthea Eedes
Dean Dicks
DESIGN
Tamlin Prinsloo
Kian Eriksen
Lee-ann Harris
WEB EDITOR
Christian Eedes

PUBLISHER
Harold Eedes
EDITOR
Fiona McDonald
JOURNALISTS
Christian Eedes
Sara Steer
NATIONAL BUSINESS MANAGER
Debbie Margolis
ADVERTISING
Marina Simpson
Bridget Davidtsz
June Golze
PRODUCTION MANAGER
Shirley McDiarmid
ASSISTANTS
Celia Gilloway
Katherine Humphreys
Cari-ann Edge
Jillian Schooling

3 Howard Drive, Pinelands, Cape Town, South Africa.
PO Box 180, Howard Place 7450, South Africa.
Tel: +27 021 530-3100, Fax: +27 021 531-2212
E-mail: wine@rsp.co.za Website: www.winemag.co.za

Noteworthy newcomers

The number of wine producers in the Cape is increasing at the rate of around 30 a year, and it's always interesting to note the performance of newcomers whose wines have been greeted with "wow!" WINE magazine's **MIKE FROUD** recalls a few from the Swartland, Stellenbosch and Wellington, contemplates some "garage" wines – and wonders about a cellar that's sprung up on the west coast.

LAMMERSHOEK

It was just over a year ago, just after publication of the 2002 edition of WINE's Pocket Guide, that the wines of the Kretzel and Stephan families' farm near Malmesbury started grabbing attention. First there was a good Viognier, then a very good Cabernet Merlot blend, then an excellent wooded Chenin – one of six finalists in the last Chenin Blanc Challenge. Though established in 1971, the cellar only came about in '99. Watch this space.

ALEXANDERFONTEIN

Also in the Swartland, overlooking the town of Darling, and also new to the own-brand business after years of selling everything to others – established in 1974, first own-wines made in 2000 – the Basson family hit the mark with a 2001 Sauvignon Blanc released later than most, and well priced. Can they do it again, and then with their Cabernet?

DIEMERSFONTEIN

Wellington's becoming known for hot property in more ways than one, and the success of David Sonnenberg's somewhat controversial reds are further proof that the area isn't only good for vine nurseries. The farm was established back in 1965, producing fruit other than for winemaking until the '90s. The maiden 2001 vintage was applauded by the Pinotage Association, Shiraz and Merlot lovers – and the Fairbairn Capital SA Trophy Wine Show judges who awarded a trophy to the flagship Carpé Diem.

STARK-CONDÉ

Back in Stellenbosch, there's a small cellar in the Jonkershoek Valley that has yet to take a wrong turn. Some confusion in the market-place between José Condé's Starke label and that of the Starke family in Muldersvlei, but, as a neighbour of Neil Ellis and son-in-law of Ellis's partner Hans Schröder on the farm Oude Nektar farm, Stark-Condé has released two 5 Star Cabernets in the space of a year! Not a bad start.

GARAGISTE

Difficult to find in-store, very small quantities, and including a variety South African winemakers are still wrestling with, Topaz Wines are the result of a Somerset West venture involving Tanja Beutler and Cape Wine Master Clive Torr. A Pinot Noir and a Shiraz; a Burgundy lover and property owner there, too. Could well make for a bottle of something very interesting, at least... And then there's a group of men in Paarl, also apparently doing more than dabbling: Schoongezicht, Main Street winery, and labels proclaiming Mason's Hill, Blyde, Anthony Smook, Bergheim and William Everson. Got folks wondering who helped them make these reds...

CAPAIA

Just 'around the corner' from Havana Hills, which has really been smokin' in terms of good reviews on numerous fronts since launching in '99, and in what could be declared a new Olifantsberg ward adjacent to the Durbanville sprawl, Capaia's German owner Alexander Baron von Essen and his winemakers (also German) are expected come up with the goods. They're only planning to release the flagship in a few years time, but 2003 should see the release of the first Blouberg Hill, a Sauvignon Blanc, to be followed by a red blend. ❑

Cape Wine's strong suits

By **TONY MOSSOP CWM**

"WE MAKE THE BEST South African wines in the World" may not actually have been uttered by some blinkered winemaker in Stellenbosch or Paarl years ago, but it sure caught on as a war cry during the dismally dark pre- Walk-to-Freedom days of the eighties. The arrogant sentiment gave particular comfort to the men in power in local wine centres, those in control of a particularly control-ridden industry.

It took a much-vilified event to sound the wake-up call: our 1995 thrashing at the hands of the Aussies in the SAA sponsored Wine Test, Cape Town – having handed these traditional foes a similar thumping during the Rugby World Cup in Newlands just a few months before. While our wine politburo fulminated at the "damage" done to our "international image" in the market, just about every winemaker under 30 – and a good few *ouer toppies,* too – phoned their travel agents and headed across the world.

The story thereafter: wine exports up by over 250% since 1995, from under 15% of total production to over one third; increasing column space devoted to our wines by print media in European markets and, since the 2002 Cape Wine trade show in Paarl, in the hitherto ignorant US too. And try and buy your favourite tipple at anything like last year's price and you wish you had pounds or euros to spend.

But how do wine buyers and drinkers in our major export markets perceive our products? For a start, they are confronted with a range of brands you and I have hardly heard of: our top performers – such as Western Wines' Kumala and Vinfruco's Arniston Bay – are on almost every retail shelf in the UK, yet are unknown in South Africa. The giant Swiss-owned African Terroir (previously SAVISA) group's export-driven ranges cover the whole politically-correct spectrum, from organic to empowerment wines: just what the media need for good copy.

Cape wines moved from flavour-of-the-month status in the early nineties, then through an awkwardly spotty teenage stage towards the end of the decade, when Europeans, and particularly the attention-span-deficient Brits, switched focus to fresh pastures like Chile and Argentina. Now we seem to be doing a few things right again. And opinion-forming journalists are starting to make overtly anti-Oz statements in public – almost a deportable offence just a few years ago. "Lost the plot... Too much simple fruit... Bubblegum flavours..." Cape winemakers are being urged to explore their strengths, not try to copy the Aussies.

It's a war zone out there in the sub £5 category (roughly the R20–R30 bracket). But this is where SA wines are strongest – some feel too strong. Particularly with white wines, mostly Chenin based. Not surprising, with Chenin fruit changing hands at around R1000 per ton, and Colombar, another prolific contributor to a good, fruity dry white blend, for a mingy R670. That's about a rand a litre for the juice!

No wonder we've cleaned up just about everyone in the el-cheapo white category – it's pumped out the wine lake no one knew what to do with just a year ago. And then the rand helped by falling some more!

This, no doubt, is the reason for South Africa occupying such a low spot in the average bottle-price pecking order for retail wines sales in the UK – well below other New World countries, except perhaps for Chile. So the fact that you can't get your hands on enough Rubicon or Fusion V, even if you could afford them, means that their export figures are not significant enough to tip the average price up even one degree.

One insect on the windscreen, though, is the volume of plonk leaving these shores in bulk tankers, some to be re-baptised or mixed (not "blended") with even worse rot-gut, possibly under a "Cape" banner, by uncontrollable offshore customers in the serious volume business. At a hefty 35% of total exports, this shows no sign of dropping. Better than ending up in antisocial *papsakke* (foil bags) on the local market, though?

So what are our strong suits out there in the wine-drinking world – South Africa's "unique selling points"? Well, Pinotage, obviously: no one else makes it, really. It's our national grape, despite being vilified by such heavyweight scribes as UK Wine magazine's Robert Joseph as a "mongrel". The fact that the Pinotage Association can claim over 170 members is sign enough of how seriously our growers take the grape. Yet there still remains much internal squabbling over style and acceptable quality – even talk of legislating some wines off the map if they don't conform to the blockbuster, heavily wooded model.

There's seemingly quite a bit of unwanted Pinotage hanging around in tanks right now, and perhaps blending is the best route in the long run – at both top and bottom ends of the market. The search for the identity of a real "Cape Red Blend" occupies a good few minds in the winelands, with

The search for the identity of a real "Cape Red Blend" occupies a good few minds in the winelands, with the accepted formula these days including around a third Pinotage – the rest, as they say, is up to you – and some pretty innovative combinations can be found. The juicy, up-front mulberry fruit of the grape can rather dominate a blend – but also gives it an alluring plushness if deftly handled.

the accepted formula these days including around a third Pinotage – the rest, as they say, is up to you – and some pretty innovative combinations can be found. The juicy, up-front mulberry fruit of the grape can rather dominate a blend – but also gives it an alluring plushness if deftly handled.

Chenin Blanc is another potential winner, as the ever-escalating quality of entries in such competitions as WINE's Chenin Blanc Challenge prove. Though our national Chenin vineyard has shrunk to below 20 percent of the total spread, from over a third just a few years ago, there still remains a core of superior fruit from old vines, but probably not more than about 5000ha, or a quarter of the total. No other country, including France, has so much Chenin – and at such low prices in world money. Of course, the debate continues between the pro- and anti-oak squads, but it seem that the former hold sway if trophy results are any guide. At least a deftly-wooded Chenin doesn't resemble a wannabe Chardonnay as much as a Sauvignon can. And don't forget Chenin's contribution to those budget-priced export volumes. Both styles are very different from the high-acid wines of the Loire, homeland of the grape. The name, though, is hardly competition to Brand Chardonnay.

Which leads us to consider the world's favourite white grape – in fact, you can probably leave out the word "white". With Chardonnay plantings having rocketed from California to Australia, from the Cape to the Languedoc, predictions of a tsunami of the stuff have begun to look accurate. The cheaper, un-wooded versions tend to expose the blandness of the grape, which shows its best in company with some oak. It's encouraging to hear visiting experts' favourable comments on our progress these last few years in raising our Chardonnay quality, with more finesse, finer fruit-acid balance and more sensitive oak treatment. Though we still have a way to go to overhaul Brand Australia in the Chardonnay image stakes – even if we have the quality, which we definitely do.

Even more striking is the international recognition for our Sauvignons, in a world where New Zealand has virtually re-written the taste manual of the variety. These are pricey wines, however, from one of the world's smaller national vineyards. The steep learning curve in our top Sauvignon cellars like Vergelegen, Mulderbosch and Thelema, has guar-

Even more striking is the international recognition for our Sauvignons, in a world where New Zealand has virtually re-written the taste manual of the variety. These are pricey wines, however, from one of the world's smaller national vineyards.

anteed sell-outs, and a locomotive effect for more moderately priced but almost-as-good regions such as the unlikely Robertson, as well as Durbanville and Darling. This wine must surely be a credible calling card in international markets. The more ordinary stuff can add some welcome zip to cheaper blends, too.

Oz winemakers seem to have made Shiraz their own in recent years – a far cry from the seventies, when loads of unwanted fruit found its way into jam and muffins! The "S" word has become so twinned with Australia that the EU, in a similar knee-jerk reaction to the use of the word "port" by anyone other than the Portuguese, are trying to legislate the name off their shelves in favour of "Syrah", knowing the confusion it will cause among punters looking for their favourite "Barossa drop".

In the Cape, our initial high hopes for Shiraz were based on older vineyards, some of which, like that supplying the early Boekenhoutskloofs, have been grubbed up. The recent dearth of gold bullion for Shiraz on shows such as Veritas and the SA Trophy Wine Show probably reflect the youth of our vineyards – Shiraz has been the most prolifically planted variety these past few years. As these vines mature, get ready for some super wines, in a style somewhere between the warmer region Australian model and those of the Rhône valley.

Which brings me to the fact that there are precious few regions in the country regarded as genuine "cool climate" spots – and those that are have hardly been tapped. The Cederberg, parts of the Elgin/Grabouw axis, and the windswept Southern Cape around Agulhas are still proving themselves, but will probably deliver unique fruit, be it Pinot Noir, Chardonnay or Sauvignon. But does it really matter to a foreign buyer? Apart from Stellenbosch, hardly any of our regions arouse any recognition when looking cursorily at labels. In this regard Europe still has the edge, with the possible exception of the word 'Napa'. In fact, the EU's desire to enforce the intellectual property of its appellations is its natural response to the cultivar-driven approach of the

New World. It will be a long time before our wines are sought out for being from Walker Bay or Elgin, rather than on the reputation of producer or variety.

Just on sheer critical mass, Cabernet and Chardonnay, followed by Merlot, will continue to exist as brands in their own right, and South Africa with less than 20% of its vineyards planted to these three, must compete on this rather un-level playing field, whether we like it or not. There is no doubt our Kanonkops, Jordans and Veenwoudens can do it. At the mid-priced levels, however, inhabited by the Winecorps, Distells and Stellenbosch Vineyards of the industry, expect some serious slugging for shelf space.

And what of the niche varieties? The tiny pockets of Viognier, for example – hardly a force to reckon with just now on export markets. Yet grapes such as this, Mourvèdre,

The EU's desire to enforce the intellectual property of its appellations is its natural response to the cultivar-driven approach of the New World. It will be a long time before our wines are sought out for being from Walker Bay or Elgin, rather than on the reputation of producer or variety.

Pinot Noir, Tempranillo, Touriga Naçional, Sangiovese and Malbec all have devotees – members of the ABC (Anything But Chardonnay/Cabernet) clubs, the wine geeks, or 'train spotters' of the wine world. The more wine made from these varieties, the more spice is added to the national brew, and the more interest they will conjure up on the rest of the menu. Charles Back and his Fairview crew have shown how it's done – probably our most high-profile exporters.

Only a few of our wine categories can be regarded as approaching the best of the Old World. Cap Classique, in common with most other bubblies from warmer climes, hasn't cracked it against the top Champagnes – though many are more desirable than the very ordinary wines from this region. The same can be said for Pinot and Chardonnay vs Burgundy. Bordeaux, on the other hand, apart from the great growths, can play second fiddle to some of the Cape's more serious Cabernets, and the same can be said of some generic Rhône reds vs SA Shiraz in the price/quality matrix. One niche category which has probably crept closer to its Old World role model than most other wines is port, with great attention being paid to varietal authenticity and style. But then, fortifieds have always been one of the Cape's strong suits.

Picking a national team to represent the true uniqueness of our winelands must, in the end, revolve largely around stickier numbers: our top natural sweet and botrytised wines, muscadels and ports are shoe-ins. Lock the scrum with a couple of well-sculpted Pinotages or Pinotage blends, with a few subtly-oaked Chenins in the backline, and, while being a bit underweight, we may just have the element of surprise... The other day I tasted a delicious "Mass wine" made for the Catholic church by Citrusdal Winery – a dead ringer for an Italian Vinsanto; made me want to convert! ❑

Benchmarks

Many Cape wine cellars have walked off with international accolades from time to time. But a good long-term track record is harder to attain – and doesn't always involve winning major gongs. Often it revolves around price and demand – in other words, reputation built up over many years.

A shortlist of 10 benchmarks, in no particular order:

Meerlust Rubicon
Kanonkop Paul Sauer... and Pinotage
Nederburg Edelkeur
Klein Constantia Vin de Constance
Hamilton Russell Chardonnay... and Pinot Noir
Thelema Chardonnay... and Merlot
The Charles Back/Fairview collection... particularly volume wines like Goats do Roam
Veenwouden Classic
Mulderbosch Sauvignon Blanc
Rust en Vrede

If you wonder where some of the other stars are, like the Vergelegens and De Traffords – well, they're still a bit wet behind the ears compared with these seasoned campaigners. And some others, like Rustenberg, are still in come-back mode. But expect this list to change annually! ❑

Wine competitions, talent scouts

The introduction into the South African wine industry calendar of the Fairbairn Capital Trophy Wine Show has added considerable fuel to the debate around wine show judging. As the chairman of the inaugural competition, **MICHAEL FRIDJHON** doesn't pretend to be a neutral voice on the subject, but with his extensive wine show-judging experience both locally and around the world he can speak with some authority on the subject.

THE BUSINESS of show judging is fraught with countless difficulties: the wines themselves are not consistent – they vary with bottle age, minimal cork taint, the so-called halo effect, the circumstances of their service, their quality on the day. Stirling Moss was "the best driver never to have won a world championship" is an adage that might just as easily be applied to show wines. At best you can hope for a pattern, some way of tracking the performance of the consistently highflying players. This does not mean that show judging is a hit-and-miss affair: simply that there are variables and it would be naïve to disregard them.

Good shows tend to identify a core group of top performers. This means that some unanimity is possible between tasters, and the definition of good is, in part, a function of coherence to the trend. Here is where the territory becomes a potential minefield: if the majority of people think the earth is flat, this kind of consistency is no guarantee that it is not a sphere. Agreement – under these circumstances – merely reflects the depth of ignorance (or a cavalier disregard of the evidence) of a sufficient number of qualified voters. Hence shows are judged by the quality of their judges, and by the transparency – statistically and aesthetically – of the judging procedures.

The former is more easily assessed than the latter. Any

judging methodology that deprives the tasters of the control of the results of their panel is essentially flawed. Whether this means a system that discounts the top and bottom scores, or simply one where the result is the arithmetical outcome of summing and averaging equally weighted totals is actually immaterial. The moment a judging panel surrenders control of its decision-making process to an anonymous machine, the outcome is no longer in the hands of those who are nominally in control of it, who are expected to assume responsibility for it.

The aesthetic issues are more difficult to track. Stylistic preferences between tasters can vary by very small amounts and these can still have a dramatic effect on a panel's chosen wines. What constitutes "ripe" fruit, when is "high" alcohol simply "too much" alcohol, how "dry" should "dry wine" be, when is oak excessive, when is the fruit overwhelmed by wood, what constitutes an appropriate level of integration, oxidation, even volatile acidity are questions for which there are no "objective" answers and which can substantially influence the decisions taken by a panel.

In part, the show's organisers can resolve these issues simply in the judges they invite to become part of their panels and by the directions they give them when it

A well-run wine-judging event pays careful attention to the comments of all of the tasters, most particularly those whose views diverge most significantly (and most consistently) from the norm. A good business is not managed by yes-men, and quality wine competitions listen carefully to dissent.

comes to judging a particular class. In a way, this should also be the responsibility of the show chairman: he – or she – should work with the panels, tweaking rather than bulldozing the course they move along. Serious tasters – people who have done this kind of work often, and in many countries and for many different chairmen – swiftly recognise the criteria, implicit or explicit, which are the defining features of a particular judging environment. They do not insist on neutralising the good a show can achieve by imposing their own agenda upon that of the competition. By the same token, a well-run wine-judging event pays careful attention to the comments of all of the tasters, most particularly those whose views diverge most signifi-

The broad guidelines are usually clear. It is the wines that are on the cusp, teetering at the brink of entry (or exit) from a particular medal class, that pose the greatest difficulties.

cantly (and most consistently) from the norm. A good business is not managed by yes-men, and quality wine competitions listen carefully to dissent.

A key problem area remains thresholds: what constitutes a good wine, what a fine wine, what a world-class example, is an exercise in hair splitting. The broad guidelines are usually clear. It is the wines that are on the cusp, teetering at the brink of entry (or exit) from a particular medal class, that pose the greatest difficulties. For the panellists this is not always important, but for the competitor the difference between silver and gold, between gold and a trophy, is a matter of deadly importance. When the Shiraz panel at the inaugural SA Trophy Wine Show spent nearly 12 hours working their way through the 90+ submissions, no one could accuse the group of being cavalier with the commercial future of the contestants. Names can be made (more easily than they can be destroyed) at wine competitions and it behoves those who sit in judgement to treat the grading of each and every wine as a life-and-death matter.

Nevertheless, mistakes do happen. I have seen a panel – suspecting a corked bottle – call for a second sample and, finding it in the same unfragrant condition, assume a fault in the winemaking. Only after the show chairman called for a third bottle was the true quality of the wine revealed. Even here, however, a doubt lingers about the overall quality of the submission. When a high percentage of the entry suffers from cork taint, what kind of consistency can the average wine drinker expect from his purchases?

If shows and show judging are fraught with difficulties, is the whole enterprise simply driven by hype and the commercial needs of a wine industry, or its key protagonists: purists may suspect 'yes' but even the most insistent among them would be forced to acknowledge that the system does serve to scout out talent. There is also the issue of the limited downside: when a good wine under-performs in a show-judging environment consumers are inclined to blame the judges, or the judging methodology. Only once the wine has failed dismally on several occa-

When a good wine under-performs in a show-judging environment consumers are inclined to blame the judges, or the judging methodology. Only once the wine has failed dismally on several occasions will they begin to suspect a fault with the wine, or the winery.

sions will they begin to suspect a fault with the wine, or the winery. On the other hand, a good performance at a credible show thrusts a producer into the limelight. For those who are well established it serves as a confirmation of their branding; for those still seeking to make their reputation, a well-run wine show provides the golden opportunity to prove their worth. ❑

Prior to chairing the 2002 South African Trophy Wine Show, Michael Fridjhon was, for many years, convenor of the South African Airways winelist selection panel. He is a senior judge at the International Wine Challenge (UK), and has judged at the International Wine and Spirit Competition in London, the Australian National Wine Show and at shows and competitions in the United States, Chile, and France.

The judging panels

TASTINGS CONDUCTED BLIND

Blind-tastings are the only fair means of rating and comparing wines... Reviews included in this guide are those of South Africa's ten most reputable panels:

WINE MAGAZINE

WINE magazine's tasting panels conduct annual category overviews as well as monthly assessments of new releases – see page 21 for more details. WINE also presents awards in two of the categories most important to the SA wine industry:

The WINE magazine Chenin Blanc Challenge, an annual competition, with one of the objectives being to help realise and promote a new world-class benchmark from the Cape.

The WINE magazine TOPS at SPAR Pinotage Champion of the Year award, in search of world-class wines made from South Africa's indigenous variety, vintage by vintage.

TROPHY WINE SHOW

The annual Fairbairn Capital SA Trophy Wine Show, convened by Michael Fridjhon and WINE magazine, seeks to identify the best wines in the country and to award trophies to the top examples in each of the major classes as well as to the producer achieving the best results overall. Judges with extensive international tasting experience – including at least one overseas judge per panel – use the 100-point scoring system in striving for consensus-driven results, where wines scoring 90 and over win gold medals (5 Stars), 80 to 89 silver medals (4 to 4½ Stars) and 70 to 79 bronze medals (3 to 3½ Stars).

VERITAS

The SA wine producers' major competition of the year involves qualified tasters, including non-South African judges. A Veritas double gold medal is awarded when at least five of the adjudicators on a panel of seven score a wine 17 or more points out of 20. Also awarded are golds, silvers and bronzes.

DINERS CLUB

The annual Diners Club (SA) Winemaker of the Year award focuses on a different category of wine each time, with the jury including various local experts and at least one judge from overseas. Besides the overall winner, the Top Ten wines entered are also lauded.

JOHN PLATTER

Seasoned tasters, some with international experience, are involved in assessing the country's latest and future releases for the annual *John Platter South African Wine Guide*. For a 5 Star rating, the judges are presented with a selection of wines, each considered by at least one of the tasters as "superlative, a Cape classic", which are then tasted blind by all and a vote taken as to which are deserving of the guide's ultimate accolade.

MICHELANGELO

Michelangelo International grand gold, gold and silver medals are awarded annually by panels of local and overseas-based judges. The Sandton hotel's tastings are approved by the Concours Mondial du Vin of Brussels, and conducted according to the principles championed by the Office International de la Vigne et du Vin (OIV).

SOUTH AFRICAN AIRWAYS

The annual SAA on-board winelist selections involve leading local and international judges. Trophies are awarded to top-scorers in the process of choosing what will be served to First- and Business-class passengers.

PINOTAGE ASSOCIATION TOP TEN

The annual Pinotage producers' Top 10 awards are geared towards establishing benchmarks of international quality in respect of the country's indigenous variety. Judges include specialists in this field as selected by the Pinotage Association, who are charged with assessing wines of various vintages on sale at the time that the competition results are announced each year.

PETER SCHULZ

Recipients of the Peter Schulz Excellence Awards and trophy, sponsored by importer-wholesaler NMK Schulz, are chosen on the basis of a competition hosted by the SA Port Producers Association in association with WINE magazine, whose panel conducts annual reviews of the various styles of port.

APB CLASSIC

The Air France Preteux Bourgeois Classic trophy, category and gold medal winners are adjudged by a panel of "French palates", some based in South Africa and some flown in from Europe every year.

JULIET CULLINAN

The wines on offer at the annual Juliet Cullinan SCMB Wine Festival, Gauteng, are judged by a panel of Cape Wine Masters. The Juliet Cullinan Cape Wine Masters Award aims at identifying the best three examples of each grape variety on show, with gold, silver and bronze medals awarded.

WINE Panel Tastings

WINE magazine conducts and reports on several panel tastings every month, all audited by chartered accountants Deloitte & Touche. Teams of experienced judges include Cape Wine Masters and Masters of Wine (UK), top winemakers and leading wine writers.

Tastings are conducted blind, with wines sourced direct from the producers in most instances to minimise the risk of storage under anything other than ideal cellaring conditions. Once the panellists have assessed the line-up of wines independently, star ratings are awarded on the basis of discussion, whereby rather than rely solely on arithmetical averaging, subtle attributes or minor flaws associated with each wine are debated with a view to reaching consensus. A 20-point scoring system is used: 5 Stars awarded for a score of 18 or more, 4 to 4½ Stars for a score of between 16 and 17, 3 to 3½ Stars for 15 out of 20, 2 Stars for 14 points and 1 Star for 13 out of 20.

★★★★★	Superlative wine, a masterpiece. World class.
★★★★	Excellent wine of distinction. Recommended for special occasions and cellaring.
★★★	Good to very good. Fine character and style. Suitable for cellaring.
★★	Above average. Appealing.
★	Average. Acceptable.

WINE tasting reports include the full list of the wines reviewed, together with the prices, presented in a manner highlighting the best value-for-money within each quality grouping. Top-rated wines in every WINE panel tasting are assessed in the magazine by way of an overview of the tasters' comments raised in discussion and noted on their score-sheets, inclusive of the panels' advice on 'when to drink'.

WINE subscribers are first to know the results of each WINE tasting – the identity of the wines and details of the ratings awarded are withheld from all but the editorial team compiling the tasting reports until after publication thereof, at which point there is full disclosure. ❑

Pick of the Best Wines

Wines with the best records in terms of consistently good reviews in blind-tastings during recent years – as rated by the ten most authoritative tasting panels in South Africa (see pages 16 to 19)

REDS

CABERNET FRANC

Bellingham 'Spitz'
Warwick

CABERNET SAUVIGNON

Boekenhoutskloof
Cederberg
De Trafford
Delheim
Fleur du Cap
Grangehurst
Jordan
Kanonkop
Klein Constantia
L'Avenir
Laborie
Lanzerac
Le Bonheur
Le Riche
Longridge
Môreson
Morgenhof
Neil Ellis
Rustenberg
Rust en Vrede
Saxenburg
Stark-Condé
Thelema
Vergelegen
Vergenoegd
Vriesenhof
Waterford

MERLOT

Boschendal
Delaire
De Trafford
Dieu Donné
Fleur du Cap
Jordan
Môreson
Overgaauw
Spice Route
Saxenburg
Steenberg
Thelema
Veenwouden
Vergenoegd

PINOTAGE

Altydgedacht
Avontuur
Bellingham 'Spitz'
Broken Stone
Cloof
Clos Malverne

Delheim
Diemersfontein
Grangehurst
Hidden Valley
Kaapzicht
Kanonkop
L'Avenir
Môreson
Morgenhof
Neethlingshof
Rijk's
Saxenburg
Simonsig
Southern Right
Spice Route
Steytler
Uiterwyk
Villiera

RIJK'S 2000
WINE TOPS at SPAR
Pinotage Champion 2002
(Pick of the 2000 vintage)

CLOOF 1999
Pinotage Champion 2001
(WINE's pick of the 1999s)

STEYTLER 1998
Pinotage Champion 2000
(WINE's pick of the 1998s)

HIDDEN VALLEY 1997
Pinotage Champion 1999
(WINE's pick of the 1997s)

UITERWYK TOP-OF-THE-HILL 1996
Pinotage Champion 1998
(WINE's pick of the 1996s)

ALTYDGEDACHT 1995
Pinotage Champion 1997
(WINE's pick of the 1995s)

PINOT NOIR

Bouchard Finlayson
Hamilton Russell
WhaleHaven

SHIRAZ

Boekenhoutskloof
Brahms
Bredell's
Coleraine
De Trafford
Delheim
Gilga
Graham Beck
Havana Hills
Kevin Arnold
Klein Constantia
Lievland
Merindol
Neil Ellis
Nitida
Rickety Bridge
Rust en Vrede
Saxenburg
Slaley
Stellenzicht
Swartland
Vergelegen
Vergenoegd

RED BLENDS

Bertrams Robert Fuller
Beyerskloof
Cathedral Cellar Triptych
Clos Malverne
Cordoba Crescendo
De Toren Fusion V
Du Preez Polla's Red

Glen Carlou Grande Classique
Grangehurst
Jordan
Kanonkop Paul Sauer
Kanu Keystone
Klein Constantia Marlbrook
Klein Gustrouw
Le Bonheur Prima
Lievland DVB
Meerlust Rubicon
Morgenhof Première Selection
Mulderbosch Faithful Hound
Nitida Calligraphy
Rupert & Rothschild
Rust en Vrede
Rustenberg
Simonsig Tiara
Uiterwyk
Veenwouden
Vergelegen
Vergenoegd
Villiera

WHITES

CHARDONNAY

Amani
Avontuur
Boschendal
De Wetshof
Eikendal
Fairview
Glen Carlou
Groot Constantia
Hamilton Russell
Jordan
Mulderbosch
Neil Ellis
Rupert & Rothschild
Simonsig
Stellenzicht
Thelema
Vergelegen

CHENIN BLANC

Avontuur
Cederberg
De Trafford
Jordan
Kanu
Ken Forrester
Morgenhof
Mulderbosch
Rudera
Villiera

DE TRAFFORD KEERMONT 2001
KANU LIMITED RELEASE 2000
Joint winners of WINE's 2002 Chenin Blanc Challenge

KANU LIMITED RELEASE 1999
Winner of the 2001 Chenin Blanc Challenge

DE TRAFFORD 1999
Winner of the 2000 Chenin Blanc Challenge

VILLIERA 1996
Winner of the 1999 Chenin Blanc Challenge

L'AVENIR 1997
Winner of the 1998 Chenin

Blanc Challenge

MORGENHOF P' BIN 210 1996

Winner of the 1997 Chenin Blanc Challenge

BOSCHENDAL JEAN LE LONG '93

Winner of the 1996 Chenin Blanc Challenge

SAUVIGNON BLANC

Buitenverwachting
Cape Point
Jordan
Mulderbosch
Neil Ellis
Nitida
Rickety Bridge
Steenberg
Vergelegen
Villiera

SÉMILLON

Boschendal
Constantia Uitsig
Fairview
Steenberg
Stellenzicht

VIOGNIER

Fairview

DESSERT

NATURAL SWEET

Brampton (NLH)
KC Vin de Constance
Meerendal
Signal Hill

NOBLE LATE HARVEST

Avontuur
Boland
De Wetshof
Ken Forrester
Klein Constantia
L'Avenir
Nederburg
Neethlingshof
Signal Hill
Simonsig
Slanghoek
Spier
Stellenzicht
Vergelegen
Wide River

VIN DE PAILLE (Straw Wine)

De Trafford
Fairview

SPARKLING

CAP CLASSIQUE

Avontuur
Bon Courage
Graham Beck
JC le Roux
Laborie
Morgenhof
Simonsig
Twee Jonge Gezellen
Villiera
Weltevrede

FORTIFIED

MUSCAT

Bakenskop
Boland
Bon Courage
Die Krans
Du Toitskloof
Klawer
Landzicht
Monis
Nuy
Slanghoek
Weltevrede

PORT

Allesverloren
Axe Hill
Boplaas
Bredell's
Die Krans
KWV
Landskroon
Monis
Morgenhof
Overgaauw
Swartland

SHERRY

KWV
Monis

Pick of the Cape's best cellars overall

ON QUALITY

Cellars with the best track records in terms of consistently good reviews for their (non-fortified) wines at various blind-tastings in recent years:

AVONTUUR

BEYERSKLOOF

CHARLES BACK

DE TRAFFORD

GLEN CARLOU

GRANGEHURST

HAMILTON RUSSELL

JORDAN

KANONKOP

KLEIN CONSTANTIA

LE BONHEUR

MORGENHOF

MULDERBOSCH

NEIL ELLIS

RUSTENBERG

RUST EN VREDE

SAXENBURG

STELLENZICHT

THELEMA

VEENWOUDEN

VERGELEGEN

VERGENOEGD

Pick of the Cape's best value cellars

ON PRICE

Producers of a range of wines that have, during the past two years, received good reviews and stood out as inexpensive relative to others in the same league, quality-wise:

Amani

Avondale

Bergsig

Bloemendal

Boland

Bon Courage

Cederberg

Darling Cellars

Delheim

Du Toitskloof

Excelsior

Goede Hoop

Graham Beck

Groot Constantia

Jonkheer

Jordan

Kanu

Nitida

Rhebokskloof

Rickety Bridge

Riebeek

Simonsig

Swartland

Van Loveren

Villiera

Welmoed

Windmeul

Statistics, by the numbers

The latest statistics available from SA Wine Industry Information & Systems (SAWIS), all pertaining to 2001 unless stated otherwise.

310	**Private cellars** – 91 estates, 219 non-estates (up from a total number of 277 private cellars in 2000)
67	**Co-operatives** (69 in 2000)
11	**Producing wholesalers** (9 in 2000)
388	**Total number of wine cellars** (up from 355 in 2000)
135	**Cellars crushing less than 100 tons of grapes** and/or producing less than 54 000 litres, i.e. less than 12 000 cases of 6x750ml or equivalent
91	**Bulk wine buyers** 70 wholesalers, 21 for export only
4 390	**Primary producers:** wine growers/farmers (4 501 in 2000)
106 331ha	**Area under vines:** wine grape varieties (105 566ha in 2000)
314m	**Vines** – Worcester 19%, Paarl 17%, Stellenbosch 17%, Robertson 14%, Malmesbury 12%, Orange River 9%, Olifants River 9%, Little Karoo 3%
0,98m tons	**Grapes crushed** for winemaking
530m litres	**Wine produced** (down from 540m litres in 2000)
379m litres	**Local wine sales** (up from 365m litres in 2000)
177m litres	**Wine exports** (up from 141m litres in 2000)
8	**Ranking of South Africa (2,8%) among the world's largest wine-producing countries.** According to 1999 OIV statistics, the biggest are: 1 France (22%), 2 Italy (20,5%), 3 Spain (11,5%), 4 USA (7,3%), 5 Argentina (5,6%), 7 Australia (3%). Spain has the largest percentage of total plantings at 15%

WINE GRAPE VARIETIES AS A PERCENTAGE OF TOTAL PLANTINGS:

Chenin Blanc	19,8%	Cabernet Sauvignon	9,8%
Colombar(d)	10,5%	Pinotage	6,5%
Chardonnay	5,6%	Shiraz	6,7%
Sauvignon Blanc	5,4%	Merlot	5,4%
Hanepoot	3,4%	Cinsaut	3,2%
Crouchen Blanc*	1,7%	Other reds	2,5%
Sémillon	0,9%		
Riesling	0,4%		
Other whites	16,1%		
Total white	64%	**Total red**	36%

* Crouchen Blanc referred to locally as Cape Riesling or Riesling, whereas Riesling proper is termed Weisser Riesling or Rhine Riesling

Most-planted varieties in 2001: Cabernet Sauvignon, followed by Shiraz, Merlot, Pinotage, Sauvignon Blanc

Most-uprooted varieties in 2001: Chenin Blanc, followed by Colombar(d), White French (Fransdruif), Hanepoot, Crouchen Blanc (Cape Riesling)

PACKAGING:

33% in glass (56% in 750ml), 25% in plastic, 19% bag-in-box, 22% in foil-bag, 2% other

PER CAPITA CONSUMPTION OF WINE IN SOUTH AFRICA:

7 litres a year, which compares with Luxembourg (61), France (56), Italy (51), Portugal (50), Argentina (35), Australia (20), Chile (19), USA (8). Source: World Drink Trends 2002

Registered Wine Estates

A wine estate is defined in terms of the Liquor Products Act as the smallest demarcated wine production unit and can consist of one or more bordering farms, as long as it is farmed as a unit and has its own production cellar on the estate where the wine is produced. When the term Estate Wine appears on a label, it confirms that the wine is produced from grapes which are grown on that specific estate. The following farms were registered as wine estates as at 1 August 2002:

DURBANVILLE

Altydgedacht
Bloemendal
Diemersdal
Meerendal

CONSTANTIA

Groot Constantia
Klein Constantia

ELGIN

Paul Cluver
Wildekrans

FRANSCHHOEK

Cabrière

FREE STATE

Goudveld

LITTLE KAROO

Mons Ruber

PAARL

Backsberg
De Zoete Inval
Hildenbrand
Jacaranda
Laborie
Mischa
Nelson's Creek
Onverwacht
Oude Wellington
Rhebokskloof
Seidelberg
Upland
Welgemeend

ROBERTSON

Bon Courage
De Wetshof
Excelsior
Goedverwacht
Le Grand Chasseur
Rietvallei

Springfield
Van Zylshof
Weltevrede
Zandvliet

STELLENBOSCH

Alto
Asara
Avontuur
Bellevue
Bonfoi
Elsenburg
Fort Simon
Goede Hoop
Hartenberg
Jacobsdal
Kaapzicht
Kanonkop
Klawervlei
Klein Gustrouw
L'Avenir
Le Bonheur
Lievland
Lushof
Meerlust
Middelvlei
Mooiplaas
Morgenhof
Morgenster
Mount Rozier
Muratie
Neethlingshof
Oude Nektar
Overgaauw
Remhoogte
Rust en Vrede
Simonsig
Slaley
Uiterwyk
Uitkyk
Vera Cruz
Vergenoegd
Warwick
Zevenwacht

SWARTLAND

Allesverloren

TULBAGH

De Heuvel
Kloofzicht
Lemberg
Theuniskraal
Twee Jonge Gezellen

WALKER BAY

Goedvertrouw
Hamilton Russell

WORCESTER

Bergsig
Deetlefs
Du Preez
Opstal

Single Vineyard Wines

It's arguable that the following wines are the most terroir-specific that you can find in South Africa – made from grapes derived from a single patch of land:

Alexanderfontein: all
Asara: Pinotage, S Blanc, Gamay Noir
Avondvrede: Koningshof Cab, Cuvée Marie Christine
Badsberg: Chardonnay Sur Lie
Barrydale: Tradouw
Beaumont: Hope Marguerite, Mourvèdre, S Blanc
Bergkelder: Fleur du Cap Unfiltered Cab, Fleur du Cap Unfiltered Merlot
Bilton: Merlot
Blaauwklippen: Vineyard Selection Shiraz, Vineyard Selection Cab
Bloupunt: all
Bon Courage: Inkará range
Bonnievale: Shiraz, Chardonnay, S Blanc
Boschendal: S Blanc
Brahms: Shiraz
Capaia: Blouberg Hill Cab Merlot, Blouberg Hill S Blanc
Cederberg: V Generations Cab
Clos du Ciel: Chardonnay
Daschbosch: Chardonnay, S Blanc, Merlot, Pinotage
De Wetshof: Bateleur Chardonnay, D'Honneur Chardonnay, Edeloes, Muscadel
Delheim: Vera Cruz
Dieu Donne: Cap Classique, Pinotage, S Blanc, Shiraz
Du Toitskloof: Bukettraube, Hanepoot Jerepigo
Durbanville Hills: Biesjes Craal S Blanc, Caapmans Cab Merlot, Luipaardsberg Merlot
Eikehof: Bush Vine Sémillon
Elephant Pass: all
Excelsior: Cab
Flagstone: BK5, Elim Vineyards, Heywood House, Music Room, Strata Series, Two Roads, Writer's Block
Gilga: Shiraz
Glenwood: Chardonnay, Sémillon
Graceland: Cab
Graham Beck: Old Road Pinotage; Ridge Syrah

Groot Eiland: Shiraz
Gusto: S Blanc
Hartenberg: Gravel Hill Shiraz
Hidden Valley: Pinotage
High Constantia: all
Hildenbrand: Cab, Chardonnay, Chenin, Sémillon
Hoopenburg: Chardonnay, Merlot, Pinot
Jacaranda: all
Jordan: Cobblers Hill
Joubert-Tradauw: Chardonnay
Kanu: Chardonnay, Chenin Wooded, Shiraz
Katbakkies: all
Klawervlei: all
Kleine Zalze: Vineyard Selection
Klompzicht: Merlot, Shiraz
Kloofzicht: Alter Ego
Kloovenburg: Cab, Chardonnay, S Blanc, Merlot
KWV: Abraham Perold Shiraz
L'Avenir: Chardonnay, Chenin
La Petite Ferme: Shiraz
Landau du Val: S Blanc, Sémillon
Le Chasseur: all
L'Émigré: L'Émigré Shiraz, L'Émigré Pinotage
Leidersberg: Cab
Linton Park: Cab, Chardonnay, Merlot, Shiraz
Longridge: Longridge wines
Louiesenhof: all
Louisvale: Chardonnay
Main St: Merlot, Shiraz
McGregor: Cab, Res S Blanc
Mont Destin: Chenin, Pinotage
Montpellier: all
Nederburg: Reserve wines
Neethlingshof: Lord Neethling Cab, Lord Neethling Cab Franc, Lord Neethling Pinotage
Nordale: Captains' Drift Shiraz
Onderkloof: Cab, Chardonnay, Chenin, Pinotage, S Blanc
Oranjerivier: Chenin, Muscadels, Pinotage, Ruby Cab
Oudekloof: Foregon Cinsaut Pinotage
Oude Wellington: Chardonnay, Currant Abbey, Gewürz, Ruby Cab
Paradyskloof: Talana Hill Chardonnay, Talana Hill Royale
Plaisir de Merle: Cellarmaster's Release Cab
Quando: S Blanc

Rhebokskloof: Shiraz
Rijk's: Cab, Chardonnay, Chenin, Merlot, Pinotage, S Blanc, Sémillon, Shiraz
Robertson: Almond Grove Noble Late Harvest, Wide River Reserve S Blanc
Rustenberg: Five Soldiers, Peter Barlow
Savanha: Naledi, Sejana
Saxenburg: Shiraz Select (SSS)
Signal Hill: Rebattet, Rouge De Constance, Vin de L'Empereur
Simonsig: Merindol Syrah, Redhill Pinotage
Slaley: Slaley Merlot, Slaley Shiraz, Slaley Pinotage, Broken Stone S Blanc
Slanghoek: Cab, S Blanc
Spier: Private Collection wines
Springfield: Life from Stone S Blanc, Méthode Ancienne Chardonnay, Special Cuvée S Blanc
Stark-Condé: Condé Cab
Stellenbosch Vyds: Genesis Chardonnay, Genesis Merlot, Genesis S Blanc, Genesis Shiraz
Stellenzicht: Stellenzicht Syrah
Sumaridge: Syrah
SylvanVale: all
Uiterwyk: S Blanc, Top-of-the-Hill Pinotage, Viognier
Upland: Cab
Van Loveren: Spes Bona Chardonnay
Veenwouden: Thornhill
Vergelegen: Res Chardonnay, Res S Blanc
Villiera: Bush Vine S Blanc
Von Ortloff: all
Wamakersvallei: La Cave wines
Warwick: Cab Franc, Old Bush Vines Pinotage, Merlot
Wellington: Cab, Merlot, Pinotage, Port, S Blanc, Shiraz
Weltevrede: Oude Weltevreden Chardonnay, Oude Weltevreden Merlot Cab, Weltevrede S Blanc, Woolworths Cape Riesling
Western Wines: Kumala Single Varietal Cab, Kumala Single Varietal S Blanc
Wildekrans: Barrel Selection Pinotage, Cab, Chardonnay, Merlot, S Blanc, Sémillon
Zandvliet: Kalkveld Syrah
Zevenwacht: Tin Mine Chenin

CIRCA 1693
THE OLD MATURATION CELLAR
BELLINGHAM
BELLINGHAM
EST. 1693
BELLINGHAM
Shiraz
COASTAL REGION
SOUTH AFRICA
BELLINGHAM
Classic, Modern, Exceptional
NOT FOR SALE TO PERSONS UNDER THE AGE OF 18
www.bellingham.co.za

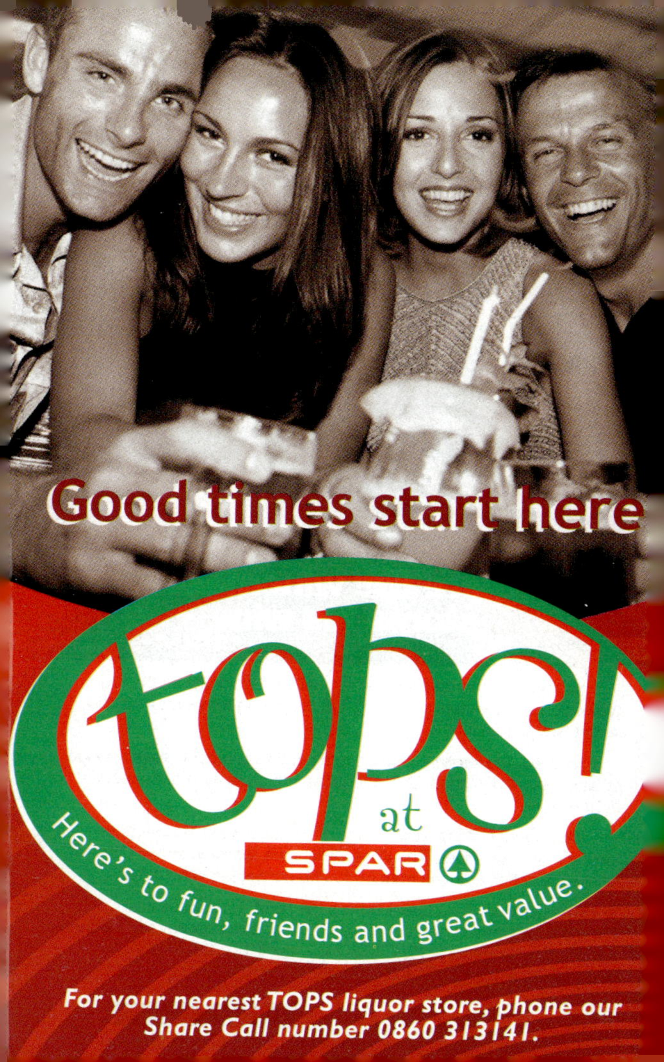
Good times start here
tops!
at
SPAR
Here's to fun, friends and great value.
For your nearest TOPS liquor store, phone our Share Call number 0860 313141.

A-to-Z directory

SA wines & cellars

A comprehensive, ready-reference encompassing everybody who's anybody in the South African wine industry...

Including a guide to the country's best: very good to world-class wines as judged by South Africa's 10 most reputable tasting panels (see page 18 to 20) during the past five years – October 1997 to September 2002.

KEY TO THE RATINGS

CC	Winner of the Chenin Blanc Challenge	
DT	Diners Club Top Ten commendation	
DW	Diners Club Winemaker of the Year award	
JC	Juliet Cullinan Cape Wine Masters Award	
JP	5 Stars in the *John Platter SA Wine Guide*	
MG	Michelangelo Int grand gold medal & trophy	
PB	Air France PB Classic trophy & gold medal	
PC	Pinotage Champion – best of the vintage	
PS	Peter Schulz Port Excellence award & trophy	
PT	Pinotage Association Top Ten award	
SA	SAA winelist selections: trophy & certificate	
TG	Gold medal at The SA Trophy Wine Show	
TT	Trophy at The SA Trophy Wine Show	
VD	Veritas Double Gold medal	
W5	5 Stars in WINE magazine	★★★★★
W4½	4½ Stars in WINE magazine	★★★★☆
W4	4 Stars in WINE magazine	★★★★
W3½	3½ Stars in WINE magazine	★★★☆

Abbreviated dates given after each rating pertain to the year it was awarded (in the case of SAA, the year the wine was offered to airline passangers)

AAN DE DOORNS (co-op)

WORCESTER MAP

R43, between Worcester & Villiersdorp
PO Box 235, Worcester 6849
TEL: 023 347-2301
FAX: 023 347-4629
aded@intekom.co.za
DIRECTORS: Andries du Toit (chairman), HJ Griessel, GC Hugo, SF Hugo, HE Ungerer
GM: Johan Morkel
C'MASTER: Gert van Deventer
V'CULTURIST: Nevin du Toit
SALES: Mon–Fri 08:00–17:00
TASTINGS: no fee
TOURS: by appointment
PICNICS: welcome to bring
EST 1954. Area under vine: 1 182ha (78% white). Annual production: 17 719 tons, 13m litres
FLAGSHIP: Doornroodt

LISTED WINES

D WHITE: Colombar (U); Colombar Chardonnay (U)
SS PINK: Blanc de Noir (U, Muscadel)
RED: Doornroodt (W, Ruby Cab & Merlot); Pinotage (W)
FORTIFIED: Muscadel (U); Port (U, Tinta B)

ABENDSGLÜCK ▷ BERNHEIM
ABOVE ROYALTY ▷ AVONTUUR
ABRAHAM PEROLD ▷ KWV
ADELBERG ▷ SIMONSIG
ADELBLANC ▷ SIMONSIG
ADELPRACHT ▷ DROSTDY
A FEW GOOD MEN ▷ RIEBEEK
AFRICAN RED ▷ AFRICAN TERROIR
AFRICAN RED ▷ LONG MOUNTAIN
AFRICAN SUNSET ▷ LADISMITH

AFRICAN TERROIR

STELLENBOSCH MAP

P CELLARS: Sonop (Paarl), Cilmor (Worcester)
HEAD OFFICE: Distillery Rd, KWV Park, Stellenbosch 7599
TEL: 021 869-8723; 021 863-8534 (Sonop cellar); 021 887-2409 (Cilmor cellar)
FAX: 021 886-4838
office@africant-terroir.co.za
www.africanterroir.co.za
OWNER: Jacques Germanier Group (SAVISA)
MD: Sebastian Dayer
C'MASTER: Alain Cajeux
V'CULTURIST: Thys Greeff
VISITS: by appointment only
EST 1994 (previously trading as SAVISA). Area under vine: 74ha (African Terroir). Annual production: 600 tons (African Terroir), 1 300 tons (Sonop), 6 600 tons (Cilmor)
ORGANIC WINE: Sonop Organic Terroir

LISTED WINES

Azania

RED: Shiraz (W) Out of Africa
D WHITE: Chardonnay (W);

S Blanc (U)
RED: Cab (W); Pinotage (W); Shiraz (W)

Sonop Organic Terroir
D WHITE: Chardonnay (W); S Blanc (U)
RED: Pinotage (W); Shiraz (W)

Tribal
D WHITE: African White (U, Chenin & Sémillon & S Blanc)
RED: African Red (W, Cinsaut & Pinotage & Ruby Cab & Cab)

Winds of Change
D WHITE: Chardonnay (W); S Blanc (U)
RED: Merlot Pinotage (W); Pinotage Cab (W)

AFRICAN WINES & SPIRITS

NÉGOCIANT: African Wines & Spirits – not open to the public
PO Box 389, Constantia 7848
TEL: 021 794-6697
FAX: 021 794-3990
aws@awsomedrinks.com
DIRECTORS: James McLachlan (MD); Directors Mike Cox, Chris Weeden, André van Wyk
SALES: at retail stores only
EST 1992

LISTED WINES

Bertrams
RED: Cab (W); Cab Merlot (W); Shiraz (W)

Craighall
D WHITE: Chardonnay S Blanc (W); S Blanc (U)
RED: Cab Merlot (W)

South Southwest
D WHITE: Chardonnay Chenin (U)
RED: Ruby Cab Pinotage (U)

HIGHLY RATED

Bertrams Cabernet Sauvignon
1996 – W3½ 98
Bertrams Robert Fuller Reserve
2001 – W3½ 02
1997 – W3½ 01
1996 – VD 98, W3½ 98
Bertrams Shiraz
1996 – W4 98

AGTERKLIPHOOGTE

ROBERTSON MAP

CO-OP: Robertson McGregor Rd, 25km from Breede River
PO Box 267, Robertson 6705
TEL: 023 626-1103
FAX: 023 626-3329
akhwyn@lando.co.za
GM: Helmard Hanekom
C'MASTER: Helmard Hanekom
V'CULTURIST: Brian Stipp
SALES: Mon–Fri 8:00–17:00
TASTINGS: no fee
TOURS: by appointment
EST 1965. Area under vine: 308ha (75% white). Annual production: 4 500 tons, 324 000 litres

LISTED WINES

Wandsbeck
D WHITE: Chenin (U); Sunay (U, Colombar & Chenin)
SS WHITE: Chenay (U,

Chenin & Chardonnay)
RED: Ruby Cab (U);
Shiraz (W)
DESSERT: Muscadel (U)

AGULHAS ▷ MERWESPONT

AGUSTA

FRANSCHHOEK MAP
ESTATE: Grande Provence, Main Rd, Franschhoek
PO Box 393, Franschhoek 7690
TEL: 021 876-3195
FAX: 021 876-2652
orders@agustawines.co.za
www.agustawines.co.za
OWNER: Count Riccardo Agusta
MD: Wolf Peter Berthold
GM: Christian von Palace
C'MASTER: Alastair Rimner
V'CULTURIST: Anton Olivier
SALES: in summer Mon–Sun 10:00–18:00; in winter Mon–Sun 09:00–17:00
TASTINGS: fee of R10
TOURS: by appointment
FUTURES (EN PRIMEUR): yes
N MONUMENTS: La Provence manor house & Grande Provence guest house (1756)
ACCOM: Grande Provence Guest House, Tel 021 896-2163, Fax 021 876-4204, guesthouse@agustawines.co.za
RESTAURANT: La Brasserie, Tues–Sun 11:30 until late, Tel 021 876-4143
PICNICS: available for purchase
EST 1975 (producing wine since 1991). Area under vine: 40ha (65% white).
Annual production: 145 tons, 700 000 litres
FLAGSHIPS: Cab, Chardonnay

LISTED WINES

Count Agusta
D WHITE: Chardonnay (W); S Blanc (U)
RED: Cab (W); Merlot (W)

Angels' Tears
OD WHITE: Angels' Tears (U, Chenin & Hanepoot)

HIGHLY RATED

Count Agusta Chardonnay
1998 – W4 99
Count Agusta Sémillon
1999 – W3½ 01
Count Agusta Cabernet Sauvignon
1998 – W4 01
1997 – W3½ 98
Haute Provence Blanc Fumé
1998 – W3½ 99
Haute Provence Chenin Blanc
1997 – W4 98
Haute Provence Reserve Chardonnay
1998 – W4 99, VD 98
Haute Provence Cabernet Sauvignon
1996 – VD 97

AKKERBOS ▷ CHARLES BACK
ALBENET ▷ NELSON'S CREEK

ALEXANDERFONTEIN

SWARTLAND MAP

P CELLAR: Outside Darling
PO Box 201, Darling 7345
TEL: 022 492-3219 (sales), 082 655-2142 (cellar)
FAX: 022 492-3171
darlrome@iafrica.com
OWNER: Bassenberg Pty Ltd
DIRECTORS: T Basson, NJ Basson
W'MAKER: Wouter Pienaar (consulting)
V'CULTURIST: T Basson
VISITS: by appointment only
SINGLE VYD WINES: all
EST 1974 (producing wine since 2000). Area under vine: 292ha (50% red). Annual production: 40 tons, 25 000 litres

LISTED WINES

D WHITE: Chardonnay (W); S Blanc (U)
RED: Cab (W, Cab & Merlot)

HIGHLY RATED

Alexanderfontein Sauvignon Blanc
2001 – W4 02

ALEXIS ▷ AMBELOUI

ALLESVERLOREN

SWARTLAND MAP

ESTATE: R311, Riebeek West
PO Box 23, Riebeek West 7306
TEL: 022 461-2589 (tastings, sales); 022 461-2320 (cellar, manor house)
FAX: 022 461-2444
dmalan@wcaccess.co.za
www.allesverloren.co.za
OWNER: Malan Boerdery Trust (Fanie & Danie Malan)
DIRECTORS: Daniel Francois Malan (MD), Stephanus Francois Malan
C'MASTER: Daniel Francois Malan
SALES: Mon–Fri 08:30–17:00, Sat 08:30–12:30
TASTINGS: no fee
TOURS: by appointment
FUTURES (EN PRIMEUR): yes
PICNICS: welcome to bring
EST pre-1872. Area under vine: 180ha (97% red). Annual production: 1 600 tons, 750 000 litres

LISTED WINES

RED: Cab (W); Shiraz (W); Tinta B (W)
PORT: LBV (W, Tinta B & Souzão & Pontac & Tinta R & Tinta F & Touriga N & Malvasia Rey)

HIGHLY RATED

Allesverloren Cabernet Sauvignon
1996 – W4 99
Allesverloren Shiraz
1996 – DW 98
Allesverloren Port
1996 – W3½ 02
1995 – W3½ 02
1993 – W4 PS Trophy 00

ALMOND GROVE ▷ ROBERTSON
ALPHEN ▷ AFRICAN WINES
ALPHEN HILL ▷ TOMORROW
ALTER EGO ▷ KLOOFZICHT

ALTO

STELLENBOSCH MAP

ESTATE: Annandale Rd, off R44 between Somerset West & Stellenbosch
PO Box 15, Lynedoch 7603
TEL: 021 881-3884
FAX: 021 881-3894
nee@mweb.co.za
www.alto.co.za
OWNER: Lusan Holdings (Pty) Ltd
DIRECTORS: R Jones (MD); A Diedericks, HA Hëyl, BCM Tan, Prof CJ van Wyk
C'MASTER: Schalk van der Westhuizen
V'CULTURISTS: David Saayman, Eben Archer
SALES: Mon–Fri 09:00–17:00, Sat 09:00–12:30
TASTINGS: fee of R5
TOURS: by appointment
FUTURES (EN PRIMEUR): yes
EST 1918. Area under vine: 94ha (100% red). Annual production: 300 tons, 180 000 litres
FLAGSHIP: Cabernet

LISTED WINES

RED: Cab (W); Rouge (W, Merlot & Cab & Shiraz)

HIGHLY RATED

Alto Cabernet Sauvignon
1996 – W4 02
1995 – W4 98

ALTYDGEDACHT

DURBANVILLE MAP

ESTATE: corner of Tygerberg Valley Rd (R31) & Racecourse Rd (M13), Durbanville
PO Box 213, Durbanville 7551
TEL: 021 976-1295
FAX: 021 976-4318
altydgedacht@mweb.co.za
OWNERS: Parker family
C'MASTER: Oliver Parker
V'CULTURIST: John Parker
SALES: Mon–Fri 09:00–17:00, Sat 09:00–13:00
TASTINGS: no fee
TOURS: by appointment
FUTURES (EN PRIMEUR): yes
EST 1730. Area under vine: 150ha (50% red). Annual production: 90 tons, 70 000 litres

LISTED WINES

D WHITE: Chardonnay (W); Chenin S Blanc (U); Gewürz (U); S Blanc (U)
SS WHITE: Chatelaine (U, Gewürz & Chenin)
RED: Barbera (W); Cab (W); Merlot (W); Pinotage (W); Shiraz (W)

HIGHLY RATED

Altydgedacht Barbera
1999 – W3½ 01
Altydgedacht Cabernet Sauvignon

1998 – W4 02
Altydgedacht Pinotage
2000 – W3½ 02
1999 – W4 02, JC 02

AMADÉ ▷ WELGEMEEND

AMANI

STELLENBOSCH MAP

P CELLAR: Polkadraai Rd (M12), ±10km outside Stellenbosch
PO Box 12422, Die Boord 7613
TEL: 021 905-1126
FAX: 021 905-4404
wine@amani.co.za
OWNER: Jim L Atkinson
MD: Alphonso H Bowe
GM: Fanie Joubert
W'MAKERS: consultants Rod Easthope & Emma Williams
SALES: Tues–Fri 09:30–16:00, Sat 09:30–13:30
TASTINGS: no fee
TOURS: by appointment
PICNICS: welcome to bring
EST 1997. Area under vine: 20ha (60% white). Annual production: 100 tons, 65 000 litres
FLAGSHIP: Chardonnay

LISTED WINES

D WHITE: Brushed Chardonnay (W); Chardonnay (W); Chenin (W); S Blanc (U)
RED: Merlot (W)
DESSERT: Charlotte NLH (W, Chenin)

HIGHLY RATED

Amani Chardonnay
2001 – JC 02
2000 – W3½ 01
1999 – W4 00
1998 – W4 00
Amani Sauvignon Blanc
1998 – VD 98
Amani Merlot
1999 – W4 01

AMBELOUI

CONSTANTIA MAP

P CELLAR: 4075 Valley Rd, Hout Bay
PO Box 26800, Hout Bay 7873
TEL: 082 441-6039
FAX: 021 790-7873
ambeloui@icon.co.za
OWNERS: Nick & Ann Christodoulou
C'MASTER: Nick Christodoulou
VISITS: by appointment only
EST 1995 (producing wine since 1998). Area under vine: 0,6ha (50% white)

LISTED WINES

MCC SPARKLING: Alexis (Chardonnay & Pinot); Christo (Chardonnay & Pinot); Miranda (Chardonnay & Pinot)

HIGHLY RATED

Ambeloui Lisa
1998 – W4 00

AMBER ▷ MURATIE

AMERICANA ▷ MISCHA
AMITY ▷ WELVANPAS
AMOS ▷ CHARLES BACK
ANCIENNE ▷ SPRINGFIELD
ANDREW BAIN ▷ CAPE VINEYARDS
ANDREW'S HOPE ▷ CHARLES BACK
ANGELS ▷ VREDENHEIM
ANGELS' TEARS ▷ AGUSTA
ANNA ▷ DOUGLAS GREEN

ANNANDALE

STELLENBOSCH MAP

P CELLAR: Annandale Rd, off the R44 between Stellenbosch & Somerset West, 5km outside Stellenbosch
PO Box 12681, Stellenbosch 7613
TEL: 021 881-3562 (cellar, tastings); 021 881-3560 (sales, manor house)
FAX: 021 881-3562
annandale@telkomsa.net
www.annandale.co.za
OWNER: Hempies du Toit
C'MASTER: Pieter Gerhardus du Toit
SALES: Mon–Sun 10:00–15:00
TASTINGS: fee of R5
TOURS: by appointment
N MONUMENT: cellar (1688)
PICNICS: welcome to bring
EST 1995. Area under vine: 45ha (100% red). Annual production: 150 tons, 50 000 litres
FLAGSHIP: Cabernet

LISTED WINES

RED: Cab (W); Cavalier (W, Shiraz & Cab); Res (W, Shiraz)

HIGHLY RATED

Annandale Cabernet Sauvignon 1996 – W3½ 00

ANNA SIMON ▷ FORT SIMON
ANNIE ROZIER ▷ MOUNT ROZIER
ANSELA ▷ MURATIE

ANT HILL

NÉGOCIANT: wines made at Topaz Wines CC
POSTAL: 14 Somerset Street, Somerset West 7130
TEL: 082 895-9008
FAX: 021 885-5086
anthill@absamail.co.za
OWNERS: Mark Howell & Hylton Schwenk
W'MAKER: Mark Howell
VISITS: by appointment only
FUTURES (EN PRIMEUR): yes
EST 2002. Annual production: 4,2 tons, 2 470 litres

LISTED WINES

RED: Pinotage (W)

ANTHONY SMOOK ▷ JA CLIFT
ANTICA ▷ SIGNAL HILL
ANTONIO BIANCO ▷ DE HEUVEL

ANURA

PAARL MAP

P CELLAR: Simondium Rd (connecting R45 & R44),

Simondium
PO Box 192, Simondium 7670
TEL: 021 875-5360
FAX: 021 875-5657
info@anura.co.za
www.anura.co.za
OWNER: Tymen Bouma
W'MAKERS: Tymen Bouma, Carla van der Mescht
V'CULTURIST: Hannes Kloppers
SALES: Mon–Sun 09:30–16:00
TASTINGS: fee of R5
TOURS: by appointment
RESTAURANT: Mon–Sun 09:30–16:00
FROMAGE cheese factory
EST 1990 (producing wine since 2001). Area under vine: 109ha (70% red). Annual production: 130 tons, 91 000 litres under own label
FLAGSHIP: Anura Reserve wines

LISTED WINES

Anura

D WHITE: Chardonnay (W); Chenin (W); S Blanc (U)
RED: Cab (W); Merlot (W); Pinotage (W); Res Cab (W); Res Syrah (W); Shiraz (W)

Froghill

RED: Cab Merlot (W); Merlot (W); Pinotage (W)

HIGHLY RATED

Anura Merlot
2001 – W3½ 02

APOSTLES FALLS ▷ INT W SERVICES

APPELSDRIFT

ROBERTSON MAP
P CELLAR: Clairvaux Winery, Voortrekker Rd, Robertson
PO Box 179, Robertson 6705
TEL: 023 626-3842
FAX: 023 626-1925
appelsdrift@lando.co.za
OWNER: Wouter de Wet
C'MASTER: Kobus van der Merwe
V'CULTURIST: Brian Stipp
SALES: Mon–Fri 08:30–17:30, Sat 09:00–12:30
TASTINGS: no fee
TOURS: by appointment
PICNICS: welcome to bring
EST 1999. Area under vine: 150ha (70% red). Annual production: 1 400 tons, 500 000 litres
FLAGSHIP: Clairvaux Cab Shiraz

LISTED WINES

Clairvaux

D WHITE: Chardonnay Colombar (U)
NATURAL SWEET: Rosé (U, Colombar & Pinotage); Soleil (U, White Muscadel & Colombar)
RED: Cab Shiraz (W); Sandberg Purple (U, Pinotage & Ruby Cab)
DESSERT: Madonna's Kisses (U, White Muscadel)
FORTIFIED: Port (W, Ruby Cab)

ARIANE ▷ BEAUMONT
ARNIM ▷ CABRIÈRE
ARNISTON BAY ▷ VINFRUCO
ARTEMIS ▷ MONTPELLIER

ASARA

STELLENBOSCH MAP

ESTATE: Polkadraai Rd (R310), Stellenbosch
PO Box 79, Vlottenburg 7604
TEL: 021 888-8018 (sales); 021 888-8040 (tastings); 021 888-8020 (cellar); 021 888-8000 (office)
FAX: 021 888-8001
info@AsaraWine.com
www.AsaraWine.com
OWNERS: Markus & Christiane Rahmann
GM: Twiggy Tolken
C'MASTER: Jan van Rooyen
V'CULTURIST: Pieter Rossouw
SALES: Mon–Fri 09:00–17:00, Sat 10:00–17:00 (Nov–Apr) 10:00–14:00 (May–Oct)
TASTINGS: fee of R10
TOURS: by appointment
FUTURES (EN PRIMEUR): yes
PICNICS: welcome to bring
EST 1772 (producing wine since 1953, as Verdun until 2001). Area under vine: 130ha (70% red). Annual production: 300 tons
FLAGSHIP: Belltower Collection
SINGLE VYD WINES: Pinotage, S Blanc, Gamay Noir

LISTED WINES

D WHITE: Chardonnay (W); Chenin (W); Ivory (U, S Blanc & Chardonnay); S Blanc (U)
PINK: Rosé (U, Merlot & Gamay & Cab & Pinotage)
RED: Belltower Collection (W, Cab & Merlot); Cab (W); Carillon (W, Merlot & Cab); Ebony (W, Shiraz & Pinotage & Gamay); Gamay Noir (U); Merlot (W); Pinotage (W); Shiraz (W)
DESSERT: Noble Late Harvest (W, Chenin)
FORTIFIED: White Port (W, Chenin)

HIGHLY RATED

Asara Chardonnay 2000 – W3½ 02
Asara Sauvignon Blanc 2001 – W4 02
Asara Belltower Collection 1998 – W3½ 02
Asara Cabernet Sauvignon 1998 – W4½ 02

ASHANTI

PAARL MAP

ESTATE: Sonstraal Rd, Klein Drakenstein, Paarl
PO Box 934, Huguenot 7645
TEL: 021 862-0789
FAX: 021 862-2864
ashanti@icon.co.za
www.ashantiwines.com
OWNERS: Ashanti Estates (Pty) Ltd
GM: Michael Grabowski

C'MASTER: Sydney Burke
SALES: Mon–Fri 09:30–17:00, Sat 09:30–16:00 (Sun as per Sat in season)
TASTINGS: fee of R5
TOURS: by appointment
RESTAURANT: Il Casale, Tues–Sat 11:00–23:00 (Sun lunch in season), Tel 021 862-6288
EST 1987. Area under vine: 195ha (70% red). Annual production: 500 tons, 300 000 litres.
FLAGSHIP: Ashanti Chiwara

LISTED WINES

Ashanti
RED: Cab (W); Chiwara (W, Pinotage & Cab & Shiraz); Pinotage (W)

Joseph's Hat
RED: Cab Pinotage (W)

Nicole's Hat
D WHITE: Chardonnay (U); Chenin (U)

HIGHLY RATED

Ashanti Malbec
1999 – W3½ 00
1998 – W3½ 99
Ashanti Bin 98 Vin de Paille
NV – W3½ 00

ASHER ▷ KLOOFZICHT

ASHTON

ROBERTSON MAP
CO-OP: Main Rd, Ashton
PO Box 40, Ashton 6715
TEL: 023 615-1135
FAX: 023 615-1284
ashkel@mweb.co.za
GM: Willem Joubert
W'MAKERS: Morno Brink, Philip Louw
V'CULTURIST: Hennie Visser
SALES: Mon–Fri 08:00–17:00, Sat 08:30–12:30
TASTINGS: no fee
TOURS: by appointment
PICNICS: welcome to bring
EST 1962. Area under vine: 1 350ha (50% red). Annual production: 18 000 tons
FLAGSHIP: Satynrooi

LISTED WINES

D WHITE: Chardonnay (W); Colombar (U); Colombar Chardonnay (U); S Blanc (U)
SEMI-SWEET: Blanc de Noir (U, Red Muscadel); Late Harvest (U, Colombar & White Muscadel); Petillant Blanc (U, Colombar & White Muscadel, Perlé); Special Late Harvest (U, White Muscadel)
RED: Cab (W); Satynrooi (U, Ruby Cab & Pinotage); Shiraz (W)
FORTIFIED: Port (U, Pinotage); Red Muskadel (U); White Muskadel (U)

ASHWOOD ▷ WINECORP
ASSEGAI ▷ WOODLANDS
ASTONVALE ▷ ZANDVLIET

@ ▷ MAKRO
ATTICUS ▷ BELLEVUE

AUDACIA

STELLENBOSCH MAP

P CELLAR: Off the R44 Rd between Stellenbosch & Somerset West
PO Box 12679, Die Boord, Stellenbosch 7613
TEL: 021 881-3052
FAX: 021 881-3137
audacia@icon.co.za
OWNER: Trevor Strydom
C'MASTER: Elsa Carstens
SALES: Mon–Fri 09:00–17:00, Sat 12:00–17:00 (Sun 12:00–16:00 in season)
TASTINGS: fee of R5, redeemable on purchase of wine
EST 1930. Area under vine: 15ha (100% red). Annual production: 114 tons, 80 000 litres
FLAGSHIP: Cabernet

LISTED WINES

RED: Cab (W); Merlot (W); Shiraz (W)

HIGHLY RATED

Audacia Cabernet Sauvignon 1998 – W3½ 01
Audacia Merlot 2000 – W3½ 02

AUGUSTINE ▷ DOUGLAS GREEN
AURET ▷ CLOS MALVERNE
AUSTIN ▷ BODEGA

AUTUMN HARVEST ▷ DISTELL
AVOCA ▷ DOUGLAS

AVONDALE

PAARL MAP

P CELLAR: Lustigan Rd (R303), South Drakenstein, South Paarl
PO Box 602, S Paarl 7624
TEL: 021 863-1976
FAX: 021 863-1534
wine@avondalewine.co.za
www.avondalewine.co.za
OWNER: The Avondale Trust
TRUSTEES: John Grieve (managing), VR Grieve, EH Grieve, B Robertson, RE Robertson, JM Welham
C'MASTER: Dewaldt Heyns
V'CULTURIST: Jonathan Grieve
SALES: Mon–Fri 09:00–17:00, Sat 09:00–13:00
TASTINGS: no fee
TOURS: by appointment
EST 1693 (producing wine since 1999). Area under vine: 75ha (70% red). Annual production: 250 tons, 187 000 litres
FLAGSHIP: Avondale Cab

LISTED WINES

Avondale

D WHITE: Chardonnay (W); Chenin (U); Eclipse (U, Chenin & S Blanc & Chardonnay); S Blanc (U)
RED: Cab (W); Julia (W, Cab & Ruby Cab &Cab Franc & Merlot); Merlot Le Pleurs (W);

Reserve (W, Cab & Cab Franc & Merlot & Shiraz); Shiraz (W, Shiraz & Cab)
FORTIFIED: Muscat Blanc (U, Muscat de F); Muscat Rouge (W, Muscat de F)

Ploughman

D WHITE: Chenin (U, Chenin S Blanc)
RED: Cab Shiraz (U)

HIGHLY RATED

Avondale Sauvignon Blanc
1999 – W3½ 00
Avondale Cabernet Sauvignon
1999 – W3½ 02
Avondale Dry Red
1999 – VD 00
Avondale Julia
2000 – W3½ 02
1999 – W3½ 01
Avondale Shiraz
2000 – W3½ 02
Ploughman Chenin Blanc
2001 – W3½ 02

AVONDVREDE

PAARL MAP

P CELLAR: Valley Rd (R44), Klapmuts
PO Box 152, Klapmuts 7625
TEL: 083 658-0595
FAX: 021 875-5609
OWNERS: John & Christine Enthoven
C'MASTER: JFP Enthoven
VISITS: by appointment only
EST 1996 (wine produced since 1999). Area under vine: 8,6ha (100% red). Annual production: 5 tons, 2 700 litres
FLAGSHIP: Koningshof Noble Red
SINGLE VYD WINES: Koningshof Cab, Cuvée Marie Christine

LISTED WINES

Koningshof

RED: Cab (W)

Marie Christine

RED: Merlot (W)

AVONTUUR

HELDERBERG MAP

ESTATE: R44 Rd between Stellenbosch & Somerset West
PO Box 1128, Somerset West 7129
TEL: 021 855-3450
FAX: 021 855-4600
avonwine@mweb.co.za
www.avontuurestate.co.za
OWNER: Tony Taberer
GM: Pippa Mickleburgh
C'MASTER: Willie Malherbe
SALES: Mon–Fri 08:30–17:00, Sat 09:00–13:00
TASTINGS: fee of R10
TOURS: by appointment
RESTAURANT: lunch Mon–Fri (Sat by appointment for groups)
EST 1982 (producing wine since 1987). Area under vine: 50ha (60% red). Annual production: 370 tons, 240 000 litres

FLAGSHIPS: Above Royalty, Luna De Miel, Baccarat

LISTED WINES

MCC SPARKLING: Avon Brut (U, Pinot Chardonnay)
D WHITE: Chardonnay (W); Chenin (U); Chenin Barrel Select (W); Luna de Miel (W, Chardonnay); S Blanc (U)
OD RED: Ruby Bush (U, Ruby Cab)
D RED: Baccarat (W, Cab & Merlot & Cab Franc); Cab (W); Cab Franc (W); Frantage (W, Cab Franc & Pinotage); Merlot (W); Pinot (W); Pinotage (W); Avon Rouge (W, Cab & Merlot)
DESSERT: Above Royalty (W, Riesling, NLH)
BRANDY: Estate 5 Year Old (W, Chenin)

HIGHLY RATED

Avontuur Brut
NV – W4 00, W3½ 98
Avontuur Barrel Select Chenin Blanc
2001 – TT 02, W5 02, DT 01
Avontuur Le Chardon
1995 – PB 98
Avontuur Luna de Miel Chardonnay
2000 – W4 02
Avontuur Reserve Chardonnay
1999 – JP 01, SA Trophy 01, W4 00
Avontuur Sauvignon Blanc
1999 – W3½ 00
Avontuur Baccarat
1999 – W4 02
Avontuur Cabernet Franc
2000 – W4 01
1996 – W4 98
Avontuur Cabernet Sauvignon
1999 – W4 01
Avontuur Pinotage
2000 – W3½ 02
1999 – MG 01, W4½ 01, SA Trophy 01, PT 00
1997 – W4 98
Avontuur Pinot Noir
2000 – W4 02
Avontuur Above Royalty
2000 – W4 02, JP 02
1999 – JP 01, W3½ 00

AXE HILL

LITTLE KAROO MAP

P CELLAR: Wes-Oewer, Calitzdorp
PO Box 43942, Scarborough 7975
TEL: 021 278-1051 (sales); 082 490-4248 (c'master); 044 213-3585 (cellar)
FAX: 021 780-1178
tony@axehill.co.za
OWNER: Axe Hill Trust
TRUSTEES: Tony & Lyn Mossop
C'MASTER: Tony Mossop
VISITS: by appointment only
EST 1994 (producing wine since 1997). Under vine: 1,3ha (100% red). Annual production: 618 tons, 3 500 litres

LISTED WINES

Axe Hill

PORT: Cape Vintage (W, Touriga N & Tinta B)

Masters (made by Boplaas, Die Krans, Axe Hill)

PORT: Cape Vintage Res (W, Touriga N & Tinta B & Souzão)

HIGHLY RATED

Axe Hill Cape Vintage Port
2000 – TT 02, W5 02
1999 – JP 02, W4 01
1998 – JP 01
1997 – JP 00, W4 99

Masters Cape Vintage Reserve
1998 – W4 00
1997 – W4 99

AZANIA ▷ AFRICAN TERROIR
AZURE ▷ L'ÉMIGRÉ

BABBLING BROOK ▷ MAKRO
BABYLONSTOREN ▷ BACKSBERG
BACCARAT ▷ AVONTUUR
BACCHANTE ▷ BRANDVLEI
BACCHUS ▷ VEENWOUDEN

BACKSBERG

PAARL MAP

ESTATE: Simondium Rd, off R44 (Franschhoek turn-off) between N1 (Exit 47) and Stellenbosch
PO Box 537, S Paarl 7624
TEL: 021 875-5141
FAX: 021 875-5144
info@backsberg.co.za
www.backsberg.co.za
OWNER: Michael Back
MD: Anthony John Vann
C'MASTER: Alicia Rechner
V'CULTURIST: Clive Trent
SALES: Mon–Fri 08:30–17:00, Sat 08:30–14:00, Sun 11:00–14:00
TASTINGS: fee of R10
TOURS: self-guided, no appointment necessary, no charge
RESTAURANT: Tables at Backsberg, Tues–Fri 10:00–15:00, Sat–Sun 11:00–15:00, Tel 021 875-5952
EST 1916 (producing wine since early 1920s). Area under vine: 130ha (75% red). Annual production: 1 000 tons, 700 000 litres
FLAGSHIP: Klein Babylonstoren
ORGANIC WINE: Chardonnay Organic

LISTED WINES

MCC SPARKLING: Brut (U, Chardonnay & Pinot)
D WHITE: Chardonnay (W); Chardonnay Kosher U); Chardonnay Organic (U); Chardonnay Unwooded; Chenin (U); John Martin (W, S Blanc & Sémillon); S Blanc (U); Sémillon (U)
OD WHITE: R Riesling (U)
S WHITE: Special Late Harvest (U, Chenin)
OD PINK: Rosé (U, Pinotage)

RED: Cab (W); Dry Red (U, various cultivars); Klein Babylonstoren (W, Cab & Merlot & Petit Verdot); Malbec (W); Merlot (W); Pinotage (W); Shiraz (W)
FORTIFIED: Hanepoot (U)
BRANDY: Sydney Back

HIGHLY RATED

Backsberg Merlot 1994 – W4 97

BADSBERG

WORCESTER MAP

CO-OP: Grootvlakte, Rawsonville – 800m from Goudini Spa Resort
PO Box 72, Rawsonville 6845
TEL: 023 349-3021
FAX: 023 349-3023
badsberg@lando.co.za
DIRECTORS: FR Botha (MD), PP van der Merwe, DL Brink, T Griessel, CW Botha, JW Pieterse
GM: Willie Burger
W'MAKERS: Willie Burger, Hugo Conrad
V'CULTURIST: Jurie du Plessis
SALES: Mon–Fri 08:00–17:00, Sat 10:00–13:00
TASTINGS: no fee
TOURS: by appointment
PICNICS: welcome to bring
EST 1957. Area under vine: 1 000ha (75% white). Annual production: 14 500 tons, 10,9-m litres
FLAGSHIP: Chardonnay Sur Lie
SINGLE VYD WINE: Chardonnay Sur Lie

LISTED WINES

S SPARKLING: Vin Doux (U, Hanepoot)
D WHITE: Chardonnay Sur Lie (W); Chenin (U); S Blanc (U)
S SWEET: Rosé (U, Merlot); Special Late Harvest (U, Chenin & Hanepoot)
RED: Merlot (W); Pinotage (W)
FORTIFIED: Hanepoot (U); Port (U, Shiraz & Ruby Cab); Red Jerepigo (U, Pinotage)

BAIN'S WAY ▷ WAMAKERSVALLEI
BAKENSKOP ▷ JONKHEER
BALEINE NOIR ▷ WHALEHAVEN
BALM OF THE NIGHT ▷ TJG
BALTHAZAR ▷ ROODEZANDT

B & E NÉGOTIANTS'

NÉGOCIANT: 13 Swartdam Rd, Hermanus – visits strictly by appointment
POSTAL: 293 9th Street, Hermanus 7200
TEL: 082 920-7108 & 072 374-0246 (sales); 028 313-2047 (tastings)
FAX: 028 312-4029
gerhardbritz@yahoo.com
OWNERS: Bartho Eksteen, Gerhard Britz

W'MAKER: Bartho Eksteen
TASTINGS: at Wine & Company, 7 High St, Hermanus
EST 2002

LISTED WINES

Urbane
D WHITE: Chardonnay (U); S Blanc (U)
RED: Shiraz (U)

BANGSHOOT ▷ FLAGSTONE
BAREFOOT ▷ BWC
BARON ▷ RUPERT & ROTHSCHILD
BARONESS ▷ RUPERT & ROTHSCHILD
BARONNE ▷ NEDERBURG
BAROUCHE ▷ BLAAUWKLIPPEN

BARRYDALE

LITTLE KAROO MAP

CO-OP: 1 Van Riebeeck St (R62), Barrydale
PO Box 59, Barrydale 6570
TEL: 028 572-1012
FAX: 028 572-1541
barrywine@dorea.co.za
www.barrydalewines.co.za
GM: Riaan Marais
C'MASTER: Riaan Marais
V'CULTURIST: Willem Botha
SALES: Mon–Fri 08:00–17:00, Sat 09:00–13:00
TASTINGS: no fee
TOURS: by appointment, at R17
EST 1941 (producing wine since 1976). Area under vine: 201ha (80% white). Annual production: 2 500 tons, 1,8m litres
FLAGSHIPS: Tradouw
SINGLE VYD WINES: Tradouw

LISTED WINES

Misty Point
OD WHITE: White (U)
RED: Pinotage (U); Red (W)

Seven Falls
D WHITE: Chardonnay
S Blanc (U)
N SWEET: Blanc de Noir (U, Muscadel)
RED: Shiraz (W)

Tradeaux
RED: Reserve (W, Cab)

Tradouw
D WHITE: Chardonnay (W); S Blanc (U)
RED: Merlot (W)

BARTHO EKSTEEN

WALKER BAY MAP

P PRODUCER: Wine & Company, 7 High St, Hermanus
Suite 47, Private Bag X15, Hermanus 7200
TEL: 028 313-2047
FAX: 028 312-4029
winenco@telkomsa.co.za
OWNERS: Bartho Eksteen, Ailsa Butler
W'MAKER: Bartho Eksteen
V'CULTURISTS: Jaori Lötter, Clive Marais
SALES: Mon–Fri 9:00–17:00 (Sat 09:00–17:00 in season)
TASTINGS: no fee

FUTURES (EN PRIMEUR): yes
RESTAURANT: Cheese & Wine Garden
PICNICS: available to purchase
EST 1998. Annual production: 30 tons

LISTED WINES

D WHITE: S Blanc (U)
RED: Shiraz (W)

HIGHLY RATED

Bartho Eksteen Sauvignon Blanc
2001 – W4 01
2000 – W3½ 00
Bartho Eksteen Shiraz
2000 – VD 01, W4 01

BATELEUR ▷ DE WETSHOF
BAYETE ▷ SPAR
BAY VIEW ▷ LONGRIDGE
BEACON ▷ CHARLES BACK
BEAUKETT ▷ ROBERTSON

BEAUMONT

WALKER BAY MAP

P CELLAR: Compagnes Drift Farm, Main Rd, Bot River
PO Box 3, Bot River 7185
TEL: 028 284-9450 (cellar); 028 284-9194 (sales); 028 284-9370 (manor house)
FAX: 028 284-9733
beauwine@netactive.co.za
www.beaumont.co.za
OWNERS: Raoul & Jayne Beaumont
C'MASTER: Niels Verburg
V'CULTURIST: Sebastian Beaumont
SALES: Mon–Fri 08:30–12:30 & 13:30–16:30, Sat am by appointment
TASTINGS: fee of R20pp for large groups
TOURS: by appointment
N MONUMENT: Water Mill (1750)
ACCOM: Mill House and Peppertree Cottages (self-catering)
EST 1972 (producing wine since 1993). Area under vine: 46ha (70% red). Annual production: 200 tons, 140 000 litres
SINGLE VYD WINES: Hope Marguerite, Mourvèdre, S Blanc

LISTED WINES

Beaumont

D WHITE: Chardonnay (W); Chenin (W); Hope Marguerite (W, Chenin); S Blanc (U)
RED: Ariane (W, Cab & Merlot & Cab Franc); Mourvèdre (W); Pinotage (W); Shiraz (W)
DESSERT: Goutte d Or (W, Sémillon & S Blanc, NLH)
PORT: Cape Vintage (W, Pinotage & Tinta B)

Raoul's Old Basket Press

RED: Rustic Red (U, Tinta B)

HIGHLY RATED

Beaumont Ariane

2000 – W3½ 01
Beaumont Shiraz
1999 – W3½ 00
Beaumont Goutte d'Or
1997 – W4 98

BELLE ROSE ▷ CABRIÈRE

BELLEVUE

STELLENBOSCH MAP
ESTATE: Bottelary Rd, Stellenbosch
PO Box 33, Koelenhof 7605
TEL: 021 882-2055/4
FAX: 021 882-2899 (office); 021 865-2779 (cellar)
bellevue1@global.co.za
OWNER: Houdamond Trust
TRUSTEE: DC Morkel (GM)
W'MAKERS: DC Morkel, W Kritzinger
SALES: Mon–Fri 10:00–16:00, Sat 10:00–15:00
TASTINGS: no fee
TOURS: by appointment
N MONUMENT: Cape Dutch Manor House (1803)
PICNICS: welcome to bring
EST 1852 (bottling wine since 1999). Area under vine: 190ha (75% red). Annual production: 1 180 tons, 500 000 litres
FLAGSHIPS: Tumara, PK Morkel Pinotage

LISTED WINES

D WHITE: S Blanc (U)
RED: Atticus (W, Pinotage & Cab); Malbec (W); Pinotage (W); PK Morkel Pinotage (W); Shiraz (W); Tumara (W, Cab & Cab Franc & Merlot & Malbec)

HIGHLY RATED

Bellevue Malbec
2000 – W4 01
Bellevue Tumara
2000 – W4 01
1999 – W4½ 01

BELLINGHAM

CELLAR in Wellington – not open to the public
PO Box 79, Groot Drakenstein 7680
TEL: 021 874-1011 (office); 021 873-1011 (production)
FAX: 021 551-1744
bellingham@dgb.co.za
www.bellingham.co.za
OWNER: DGB (Pty) Ltd
MD: Tim Hutchinson
W'MAKER: Graham Weerts
SALES: at retail stores only
EST 1947. Annual production: 3m litres
FLAGSHIPS: Spitz wines

LISTED WINES

D WHITE: Chardonnay (W); Chenin (W); Premier Grand Cru (U); Sauvenay (U, S Blanc & Chardonnay); S Blanc (U); S Blanc Chardonnay (U); Viognier (W)
SS WHITE: Johannisberger (U)

PINK: Rosé (U, Cinsaut & Pinotage)
RED: Cab (W); Merlot (W); Pinotage (W); Shiraz (W); Spitz Cab Franc (W); Spitz Pinotage (W)

HIGHLY RATED

Bellingham Chardonnay
2000 – W3½ 01
Bellingham Merlot
1995 – VD 97
Bellingham Shiraz
2000 – W3½ 01
Bellingham 'Spitz' Cabernet Franc
1998 – W4 02
1996 – W4 99
1995 – VD 97
Bellingham 'Spitz' Cabernet Sauvignon
1997 – W3½ 99, VD 98
Bellingham 'Spitz' Pinotage
1998 – JC 00
1997 – W4 99, PT 98, VD 98
1996 – VD 97, PT 97
Bellingham 'Spitz' Shiraz
1996 – W3½ 99

BELLTOWER ▷ ASARA
BERGHEIM ▷ JA CLIFT

BERGKELDER

STELLENBOSCH MAP

WINERY: cnr Adam Tas & Plankenburg Rd, Stellenbosch
PO Box 184, Stellenbosch 7599
TEL: 021 809-8492 (tastings, cellar); 021 809-8582 (sales)
FAX: 021 809-8492 (visitor's centre); 021 887-9081 (sales)
tours@bergkelder.co.za
www.bergkelder.co.za
OWNER: Distell (Pty) Ltd
MD: Jan Scanell
GM: Callie van Niekerk
C'MASTER: Linley Shultz
V'CULTURIST: Bennie Liebenberg
SALES: Mon–Fri 08:00–17:00, Sat 09:30–13:00
TASTINGS: fee of R12
TOURS: Mon–Fri 10:00, 10:30, 15:00; Sat 10:00, 10:30, 12:00 – Nov–Jan 7 tours a day on the hour 10:00–16:00
TOUR FEE: R12
FUTURES (EN PRIMEUR): yes, at the Vinoteque
RESTAURANT: Bistro, Mon–Sat 12:00–14:00 (vintner's platters in summer & soups and breads in winter); Tel 021 809-4892
EST 1968. Annual production: 15 000 tons, 10m litres
FLAGSHIPS: Fleur du Cap Unfiltered Merlot, Fleur du Cap Unfiltered Sémillon
SINGLE VYD WINES: Fleur du Cap Unfiltered Cab, Fleur du Cap Unfiltered Merlot

LISTED WINES

Fleur du Cap

D WHITE: Chardonnay (W); Chenin (W); S Blanc (U);

Unfiltered Chardonnay (W);
Unfiltered S Blanc (U);
Unfiltered Sémillon (W)
OD WHITE: Natural Light (U, Chenin & White Muscadel); Riesling (U)
RED: Cab (W); Merlot (W, Merlot & Cab); Pinotage (W); Shiraz (W)
DESSERT: Noble Late Harvest (W, Chenin)

Grünberger
SS SPARKLING: Spritzenlese (U, S Blanc & Chenin); Spritziger (U, S Blanc & Riesling & Sémillon)
SS WHITE: Freudenlese (U, S Blanc & Gewürz)
OD WHITE: Stein (U, S Blanc & Gewürz & Riesling)
SS RED: Rosenlese (U, S Blanc & Ruby Cab)

Kupferberger
SS WHITE: Auslese (U, Chenin & S Blanc)

Stellenryck
D WHITE: Chardonnay (W); S Blanc (U)
RED: Cab (W)

Two Oceans
D WHITE: S Blanc (U)
RED: Cab Merlot (W)

HIGHLY RATED

Fleur du Cap Chenin Blanc
2000 – JC 01
Fleur du Cap Unfiltered Chardonnay
2001 – W4 02
Fleur du Cap Unfiltered Sauvignon Blanc
2001 – JC 01
Fleur du Cap Unfiltered Sémillon
2001 – W4 02
Fleur du Cap Cabernet Sauvignon
1998 – W3½ 01, JC 02
1997 – W4 00
Fleur du Cap Reserve Cabernet Sauvignon
1996 – W3½ 99
Fleur du Cap Unfiltered Cabernet Sauvignon
1997 – W4 00
Fleur du Cap Merlot
2000 – W4 02
1999 – JC 01
1997 – W3½ 99, VD 98
1995 – VD 97
Fleur du Cap Unfiltered Merlot
1997 – W4½ 01
Fleur du Cap Noble Late Harvest
1996 – W4½ 02
Stellenryck Chardonnay
1999 – W4 00, JC 00
Stellenryck Sauvignon Blanc
2000 – JC 00
Stellenryck Cabernet Sauvignon
1998 – W3½ 01
1992 – VD 97

BERGSIG

WORCESTER MAP

ESTATE: R43, off the N1, towards Ceres (25km from

N1), Breede River
PO Box 15, Breede River 6858
TEL: 023 355-1603
FAX: 023 355-1658
wine@bergsig.co.za
www.bergsig.co.za
OWNERS: Lategan family
DIRECTORS: De Wet Lategan (MD), Prop Lategan, Plum Lategan, Louis Lategan
GM: Deidré Labuschagne
C'MASTER: De Wet Lategan
SALES: Mon–Fri 08:00–17:00, Sat 09:00–13:00
TASTINGS: no fee
TOURS: by appointment
EST 1843 (producing wine since 1938). Area under vine: 250ha (70% white). Annual production: 3 300 tons, 2m litres
FLAGSHIPS: Chardonnay, Chenin, Port

LISTED WINES

Bergsig

D SPARKLING: Brut (U, S Blanc)
D WHITE: Chardonnay (W); Chenin (U); S Blanc (U)
OD WHITE: Gewürz (U)
S WHITE: Chenin Blanc SLH (U)
PINK: Pinotage Blanc de Noir (U)
RED: Cab (W); Pinotage (W); Ruby Cab Merlot (W)
DESSERT: Noble Late Harvest (U, Chenin & S Blanc & Gewürz); Sweet Hanepoot (U, Morio Muscat)
PORT: Cape Ruby (W, Tinta B); Cape Vintage (W, Tinta B); LBV (W, Tinta B)

Painted Lady

OD SPARKLING: Demi Sekt (U, Pinotage)

HIGHLY RATED

Bergsig Chardonnay
2000 – W3½ 01
Bergsig Pinotage
2000 – W3½ 01
1999 – W4 00
Bergsig Cape LBV
1997 – W4½ PS 02, DT 00

BERNHEIM

PAARL MAP

P CELLAR: off R45, N Paarl
PO Box 7274, N Paarl 7623
TEL: 021 872-5618 (sales); 021 872-8358
FAX: 021 872-5618
bernheim@iafrica.com
OWNER: Schwulst Family Trust
DIRECTORS: Gisela Kolar (MD), Brigitta E Schwulst, Bernice Schwulst
C'MASTER: Gisela Kolar
V'CULTURIST: George Schwulst
SALES: Mon–Fri 09:00–17:00, Sat 09:00–13:00
TASTINGS: no fee
TOURS: by appointment
RESTAURANT: Pot Luck Restaurant to be established in 2003

PICNICS: welcome to bring
EST 1998. Area under vine: 10ha (60% red). Annual production: 80 tons, 100 000 litres
FLAGSHIPS: Cab, Pinotage

LISTED WINES

D WHITE: Reminisce (U, Chenin); Saint Omer S Blanc (U)
SS WHITE: Morgenthau (U, Chenin)
SS PINK: Pino Lite (U, Pinotage, Blanc de Noir); Serendipity (W, Chenin & Pinotage)
RED: Abendsglück (W, Pinotage); Moulin Rouge (W, Cab); Phillips Red (W, Pinotage); Pinaz (W, Shiraz)

BERRIO ▷ FLAGSTONE
BERTRAMS ▷ AFRICAN WINES
BERYL ▷ CHARLES BACK

BEYERSKLOOF

STELLENBOSCH MAP
P CELLAR: R304, Koelenhof, Stellenbosch
PO Box 107, Koelenhof 7605
TEL: 021 865-2135
FAX: 021 865-2683
wine@beyerskloof.co.za
www.beyerskloof.com
OWNERS: Beyers Truter (MD), Simon Halliday, Johann Krige, Paul Krige
C'MASTER: Beyers Truter
SALES: Mon–Fri 09:00–16:30, Sat 10:00–14:00
TASTINGS: no fee
EST 1989. Area under vine: 155ha (100% red). Annual production: 700 tons, 500 000 litres
FLAGSHIP: Beyerskloof (Cab Merlot)

LISTED WINES

Beyerskloof
RED: Beyerskloof (W, Cab & Merlot); Beyerskloof Pinotage (W)

Bouwland
See separate listing

HIGHLY RATED

Beyerskloof
1999 – W4 02
1998 – W4 01, PB 00, VD 00
1997 – W4 00, PB Trophy 99
1996 – W4½ 99, PB 98, VD 98
Beyerskloof Pinotage
1996 – PT 97

BIANCA ▷ WOOLWORTHS
BIANCO ▷ DE HEUVEL
BIESJES CRAAL ▷ DURBANVILLE HILLS

BILTON

STELLENBOSCH MAP
P CELLAR: Annandale Rd, Lynedoch
PO Box 60, Lynedoch 7603
TEL: 021 881-3714 (cellar, tastings, sales); 021 881-3759
FAX: 021 881-3721
blyhoek@mweb.co.za

www.bilton.co.za
OWNER: Mark Bilton
W'MAKER: Adrean Naudé (GM)
SALES: Mon–Fri 08:30–16:30, Sat 10:00–16:00
TASTINGS: fee of R3
TOURS: by appointment
PICNICS: welcome to bring
EST 1987 (producing wine since 1998). Area under vine: 80ha (75% red). Annual production: 50 tons, 31 500 litres
FLAGSHIP: Merlot
SINGLE VYD WINES: Merlot

LISTED WINES

RED: Cab (W); Merlot (W); Shiraz (W)

HIGHLY RATED

Bilton Cabernet Sauvignon 2000 – W4 02
Bilton Merlot 2000 – W4½ 02

BIRDFIELD ▷ KLAWER
BIZWENI ▷ HAVANA HILLS
BK5 ▷ FLAGSTONE

BLAAUWKLIPPEN

STELLENBOSCH MAP

P CELLAR: Strand Rd (R44), Stellenbosch
PO Box 54, Stellenbosch 7599
TEL: 021 880-0133/4/5
FAX: 021 880-0136
mail@blaauwklippen.com
www.blaauwklippen.com
OWNER: Markt Landhandel Gmbh
DIRECTORS: Hörst Brünner (MD), Hennie Pretorius
GM: Frank Meaker
C'MASTER: Paul Engelbrecht
V'CULTURIST: Kowie Kotze
SALES: Mon–Sun 08:30–17:00
TASTINGS: fee of R15
TOURS: Mon–Sun 08:30–17:00, no charge
FUTURES (EN PRIMEUR): yes
N MONUMENT: 300-yr-old 'werf', H-shaped Cape homestead (1789), Jonkershuis with 'holbol' gable, museum, antique coach collection
RESTAURANT: Mon–Sat 09:00–17:00, Sun 09:00–16:00
PICNICS: available for purchase
EST 1973 (producing wine since 1976). Area under vine: 90ha (85% red). Annual production: 500 tons, 315 000 litres
FLAGSHIPS: Vineyard Selection (VS) wines, prev labelled Barrel Selection
SINGLE VYD WINES: VS Shiraz, VS Cab

LISTED WINES

D WHITE: S Blanc (U); Sociable Dry White (U, S Blanc & Chenin); White Landau (W, Chardonnay & Sémillon)
RED: Barouche (W, Cab & Shiraz & Pinotage); Cab (W);

Pinotage (W); Red Landau (W, Cab Franc & Cab & Merlot & Pinotage); Shiraz (W); Sociable Dry Red (W, Cinsaut & Zin & Pinotage & Cab Franc); VS Cab (W); VS Cabriolet (W, Cab Franc & Cab & Merlot); VS Merlot (W); VS Shiraz (W); VS Zin (W)
PORT: Cape Vintage (W, Pontac & Zin)

HIGHLY RATED

Blaauwklippen Barouche
2000 – VD 01
Blaauwklippen Barrel Selection Cabernet Sauvignon
1999 – W3½ 02
Blaauwklippen Barrel Selection Merlot
2000 – W3½ 02
Blaauwklippen Pinotage
1998 – W4 99
Blaauwklippen Reserve Zinfandel
1998 – W3½ 00

BLANC DE MER ▷ BOUCHARD FINLAYSON
BLACK GRANITE ▷ DARLING CELLARS
BLANQUETTE ▷ NEDERBURG

BLOEMENDAL

DURBANVILLE MAP

P CELLAR: Tygerbergvalley Rd (M13), Durbanville
PO Box 466, Durbanville 7551
TEL/FAX: 021 976-2682
bloemendal@isoft.co.za
OWNER: Jackie Coetzee
C'MASTER: Jackie Coetzee
SALES: Mon–Fri 19:00–17:00, Sat 09:00–16:00, Sun 11:00–15:00
TASTINGS: no fee
FUTURES (EN PRIMEUR): yes
RESTAURANT: Bloemendal Restaurant – Mon–Sat buffet breakfast, lunch & dinner, Sun lunch only, Tel 021 975-7575; Die Wijnhuis – cheese platters Mon–Fri 09:00–17:00, Sat 09:00–16:00, Sun 11:00–15:00, Tel 021 976-2682
EST 1920 (producing wine since 1987). Area under vine: 150ha (80% red). Annual production: 80 tons
FLAGSHIPS: S Blanc, Merlot

LISTED WINES

D WHITE: S Blanc S Terras (U); Sémillon (W)
RED: Cab (W); Merlot (W); Shiraz (W)

HIGHLY RATED

Bloemendal Brut
NV – W3½ 00
Bloemendal Sauvignon Blanc
1998 – W3½ 98
1997 – W4 97
Bloemendal Sémillon
2002 – W3½ 02
2001 – W4 02
Bloemendal Merlot
1996 – VD 97

BLOUBERG HILL ▷ CAPAIA

BLOUPUNT

LITTLE KAROO MAP

P CELLAR: 12 Long St, Montagu 6720
TEL/FAX: 023 614-2385
bloupunt@lando.co.za
www.bloupuntwines.co.za
OWNER: Josias Martin Philip Hoffman
C'MASTER: Paul Hoffman
SALES: Mon–Fri 09:00–17:00, Sat 09:00–16:00
TASTINGS: for groups, at a fee
TOURS: by appointment
FUTURES (EN PRIMEUR): yes
N MONUMENT: residence (1862)
ACCOM: self-catering cottage
EST 1992 (producing wine since 1997). Area under vine: 10ha (100% white)
FLAGSHIP: Chardonnay
SINGLE VYD WINES: all

LISTED WINES

D WHITE: Unwooded Chardonnay; Wooded Chardonnay
RED: Merlot (W)

BLUEBERRY HILL

P PRODUCER: Middagkrans Rd, Franschhoek
PO Box 580, Franschhoek 7690
TEL: 021 876-3362
FAX: 021 876-2114
blwine@iafrica.com
OWNERS: Blueberry Hill Trust (Brian & Lindy Heyman)
W'MAKER: Nigel McNaught (Stony Brook)
V'CULTURIST: Paul Wallace
SALES: at retail stores only – wines made at Stony Brook
ACCOM: self-catering cottages
EST 1998. Area under vine: 4ha (100% red). Annual production: 4 tons, 3 500 litres
FLAGSHIP: S Blanc

LISTED WINES

D WHITE: S Blanc (U)
RED: Merlot (W)

BLUEBIRD ▷ DEVON HILL

BLUE CREEK

STELLENBOSCH MAP

P CELLAR: Blaauwklippen Rd, Stellenbosch
PO Box 3247, Matieland 7602
TEL: 021 887-6938 (cellar); 021 880-0522 (manor house); Cell 082 730-0084
FAX: 021 886-5462
blue_creek@email.com
OWNERS: Piet & Rabie Smal
C'MASTER: Piet Smal
VISITS: by appointment only
EST 1960 (producing wine since 1996).
Area under vine: 7ha (100% red).
Annual production: 30 tons, 22 500 litres
FLAGSHIP: Cabernet

LISTED WINES
RED: Cab (W); Pinotage (W)

HIGHLY RATED
Blue Creek Cabernet Sauvignon
1999 – W3½ 01
1998 – W4 01, PB 99
Blue Creek Pinotage
2000 – W3½ 01

BLUE MOUNTAIN ▷ BREEDE VALLEY
BLUE RIDGE ▷ VILLIERA
BLUE WHITE ▷ OLD VINES
BLYDE ▷ JA CLIFT

BODEGA

PAARL MAP

ESTATE: Waarburgh Rd, Joostenbergvlakte, Kraaifontein
PO Box 590, Kraaifontein 7569
TEL: 021 988-2929
FAX: 021 988-3527
austinwine@iafrica.com
OWNER: Austin Wine Farm (Pty) Ltd
DIRECTORS: Jeremy Squier (UK)
GM: Julianne Barlow
C'MASTER: Mark Carmichael-Green
V'CULTURIST: Paul Wallace
VISITS: by appointment only
EST 1974 (producing wine since 1990). Area under vine: 14ha (100% red). Annual production: 40 tons, 13 500 litres

LISTED WINES
RED: Cab (W); Merlot (W); Pinotage (W)

BOEKENHOUTSKLOOF

FRANSCHHOEK MAP

P CELLAR: Bohoek Rd, Franschhoek
PO Box 433, Franschhoek 7690
TEL: 021 876-3320
FAX: 021 876-3793
boeken@mweb.co.za
OWNERS: Boekenhoutskloof Investments (Pty)
DIRECTORS: T Rands, H Enderle, R Lascaris, J Hunt, A Teeger, M Denny, M Kent
GM: Marc Kent
C'MASTER: Marc Kent
V'CULTURIST: Pieter Siebrits
SALES: Mon–Fri 08:00–17:00
TASTINGS: no fee
TOURS: by appointment
EST 1994 (producing wine since 1996). Area under vine: 20ha (70% red). Annual production: 900 tons
FLAGSHIPS: Boekenhoutskloof wines

LISTED WINES

Boekenhoutskloof
D WHITE: Sémillon (W)
RED: Cab (W); Syrah (W)

Porcupine Ridge
D WHITE: S Blanc (U)
RED: Cab (W); Merlot (W); Syrah (W)

HIGHLY RATED
Boekenhoutskloof Cabernet Sauvignon
1999 – W4 02, JP 02
1998 – JP 01, W4 00
1997 – VD 99
Boekenhoutskloof Syrah
1999 – W4 02
1997 – W4 00, JP 00, PB 98, DT 98
Porcupine Ridge Cabernet Sauvignon
2001 – W3½ 02
Porcupine Ridge Syrah
2000 – W3½ 01

BOHLÔKWA ▷ SPAR

BOLAND

PAARL MAP

COMPANY: Drommedaris Rd, Daljosafat, N Paarl
PO Box 7007, N Paarl 7623
TEL: 021 862-6190
FAX: 021 862-5379
adris@bolandwines.co.za
www.bolandwines.co.za
DIRECTORS: Altus le Roux (MD), Jannie le Roux, Frans le Roux, Gawie Rossouw, George Eksteen, Chris du Toit, Pieter de Waal, Ernst Bester, Jacques Malan
C'MASTER: Altus le Roux
V'CULTURIST: Jurie Germishuys
SALES: Mon–Fri 08:00–17:00, Sat 09:00–13:00
TASTINGS: no fee
TOURS: by appointment
PICNICS: available for purchase Dec–Jan
EST 1947. Area under vine: 2 200ha (45% red). Annual production: 17 500 tons, 13m litres
FLAGSHIPS: Cab, Shiraz, Chardonnay, S Blanc

LISTED WINES
D SPARKLING: Brut (U, S Blanc)
S SPARKLING: Vin Doux (U, S Blanc)
D WHITE: Bon Vino (U, Chenin & Riesling & S Blanc); Chardonnay (W); Chenin (U); Riesling (U); S Blanc (U)
SS WHITE: Bon Vino (U, Chenin & Hanepoot)
RED: Bon Vino (U, Shiraz & Cab & Merlot); Cab (W); Merlot (W); Pinotage (W); Pinotage Cinsaut (W); Shiraz (W)

HIGHLY RATED
Boland Bon Vino (white)
NV – VD 01
Boland Chenin Blanc
2001 – VD 01
Boland Sauvignon Blanc
2000 – MG 00
Boland Cabernet Sauvignon
1999 – MG 01, VD 00
Boland Pinotage
1999 – W4 02, PB 00
1997 – W4 99
Boland Shiraz
2000 – W4 01

1998 – MG 01, W3½ 01
Boland Century
2000 – W3½ 02
1998 – W4 01
Boland Muskadel (white)
1997 – VD 00
1995 – W3½ 02
Boland Rooi Muscadel
1997 – W4 02, VD 00
1996 – VD 97
Boland Port
1997 – W4 02

BON CAP ORGANIC

ROBERTSON MAP

P CELLAR: Eilandia Ward Robertson, R60, Breede River
PO Box 356, 6705
TEL: 023 626-1629
FAX: 023 626-1895
boncap@intekom.co.za
DIRECTORS: Roelf du Preez (MD), Michelle du Preez
C'MASTER: Roelf du Preez
V'CULTURIST: Paul Wallace (consulting)
SALES: Mon–Sun 08:00–18:00
TASTINGS: no fee
TOURS: by appointment
ACCOM: Weltevrede Guest Farm, weltevrede.guestfarm@intekom.co.za
PICNICS: hampers for purchase by arrangement; welcome to bring
ORGANIC WINES: all
EST 1920s (producing wine since 2002). Area under vine: 60ha (90% red). Annual production: 80 tons, 60 000 litres

LISTED WINES

RED: Cab (W); Pinotage (W); Syrah (W, Shiraz)

BON COURAGE

ROBERTSON MAP

ESTATE: Bonnievale Rd (R317), 8km out of Robertson
PO Box 589, Robertson 6705
TEL: 023 626-4178
FAX: 023 626-3581
boncourage@minds.co.za
OWNER: André Bruwer
C'MASTER: Jacques Bruwer
V'CULTURIST: André Bruwer
SALES: Mon–Fri 08:00–17:00, Sat 09:30–15:00
TASTINGS: no fee
TOURS: by appointment
RESTAURANT: Café Maude, Mon–Fri 09:00–17:00, Sat 09:30–16:00
EST 1923 (producing wine since 1983). Area under vine: 170ha (60% white)
FLAGSHIPS: Inkará range
SINGLE VYD WINES: Inkará range

LISTED WINES

MCC SPARKLING: Jacques Bruere Brut (W, Pinot & Chardonnay)
OD SPARKLING: Chardonnay

Vin Sec (U)
S SPARKLING: Blush Vin Doux (U, Pinot & Muscat de F)
D WHITE: Chardonnay (W); Colombard Chardonnay (U); Prestige Cuvée Chardonnay (W); Prestige Cuvée Chenin (W); Riesling (U); S Blanc (U)
OD WHITE: Bouquet Blanc (U, Colombard & Chenin); Colombard (U); Gewürz (U)
S WHITE: Gewürz SLH (U)
SS PINK: Blanc de Noir (U, Shiraz & Muscat de F)
RED: Cab (W); Cab Shiraz (W); Pinotage (W); Rouge (U, Shiraz & Ruby Cab & Pinotage); Shiraz (W); Ltd Rel Syrah (W)
DESSERT: Noble Late Harvest (U, Riesling); Red Muscadel (U); W Riesling Natural Sweet (U); White Muscadel (U)

HIGHLY RATED

Bon Courage Jacques Bruère Brut Millennium
NV – W3½ 00
Bon Courage Reserve Jacques Bruère Brut
1998 – W4 01
Bon Courage Prestige Cuvée Chardonnay
2000 – W3½ 02
Bon Courage Reserve Sauvignon Blanc
1997 – VD 97
Bon Courage Inkara Shiraz
2001 – W3½ 02
Bon Courage Shiraz
2001 – W4 02
Bon Courage Gewürztraminer Special Late Harvest
2000 – VD 00
Bon Courage Noble Late Harvest
2000 – W4½ 01
Bon Courage Weisser Riesling Natural Sweet
2001 – W3½ 02, VD 01
Bon Courage Red Muscadel
1994 – W4 98
Bon Courage White Muscadel
1999 – W3½ 00

BONFOI

STELLENBOSCH MAP

ESTATE: Stellenboschkloof Rd, Stellenbosch
PO Box 9, Vlottenburg 7604
TEL: 021 881-3774
FAX: 021 881-3607
bonfoi@mweb.co.za
www.bonfoi.com
OWNER: Johannes van der Westhuizen
C'MASTER: J van der Westhuizen
SALES: Mon–Fri 09:00–17:00, Sat 10:00–12:30
TASTINGS: no fee
FUTURES (EN PRIMEUR): yes
EST 1934 (producing wine since 1974). Area under

vine: 98ha (62% red). Annual production: 600 tons (40 tons for own label), 30 000 litres (under own label)
FLAGSHIP: Cabernet

LISTED WINES

D WHITE: Chardonnay (W); S Blanc (U)
RED: Cab (W); Merlot (W); Overture (W, Cab & Merlot & Shiraz); Shiraz (W)

HIGHLY RATED

Bonfoi Cabernet Sauvignon
2000 – W3½ 02
1998 – W3½ 01

BONNIEVALE

ROBERTSON MAP

CO-OP: Kapteinsdrift, Bonnievale – on the R317 between Bonnievale & Stormsvlei
PO Box 206, Bonnievale 6730
TEL: 023 616-2795
FAX: 023 616-2332
winemaker@bonnievalecellar.co.za
www.bonnievalecellar.co.za
GM: Henk Wentzel
C'MASTER: Henk Wentzel
SALES: Mon–Fri 08:00–17:00, Sat 10:00–13:00
TASTINGS: no fee
TOURS: by appointment
PICNICS: welcome to bring
EST 1964. Area under vine: 800ha (65% white). Annual production: 10 000 tons, 6m litres
FLAGSHIP: Res Shiraz
SINGLE VYD WINES: Shiraz, Chardonnay, S Blanc

LISTED WINES

D WHITE: Kelkiewit (U, Colombar & Chardonnay); Res Chardonnay (W)
RED: Cab Merlot (W); CCC (U, Colombar & Chenin & Chardonnay); Kelkierooi (U, Pinotage & Cinsaut); Res Shiraz (W)
S FORTIFIED: Red Jerepigo (U)

BON VALLON ▷ DE WETSHOF
BON VINO ▷ BOLAND

BOPLAAS

LITTLE KAROO MAP

P CELLAR: Saayman St, Calitzdorp
PO Box 156, Calitzdorp 6660
TEL: 044 213-3326
FAX: 044 213-3750
boplaas@mweb.co.za
www.boplaas.co.za
OWNER: Carel Nel
C'MASTER: Carel Nel
SALES: Mon–Fri 08:00–17:00, Sat 09:00–15:00
TASTINGS: no fee
TOURS: by appointment
EST 1880 (producing wine since 1982). Area under vine: 90ha (50% red). Annual production: 700 tons,

270 000 litres
FLAGSHIP: Vintage Res Port

LISTED WINES

Boplaas
D SPARKLING: Pinot N Brut (U)
S SPARKLING: Sweet Sparkling (U, Colombar & Muscadel)
D WHITE: Chardonnay (U); Classic Dry White (U, S Blanc); S Blanc (U)
SS WHITE: Golden Harvest (U, Colombar & Muscadel); Late Harvest (U, Colombar & Muscadel)
SWEETS: Hanepoot (U); Muscadel (U); Red Dessert (U, Tinta B & Muscadel)
OD PINK: Blanc de Noir (U, Pinotage & Touriga N)
RED: Cab (W); Dry Red (U, Pinotage & Touriga N); Merlot (W); Pinotage (W); Res Cab (W); Res Merlot (W); Res Pinotage (W); Res Shiraz (W); Touriga N (W)
PORT: Ruby (W, Tinta B & Touriga N); Tawny (W); Vintage (W, Tinta B & Touriga N); Vintage Res (W, Touriga N); White (W, Colombar)
BRANDY: Carel Nel (Colombar); Potstill (Colombar)

Masters (made by Boplaas, Die Krans, Axe Hill)
PORT: Cape Vintage Res (W, Touriga N & Tinta B & Souzão)

HIGHLY RATED

Boplaas Cape Tawny
W3½ 02, W3½ 01, DT 00, W3½ 00, W3½ 98
Boplaas Cape Vintage Reserve
1999 – TG 02, W5 02
1997 – VD 00
1996 – W4½ 98
Boplaas Touriga Naçional Cape Vintage Reserve
1995 – W4 98
Bredell & Nel Cape Vintage Reserve
1997 – W4 01, SA Trophy 01
Masters Cape Vintage Reserve
1998 – W4 00
1997 – W4 99

BOREALIS ▷ TJG

BOSCHENDAL

FRANSCHHOEK MAP

P CELLAR: R310, 11/2 km from R45 T-junction, Groot Drakenstein
PO Groot Drakenstein 7680
TEL: 021 870-4210 (tastings); 021 870-4226/7 (sales); 021 870-4225 (cellar); 021 870-4281 (manor house)
FAX: 021 874-1864
boschendal@amfarms.co.za
www.boschendal.com
OWNERS: Anglo American
MD: Don Tooth
C'MASTER: JC Bekker

V'CULTURIST: Spekkies van Breda
SALES: Mon–Sat 08:30–16:30 (Sun 09:00–12:30 from Nov to Apr)
TASTINGS: fee of R6
TOURS: vineyard and cellar tours by appointment for a fee of R6
N MONUMENT: manor house (1812)
RESTAURANTS: buffet lunch at Boschendal Restaurant (guests to be seated by 13:00); light lunches and teas at Le Café, Mon–Sun 10:00–17:00
PICNICS: available for purchase at Le Pique Nique from Nov to Apr, Mon–Sun 12:15–13:30
EST 1685 (producing wine since 1715, initially under De Villiers family, and under Rhodes Fruit Farms since 1902). Area under vine: 320ha (45% red). Annual production: 3 300 tons, 1,92m litres
FLAGSHIPS: Grande Reserve, Res Chardonnay
SINGLE VYD WINE: S Blanc

LISTED WINES

MCC SPARKLING: Brut (U, Chardonnay & Pinot); Le Grand Pavillon (U, Chardonnay & Pinot)
D WHITE: Blanc de Blanc (U, S Blanc & Chardonnay & Sémillon); Chardonnay (W); Grand Vin Blanc (U, S Blanc & Chardonnay); Pinot Noir Chardonnay (W); Res Chardonnay (W); Riesling (U); S Blanc (U)
OD WHITE: Chenin (U); Le Bouquet (U, Muscat & Riesling & others); Le Pavillon Blanc (U, Chardonnay & Chenin)
D PINK: Blanc de Noir (U, Pinot & Merlot)
RED: Cab (W); Grande Reserve (W, Cab & Cab Franc & Merlot); Lanoy (W, Cab & Shiraz); Le Pavillon Rouge (W, Merlot & Cab); Merlot (W); Shiraz (W)

HIGHLY RATED

Boschendal Brut
1995 – W4 00
Boschendal Chenin Blanc
2000 – W4 01
Boschendal Grand Vin Blanc
2001 – W3½ 01
Boschendal Jean le Long Sémillon
2000 – W4 02
1999 – W4 01
Boschendal Pinot Noir Chardonnay
2001 – W3½ 02
Boschendal Reserve Chardonnay
2001 – W3½ 02
2000 – W3½ 01
1999 – W3½ 01

Boschendal Sauvignon Blanc
2001 – MG 01
Boschendal Grand Reserve
1997 – W3½ 00
Boschendal Lanoy
1998 – W3½ 00
1997 – DT 99
Boschendal Merlot
1997 – W4 00
1996 – W4 99
1995 – W4 98
Boschendal Shiraz
1998 – JC 00
1997 – W4½ 00, VD 99
Boschendal Vin d'Or
1998 – W3½ 01

BOSCHKLOOF

STELLENBOSCH MAP

P CELLAR: M12 (Polkadraai Rd)
PO Box 1340, Stellenbosch
7599
TEL: 021 881-3293
FAX: 021 881-3032
boschkloof@adept.co.za
OWNER: Dr Reenen Furter
DIRECTORS: Dr R Furter (MD), J Borman
C'MASTER: Jacques Borman
VISITS: by appointment only
EST 1995. Area under vine: 19ha (95% red). Annual production: 200 tons
FLAGSHIP: Reserve

LISTED WINES

D WHITE: Chardonnay (W)
RED: Cab (W); Cab Merlot (W); Merlot (W); Reserve (W, Cab & Merlot); Syrah (W, Shiraz)

HIGHLY RATED

Boschkloof Reserve Cabernet Sauvignon Merlot
1998 – VD 01
Boschkloof Syrah
2000 – W3½ 02

BOTELLA ▷ SWARTLAND

BOTHA

WORCESTER MAP

CO-OP: on R43 Rd between Worcester & Ceres
PO Box 30, Botha 6857
TEL: 023 355-1740
FAX: 023 355-1615
W'MAKERS: Dassie Smith, Johan Linde, Michiel Visser, Pierre Hugo
SALES: Mon–Fri 07:30–17:30, Sat 09:00–13:30
TASTINGS: no fee
TOURS: by appointment
FLAGSHIPS: Res range

LISTED WINES

SPARKLING: Chardonnay Brut (U); Chardonnay Demi-Sec (U)
D WHITE: Blanc de Blanc (U, Chenin & Colombard & Crouchen); Chardonnay Res (U)
OD WHITE: Chenin Res (U); Colombard Res (U)
SS WHITE: Special Late Harvest (U Chenin)
RED: Cab (U); Cab Res (W);

Dassies Rood (U, Cinsaut & Ruby Cab & Cab); Merlot (U); Merlot Res (W); Pinotage (U); Pinotage Res (W); Shiraz (U); Shiraz Res (W)
DESSERT: Hanepoot Jerepigo (U); Red Jerepigo (U, Pinotage)
PORT: Ruby (W)

BOTMASKOP ▷ DELAIRE

BOUCHARD FINLAYSON

WALKER BAY MAP

P CELLAR: Hemel-en-Aarde Valley (R320), outside Hermanus
PO Box 303, Hermanus 7200
TEL: 028 312-3515
FAX: 028 312-2317
info@bouchardfinlayson.co.za
www.bouchardfinlayson.co.za
OWNER: Bouchard Finlayson (Pty) Ltd
DIRECTORS: Peter Finlayson & Michael Clark (joint MDs), Victoria O'Hara Tollman, Brett Tollman, Michael Tollman
C'MASTER: Peter Finlayson
SALES: Mon–Fri 19:30–17:00, Sat 19:30–12:30
TASTINGS: fee for groups (R15pp)
TOURS: by appointment (R15pp)
EST 1990. Area under vine: 16ha (50% red). Annual production: 180 tons, 108 000 litres (80% white)
FLAGSHIPS: Galpin Peak Pinot, Missionvale Chardonnay

LISTED WINES

D WHITE: Blanc de Mer (U, Kerner & Riesling & Gewürz & Pinot Blanc & S Blanc); Kaaimansgat Chardonnay (W); Missionvale Chardonnay (W); Sans Barrique Chardonnay; S Blanc (U)
RED: Galpin Peak Pinot (W); Galpin Peak Tête de Cuvée Pinot (W)

HIGHLY RATED

Kaaimansgat Chardonnay
1998 – W3½ 99
Bouchard Finlayson Missionvale Chardonnay
1997 – W3½ 98
Bouchard Finlayson Galpin Peak Pinot Noir
2001 – JC 02
2000 – JC 01, W3½ 02
1999 – JC 00
1997 – W3½ 99, JP 99

BOUWLAND

STELLENBOSCH MAP

P PRODUCER: Stellenbosch farm, wines produced at/sold from Beyerskloof/Kanonkop
PO Box 107, Koelenhof 7605
TEL: 021 865-2135
FAX: 021 865-2683
wine@beyerskloof.co.za
OWNERS: Johann Krige, Paul Krige, Beyers Truter, Simon

Halliday
W'MAKERS: Beyers Truter, Abrie Beeslaar
V'CULTURIST: Koos du Toit
SALES: at Beyerskloof
EST 1997. Area under vine: 97ha (100% red). Annual production: 400 tons, 280 000 litres

LISTED WINES

RED: Merlot Cab (W)

HIGHLY RATED

Bouwland Reserve Pinotage 1998 – W4½ 00, SA Certificate 00

BOVLEI

WELLINGTON MAP

CO-OP: Bainskloof Rd, Wellington
PO Box 82, Wellington 7654
TEL: 021 873-1567 / 864-1283
FAX: 021 864-1483 / 873-1386
lizl@bovlei.co.za
www.bovlei.co.za
DIRECTORS: Pieter Cillie (MD), Deon Joubert, Danie Joubert, Neil Ennis, Koos Malan
C'MASTER: Johan Joubert
V'CULTURIST: Dawie le Roux
SALES: Mon–Fri 18:30–12:30 & 13:30–17:00, Sat 18:30–12:30
TASTINGS: no fee
TOURS: by appointment
EST 1907. Area under vine: 1 200ha (64% white). Annual production: 7 600 tons, 5,4m litres
FLAGSHIP: Cabernet

LISTED WINES

D SPARKLING: Brut (U, S Blanc & Sémillon); Sec (U, S Blanc)
OD SPARKLING: Demi Sec (U, S Blanc)
D WHITE: Chardonnay (U); Chenin (U); Dry White (U, Chenin); Riesling (U); S Blanc (U); Vin Blanc (U, Chenin)
SS WHITE: Bukettraube (U); Gewürz (U); Natural Sweet (U, Chenin); Special Late Harvest (U, Chenin); Stein (U, Chenin)
RED: Cab (W); Dry Red (U, Cinsaut & Ruby Cab); Grand Rouge (W, Cab Franc & Merlot); Merlot (U); Pinotage (U); Shiraz (U); Vin Rouge (U, Cinsaut, Ruby Cab)
DESSERT: Muscat d A (U)
PORT: NV (U, Grenache & Ruby Cab)

BOWE JOUBERT

STELLENBOSCH MAP

P CELLAR: M12, Polkadraai Rd, Stellenbosch
PO Box 1114, Stellenbosch 7599
TEL: 021 881-3103
FAX: 021 881-3377
info@bowejoubert.com
www.bowejoubert.com
OWNERS: Alphonso H Bowe (Nassau Bahamas),

Andre P Hilliard (Green Bay Wisconsin), Jannie JF Joubert (Stellenbosch)
MD: Jannie Joubert (MD)
GM: Fanie Joubert
C'MASTER: Jannie Joubert
V'CULTURIST: Lukas Joubert
VISITS: by appointment only
ACCOM: La Provence Guest Cottages, Tel 021 881-3716, Fax 021 881-3859
EST 1800s. Area under vine: 75ha (50% red). Annual production: 500 tons, 350 000 litres

LISTED WINES

D WHITE: Chardonnay (W); Chenin (W); Mosaic (U, S Blanc & Chenin & Colombard); S Blanc (U)
PINK: Rosé (U, Cinsaut & Chenin & Cab)
RED: Cab (W); Cab Merlot (W); Merlot (W)

BRADGATE ▷ JORDAN

BRAHMS

PAARL MAP
P CELLAR: Domaine Brahms, Windmeul
PO Box 2136, Windmeul 7630
TEL: 021 869-8555
FAX: 021 869-8555
brahms@iafrica.com
OWNERS: Judge Brahm & Adv Gesie Lategan
C'MASTER: Gesie Lategan
VISITS: by appointment only
FUTURES (EN PRIMEUR): yes
ACCOM: guest cottage
EST 1989 (producing wine since 1999). Area under vine: 25ha (98% red). Annual production: 100 tons, 650 000 litres
FLAGSHIP: Shiraz
SINGLE VYD WINE: Shiraz

LISTED WINES

Brahms
RED: Shiraz (W)
Judex
D WHITE: Chenin (W)
RED: Cab Merlot (W); Pinotage Merlot Ruby Cab (W)

HIGHLY RATED

Brahms Shiraz
2000 – W3½ 02
1999 – W3½ 02

BRAMPTON ▷ RUSTENBERG

BRANDVLEI

WORCESTER MAP
CO-OP: R43 between Worcester & Villiersdorp
PO Box 595, Worcester 6849
TEL: 023 340-4215
FAX: 023 340-4332
brandvlei@cybertrade.co.za
MANAGER: Jean le Roux
W'MAKERS: J le Roux, Jandré Human
V'CULTURIST: Pierre Snyman
SALES: Mon–Thurs 07:30–12:30

& 13:30–17:30, Fri 07:30–12:30 & 13:30–16:30
TASTINGS: no fee
TOURS: by appointment
PICNICS: welcome to bring
EST 1956. Area under vine: 1 400ha (60% white). Annual production: 21 000 tons, 15m litres

LISTED WINES

SPARKLING: Sparkling (U, Chardonnay)
D WHITE: Chardonnay (U); Chenin (U); S Blanc (U)
SS WHITE: Bacchanté (U, Chenin & Hanepoot)
RED: Ruby Cab Merlot (W)
DESSERT: Hanepoot Jerepigo (U, Hanepoot)

BREDELL'S ▷ JP BREDELL

BREEDE VALLEY

NÉGOCIANT: Breede Valley Wines – Kleinplasie, off Robertson Rd, Worcester
PO Box 59, Worcester 6849
TEL: 023 342-8710
FAX: 023 342-2294
manager@worcester-winelands.co.za
www.worcesterwinelands.co.za
OWNERS: various Breede Valley wine producers
GM: Jeanetta Marais
SALES: Mon–Fri 09:00–16:45, Sat 09:00–16:00
TASTINGS: fee of R5
N MONUMENT: Kleinplasie Open Air Farm Museum, Tel 023 342-2225
ACCOM: self-catering chalets, Tel 023 347-0091
RESTAURANT: Kleinplasie, Mon–Sat 08:00–15:30, Sun 08:00–15:00

LISTED WINES

Blue Mountain

D WHITE: Chardonnay (U); Chenin (U); S Blanc (U)
PERLÉ: Fronté (U, Muscat de F)
RED: Cab (W); Pinotage (W); Ruby Cab Merlot (W)

HIGHLY RATED

Blue Mountain Ruby Cabernet Merlot 2001 – W3½ 02

BRENTHURST

PAARL MAP

P CELLAR: off Main Rd, Paarl South
PO Box 6091, Paarl 7622
TEL: 021 863-1375
FAX: 021 424-5666
OWNER: Adv SA Jordaan
C'MASTER: Adv SA Jordaan
V'CULTURIST: Johan Wiese (consultant)
VISITS: by appointment
EST 1991 (producing wine since 1994).
Area under vine: 5ha (red). Annual production: 70 tons, 50 000 litres

LISTED WINES
RED: Cab Merlot (W, Cab & Merlot & Cab Franc & P Verdot)

BROKEN STONE ▷ SLALEY
BRULPADDA ▷ PADDAGANG

BUITENVERWACHTING

CONSTANTIA MAP
P CELLAR: Klein Constantia Rd, Constantia
PO Box 281, Constantia 7848
TEL: 021 794-5190
FAX: 021 794-1351
buiten@pixie.co.za
www.buitenverwachting.com
OWNER: S Mueller (German)
DIRECTORS: Lars Maack (MD), Meagan Maack
C'MASTER: Hermann Kirschbaum
V'CULTURIST: Peter Reynolds
SALES: Mon–Fri 09:00–17:00, Sat 09:00–13:00
TASTINGS: no fee
TOURS: by appointment
N MONUMENTS: manor house, wine cellar, slave cottage (1796)
RESTAURANT: Tues–Fri lunch & dinner, Sat dinner only, Tel 021 794-3522
PICNICS: available for purchase
EST 1985. Area under vine: 100ha (70% white). Annual production: 1 000 tons, 700 000 litres
FLAGSHIPS: Christine, S Blanc

LISTED WINES
MCC SPARKLING: Brut (U, Pinot & Chardonnay)
D WHITE: Buiten Blanc (U, S Blanc & Riesling & Chenin); Chardonnay (W); S Blanc (U)
OD WHITE: R Riesling (U)
RED: Cab (W); Christine (W, Cab & Cab Franc & Merlot & Malbec); Meifort (W, Merlot & Cab & Cab Franc); Merlot (W)

HIGHLY RATED
Buitenverwachting Brut
NV – W3½ 01
Buitenverwachting Chardonnay
2001 – W4 02
1997 – W4 98
Buitenverwachting Sauvignon Blanc
1997 – W3½ 98
1996 – W3½ 98
1995 – W3½ 98
Buitenverwachting Christine
1999 – W4 02
Buitenverwachting Merlot
1999 – W3½ 02, VD 01
Buitenverwachting Noblesse
1995 – W4 98

BUSHMAN'S CREEK ▷ TOMORROW

BWC

BAREFOOT WINE COMPANY
NÉGOCIANT: hired cellar space at Glenwood: Robertvlei Rd, Franschhoek

PO Box 13404, Mowbray 7705
TEL: 021 886-2044 (cellar); 021 887-9910 (office); 083 258-1307 (w'maker)
FAX: 021 887-9910
cathy@barefoot.co.za
www.barefoot.co.za
OWNERS: Catherine Marshall, Greg Mitchell, Peter Oxenham, Jeff Jolly
W'MAKER: Catherine Marshall
VISITS: by appointment only
FLAGSHIPS: Shiraz, Pinot
EST 1996 (year first wine made). Annual production: 20 tons, 15 000 litres

LISTED WINES

RED: BWC Pinot (W); BWC Shiraz (W); Farrago (W, Pinot & Cab)
FORTIFIED: Myriad (W, Pinot)

HIGHLY RATED

BWC Shiraz
2001 – W4 02
Barefoot Shiraz
1999 – W3½ 00

CAAPMANS ▷ DURBANVILLE HILLS
CABARET ▷ WEST CORP
CABOCHON ▷ MEERENDAL

CABRIÈRE

FRANSCHHOEK MAP

ESTATE: Clos Cabrière – Berg St, Franschhoek
P CELLAR: Haute Cabrière – Franschhoek Pass
PO Box 245, Franschhoek 7690
TEL: 021 876-2630
FAX: 021 876-3390
cabriere@iafrica.com
www.cabriere.co.za
OWNER: Clos Cabrière Ltd
DIRECTORS: Achim von Arnim (MD), Graham de Villiers, Alan Ramsay, Pierre Viljoen
C'MASTER: Achim von Arnim
V'CULTURIST: Sakkie Lourens
SALES: Mon–Fri 09:30–16:30, Sat 11:00–14:00
TASTINGS: fee of R20pp
TOURS: Sat 11:00, fee of R20pp incl tasting
RESTAURANT: Haute Cabrière, lunch daily, dinner seasonal, Tel 021 876-3688, hautecab@iafrica.com
PICNICS: welcome to bring
EST 1694 (producing wine since 1984). Area under vine: 27ha (60% white). Annual production: 350 tons, 230 000 litres
FLAGSHIPS: Pierre Jourdan

LISTED WINES

Arnim

RED: Sauvignon Rouge Fine de Jourdan
BRANDY: pot-still (W, Chardonnay)

Haute Cabrière

D WHITE: Chardonnay Pinot (U)
RED: Pinot (W)

Pierre Jourdan

MCC SPARKLING: Blanc de Blancs (W, Chardonnay); Brut (U, Chardonnay & Pinot); Brut Sauvage (U, Chardonnay & Pinot); Belle Rose (U, Pinot)
D WHITE: Tranquille (U, Chardonnay & Pinot)
FORTIFIED: Ratafia (U, Chardonnay & pot-still brandy)

HIGHLY RATED

Pierre Jourdan Cuvée Belle Rosé NV – W3½ 01

CABRIOLET ▷ BLAAUWKLIPPEN

CALITZDORP

LITTLE KAROO MAP

CO-OP: Andries Pretorius St, Calitzdorp
PO Box 193, Calitzdorp 6660
TEL: 044 213-3301
FAX: 044 213-3301
DIRECTORS: Koos Geyser (MD), Joos Nel, Paul Theron, Okkie Scholtz, Dirk Brink, Tienie Becker
C'MASTER: Alwyn Burger (GM)
V'CULTURIST: Willem Botha (consultant)
SALES: Mon–Fri 08:00–13:00 & 14:00–17:00, Sat 08:00–12:00
TASTINGS: no fee
TOURS: by appointment
PICNICS: welcome to bring
EST 1928. Area under vine: 300ha (90% white). Annual production: 2 500 tons, 2,4m litres
FLAGSHIPS: Merlot, Ruby Port, Blanc de Noir

LISTED WINES

D WHITE: Chenin (U)
SS PINK: Blanc de Noir (U, Tinta B)
DESSERT: Golden Jerepigo (U, Hanepoot & Muscadel); Natural Sweet (U, Colombar); White Muscadel (U)
RED: Merlot (W); Pinotage (W)
PORT: LBV (W, Tinta B & Touriga N); Ruby (W, Tinta B & Touriga N); Vintage (W, Tinta B & Touriga N)

CALLIGRAPHY ▷ NITIDA

CAMBERLEY

STELLENBOSCH MAP

P CELLAR: Off R310 between Stellenbosch & Franschhoek, Banhoek, Stellenbosch
PO Box 6120, Uniedal 7612
TEL/FAX: 021 885-1176
camberleywines@hotmail.com
OWNERS: Gaël & John Nel
C'MASTER: John Nel
V'CULTURIST: Aiden Morton
SALES: Mon–Sat 09:00–17:00, Sun by appointment
TASTINGS: no fee
TOURS: by appointment
FUTURES (EN PRIMEUR): yes
ACCOM: Camberley Cottage

PICNICS: welcome to bring
EST 1996. Area under vine: 2,5ha (80% red). Annual production: 14 tons, 7 500 *l*
FLAGSHIP: Cabernet Merlot

LISTED WINES

RED: Cab (W); Cab Merlot (W); Merlot (W); Pinotage (W); Shiraz (W)

HIGHLY RATED

Camberley Cabernet Sauvignon
1997 – W3½ 99
1996 – W4 98

CAMELOT ▷ TULBAGH

CAMERADI

P PRODUCER: 48 Bain St, Wellington 7655
TEL: 021 873-1225; Cell 082 323-4244
FAX: 021 873-1225/4910
lsw@iafrica.co.za
OWNER: Steluest CC
DIRECTORS: Niel Smith, Nic Swingler, Hendrik du Preez, Pieter Laubscher, Casper Lategan
GM: Casper Lategan
C'MASTER: Casper Lategan
VISITS: not open to the public
EST 1999. Annual production: 4 tons, 2 600 litres

LISTED WINES

RED: Cab (W); Shiraz (W)

CAMERCA ▷ SLANGHOEK

CAPAIA

DURBANVILLE MAP

P CELLAR: Botterberg Rd, off N7, Philadelphia, West Coast
PO Box 25, Philadelphia 7304
TEL/FAX/E-MAIL: t.b.a.
OWNERS: A & I Baron von Essen, Tel 0949 8022 70-5290, Fax 0949 8022 705-2921
W'MAKERS: Tibor Gal, Rolf Zeitvogel (Tel 082 412-5304)
V'CULTURIST: Werner Niewoudt
VISITS: by appointment only
EST 2002/3 (first wines to be made in 2003). Area under vine: 58ha (90% red)
FLAGSHIP: Capaia – to be released after Sept 2004
SINGLE VYD WINES: Blouberg Hill Cab Merlot, Blouberg Hill S Blanc

LISTED WINES

Blouberg Hill

D WHITE: S Blanc
RED: Cab Merlot

CAPE BAY

WALKER BAY MAP

P CELLAR: R320, Hemel-en-Aarde Valley, outside Hermanus
PO Box 225, Hermanus 7200
TEL: 028 312-3862
FAX: 028 312-3867
capebay@netactive.co.za
OWNERS: Johnson family

DIRECTORS: David Johnson (MD), Felicity Johnson
C'MASTER: Gordon Newton Johnson
V'CULTURIST: Johan Montgomery
SALES: Mon–Fri 08:30–16:30
TASTINGS: no fee
EST 1997. Area under vine: 3ha (60% white). Annual production: 110 tons
FLAGSHIPS: Newton Johnson Pinot, Newton Johnson Chardonnay

LISTED WINES

Cape Bay
D WHITE: Chardonnay (U); Chenin (U); S Blanc (U)
OD WHITE: Bouquet Blanc (U, Gewürz & Clairette Blanche & Colombar)
RED: Cab Merlot (U); Mellow Red (U, Ruby Cab & Cinsaut); Pinotage (U)

Newton Johnson
D WHITE: Chardonnay (W); S Blanc (U)
ROSÉ: Felicité (U, Pinot)
RED: Cab (W); Pinot (W); Pinotage (W)

Sandown Bay
D WHITE: Chardonnay (U)
RED: Cab (U); Pinotage (U)

HIGHLY RATED

Newton Johnson Chardonnay
1999 – W3½ 02
Newton Johnson Pinotage
2000 – W3½ 02
1998 – W4 00
Pour Mes Amis Pinot Noir
2000 – W3½ 01

CAPE CLASSICS

NÉGOCIANT: not open to the public – wines made at other cellars
PO Box 5421, Cape Town 8000 (based at Kanu, Goedgeloof Farm, Polkadraai Rd, Stellenbosch)
TEL: 021 881-3810
FAX: 021 881-3814
info@capeclassics.co.za
www.capeclassics.co.za
OWNER: The Natural Corporation
MD: Gary Shearer
GM: Herman du Preez
W'MAKERS: Barry Kok (production mgr), Mzokhona Mvemve
SALES: at retail stores only

LISTED WINES

Indaba
D WHITE: Chardonnay (W); Chenin (U); S Blanc
SS PINK: Blush (U, Grenache)
RED: Cab (W); Pinotage (W); Pinotage Cinsaut (U); Shiraz (W)

Kronendal
PORT: Cape Vintage (W)

CAPE INDABA ▷ NATURAL CORP
CAPELANDS ▷ LONGRIDGE

CAPELL'S COURT ▷ LINTON PARK
CAPENHEIMER ▷ DISTELL

CAPE POINT

CONSTANTIA MAP

P CELLAR: Cnr Glencairn Expressway & Kommetjie Rd, Fish Hoek (Noordhoek)
PO Box 37700, Valyland 7978
TEL: 021 785-7660
FAX: 021 785-7662
info@cape-point.com
OWNER: Sybrand van der Spuy
DIRECTORS: S van der Spuy (MD), Suzanne Morton (marketing)
C'MASTER: Emmanuel Bolliger until 2002 (replacement t.b.a)
V'CULTURIST: Chris van Niekerk
VISITS: by appointment only
EST 1997 (producing wine since 2000). Area under vine: 37ha (65% red). Annual production: 100 tons
FLAGSHIP: Cape Point Sauvignon Blanc

LISTED WINES

Cape Point

D WHITE: Chardonnay (W); S Blanc (U, S Blanc & Sémillon); Sémillon (W, Sémillon & S Blanc)
DESSERT: Noble Late Harvest (W, Sémillon)

Scarborough

D WHITE: Chardonnay (W); S Blanc (U)

HIGHLY RATED

Cape Point Chardonnay
2000 – W3½ 01
Cape Point Sauvignon Blanc
2001 – JC 02
2000 – W4 00
Cape Point Sémillon
2001 – W3½ 02
Cape Point Noble Late Harvest
2000 – W4 02, JC 02
Scarborough Sauvignon Blanc
2001 – PB 01
1999 – W3½ 00

CAPE ROCK ▷ WEST CORP
CAPE TABLE ▷ RIEBEEK
CAPE VIEW ▷ KAAPZICHT

CAPE VINEYARDS

NÉGOCIANT: Sneeukop Building, Goudyn Rd, Rawsonville
PO Box 106, Rawsonville 6845
TEL: 023 349-1585/1466
FAX: 023 349-1592
admin@cape-vineyards.com
OWNERS: five co-ops & two estates – Badsberg, Goudini, Merwida, Slanghoek, Waboomsrivier, Bergsig, Opstal
DIRECTORS: Schalk van der Merwe (MD), Pieter Hugo, Stanley Louw, Dave Richter, Ettiene du Toit, Manna Botha, Prop Lategan

GM: Henriëtte Jacobs
W'MAKING: see owners
VISITS: by appointment only
EST 1994
FLAGSHIPS: Andrew Bain wines

LISTED WINES

Andrew Bain
D WHITE: Chardonnay (W); Chenin (U)
RED: Cab Merlot (W)

Rawson's
D WHITE: Chardonnay (U); Revelry (U, Chenin & S Blanc & Riesling – USA only)
RED: Pinotage (W); Ruby Cab Merlot (U); Shiraz (W)

HIGHLY RATED
Rawson's Pinotage 2001 – W3½ 02

CAPE WINE CELLARS

WELLINGTON MAP

PRODUCER: wines made at cellars in Wellington & Paarl – see owners
PO Box 508, Wellington 7654
TEL: 021 864-8690
FAX: 021 873-6909
info@sadgroup.co.za
www.capewinecellars.co.za
www.wine.co.za
OWNERS: Bovlei Co-op, Wellington Co-op, Wamakersvallei Co-op, Boland Kelder
GM: Tinus Broodryk
C'MASTER: Jeff Wedgwood
SALES: at SAD Fruit Shop, 21 Main Rd, Wellington, Mon–Fri 08:30–17:00, Sat 09:00–12:30
EST 1995. Farms under vine: 300 (70% red). Annual production: 150 000 tons (domestic market)

LISTED WINES

Kleinbosch
D WHITE: Chardonnay (W); Chenin (U)
RED: Cab (W, Cab & Shiraz); Merlot (W); Pinotage (W); Rouge (W, Cinsaut & Cab); Shiraz (W)

CAPTAIN'S ▷ V JOHNSON'S
CAPTAINS' DRIFT ▷ NORDALE
CARDOUW ▷ CITRUSDAL
CAREL NEL ▷ BOPLAAS
CAREME ▷ ASHANTI
CARESSE ▷ SPAR
CARESSE MARINE ▷ WILDEKRANS

CARISBROOKE

STELLENBOSCH MAP

P CELLAR: Stellenboschkloof Rd, Vlottenburg
PO Box 25, Vlottenburg 7604
TEL: 021 881-3798; 021 881-3034 (manor house)
FAX: 021 881-3798
wjpret@mweb.co.za
OWNER: Willem Pretorius
C'MASTER: Kowie du Toit
V'CULTURISTS: Willem Pretorius, K du Toit
SALES: Mon–Fri 09:00–17:00,

Sat 09:00–14:00
TASTINGS: no fee
EST 1992 (producing wine since 1996). Area under vine: 6ha (100% red). Annual production: 40 tons, 13 500 litres
FLAGSHIP: Cabernet

LISTED WINES

RED: Cab (W)

CARL EVERSON ▷	OPSTAL
CARLONET ▷	UITKYK
CARLTON HILL ▷	TRAVINO
CARNIVAL ▷	SPAR
CARPE DIEM ▷	DIEMERSFONTEIN
CATHARINA ▷	STEENBERG
CATHEDRAL CELLAR ▷	KWV
CAVALIER ▷	ANNANDALE
CCC ▷	BONNIEVALE

CEDERBERG

OLIFANTS RIVER MAP

P CELLAR: Dwarsrivier, Cederberg – ±28km N of Citrusdal along N7, take Algeria turn-off and travel another 48km on gravel road
PO Box 84, Clanwilliam 8135
TEL: 027 482-2827; 027 482-2825 (manor house)
FAX: 027 482-1188
cederwyn@iafrica.com
www.cederbergcellars.com
OWNERS: Nieuwoudt family
DIRECTORS: Ernst Nieuwoudt (MD), David Niewoudt
C'MASTER: David Niewoudt
V'CULTURIST: Frikkie Schutte
SALES: Mon–Sat 08:00–12:30 & 14:00–17:00
TASTINGS: fee of R8pp
TOURS: by appointment
N MONUMENT: walks to Cederberg's Maltese Cross, Wolfberg Cracks and Arch
ACCOM: Sanddrif holiday resort (self-catering), Tel 027 482-2825, Fax 027 482-1188, sandrif@mweb.co.za
PICNICS: welcome to bring
EST 1973 (producing wine since 1977. Area under vine: 35ha (70% red). Annual production: 200 tons, 135 000 litres
FLAGSHIP: Cabernet, S Blanc, V Generations Cab
SINGLE VYD WINE: V Generations Cab

LISTED WINES

D WHITE: Chenin (U); Chenin Barrel Fermented; Chardonnay Barrel Fermented; S Blanc (U)
SS WHITE: Bukettraube (U)
RED: Cab (W); Cederberger (W, Ruby Cab & Merlot & Pinotage); V Generations Cab (W); Pinotage (W); Shiraz (W)

HIGHLY RATED

Cederberg Chenin Blanc 2000 – VD 00
Cederberg Barrel Fermented

Chenin Blanc
2001 – W3½ 02
1999 – W3½ 00
Cederberg Cabernet Sauvignon
2000 – W4 02
1998 – W3½ 00
Cederberg V Generations Cabernet Sauvignon
2000 – TG 02, W5 02
Cederberger
2000 – W4 02
1998 – W4 00
1997 – VD 98
Cederberg Pinotage
2000 – W4 02
1998 – W3½ 00

CELEBRATION ▷ ELEPHANT PASS
CELINE ▷ DOUGLAS GREEN
CELLARDOOR ▷ VILLIERA
CHAMELEON ▷ JORDAN

CHAMONIX

FRANSCHHOEK MAP

P CELLAR: Uitkyk St, Franschhoek
PO Box 28, Franschhoek 7690
TEL: 021 876-2494 (cellar, sales); 021 876-3241 (tastings)
FAX: 021 876-3237
marketing@chamonix.co.za
www.chamonix.co.za
OWNER: Cape Chamonix (Pty) Ltd
DIRECTORS: CRG Hellinger (MD/GM), MI Meer, I Vukic
C'MASTER: Gottfried Mocke
V'CULTURISTS: G Mocke, Rodney Kitching
SALES: Mon–Sun 09:30–17:00
TASTINGS: fee of R5pp
TOURS: by appointment (R15pp)
FUTURES (EN PRIMEUR): yes
N MONUMENT: Old Blacksmith Cottage
ACCOM: guest houses
RESTAURANT: La Maison de Chamonix, Tues–Sun 12:30–17:00, Tel 021 876-2393
EST 1988 (producing wine since 1992). Area under vine: 49,8ha (60% red). Annual production: 230 tons
FLAGSHIP: Chardonnay

LISTED WINES

MCC SPARKLING: Chardonnay (W)
D WHITE: Chardonnay (W); S Blanc (U); Res Chardonnay (W); Res S Blanc (W)
RED: Cab (W); Merlot (W); Pinot (W); Pinotage (W)
OTHER: Aperitifs, Schnapps, Grappa

HIGHLY RATED

Chamonix Courcheval Brut
1999 – W3½ 01
Chamonix Chardonnay
1998 – W4 02
Chamonix Reserve Chardonnay
1997 – W3½ 00, PB 99, PB 98

CHANSON ▷ JC LE ROUX
CHANT DE NUIT ▷ NUY

CHARLES BACK

P CELLARS: Fairview (Paarl), Spice Route (Swartland)
HEAD OFFICE: at Fairview, Suid-Agter Paarl Rd, S Paarl
PO Box 583, S Paarl 7642
TEL: 021 863-2450
FAX: 021 863-2591
fairwine@mweb.co.za
www.fairview.co.za
OWNER: Charles Back
C'MASTERS: Anthony de Jager (Fairview), Charl du Plessis (Spice Route)
V'CULTURISTS: Johan Botha (Fairview), Dawie Krynauw & Andrew Teubus (Spice Route)
SALES: at Fairview – Mon–Fri 08:30–17:00, Sat 08:30–13:00
TASTINGS: at Fairview – fee of R10pp
FLAGSHIPS: Fairview Akkerbos Chardonnay, Fairview Beacon Shiraz, Fairview Oom Pagel Sémillon; Fairview Primo Pinotage, Fairview Solitude Shiraz, Spice Route Flagship wines

LISTED WINES

Andrew's Hope (Spice Route)
RED: Cab Merlot (W, Cab & Merlot & Cab Franc)

Fairview
D WHITE: Akkerbos Chardonnay (W); Chardonnay (W); Oom Pagel Sémillon (W); Sémillon (W); Viognier (W)
RED: Amos Pinotage (W); Beacon Shiraz (W); Cab (W); Cyril Back Shiraz (W); Malbec (W); Pegleg Carignan (W); Pinotage (W); Primo Pinotage (W); SMV (W, Shiraz & Mourvèdre & Viognier); Solitude Shiraz (W)
DESSERT: La Beryl Blanc (W, Chenin, Straw Wine)

Goat Roti
RED: Goat Roti (W, Shiraz & Viognier & Grenache & Carignan)

Goats Do Roam
D WHITE: White (U, Crouchen & Clairette Blanche & Grenache Blanc & Muscat de F)
PINK: Rosé (U, Cinsaut & Grenache & Gamay Noir)
RED: Red (W, Shiraz & Cinsaut & Pinotage & Grenache & Carignan)

Spice Route
D WHITE: Chenin (U); Sémillon (W); Terroir Trial S Blanc (U)
RED: Cab Merlot (W, Cab & Merlot & Cab Franc); Flagship Merlot (W); Flagship Pinotage (W); Flagship Syrah (W); Pinotage (W)

HIGHLY RATED

Fairview Akkerbos Chardonnay 2000 – SA Certificate 02, W4 01

1999 – W3½ 00
Fairview Chardonnay
1998 – W3½ 99
Fairview Oom Pagel Sémillon
2001 – W4 02
2000 – TT 02, W5 02, JP 02
1999 – W3½ 01
Fairview Sémillon
1999 – W3½ 01
Fairview Viognier
2001 – W4 02, PB Trophy 01
2000 – PB 00, W3½ 00
1999 – W4 99
Fairview Carignan
2001 – W4 02
Fairview Cyril Back Shiraz
2000 – W4 02
Fairview Cyril Back Zinfandel
1997 – W3½ 00
Fairview Malbec
1999 – W3½ 99
Fairview Merlot
1999 – W3½ 02
Fairview Primo Pinotage
2000 – W4 02, JP 02, MG 01
Fairview Shiraz Mourvèdre
2000 – W3½ 01
1998 – W4 99
Fairview La Beryl Vin de Paille
2000 – W4½ 01
1999 – W4 00
Spice Route Chenin Blanc
1999 – W4 01
1998 – W3½ 00
Spice Route Flagship Merlot
2000 – W3½ 02
1999 – W4½ 01, JP 01
1998 – PB 99
Spice Route Flagship Pinotage
2000 – W4 02
1999 – W3½ 01
1998 – W4½00, JP 00
Spice Route Flagship Syrah
1999 – PB 01, W4 01, PB 00
1998 – W4 00
Spice Route Shiraz
1998 – W3½ 99
Spice Route Cabernet
Sauvignon Merlot
2000 – W3½ 01
1998 – W4 99

CHARLOTTE ▷ AMANI
CHATEAU LIBERTAS ▷ DISTELL
CHATELAINE ▷ ALTYDGEDACHT
CHATILLON ▷ HARTENBERG
CHEANDRIE ▷ LIEVLAND
CHEERFUL WHITE ▷ OLD VINES
CHENAY ▷ AGTERKLIPHOOGTE
CHÊNE ▷ KWV
CHIANTI ▷ CITRUSDAL
CHIWARA ▷ ASHANTI
CHRISTINE ▷ BUITENVERWACHTING
CHRISTO ▷ AMBELOUI
CILLIERS ▷ STELLENDRIFT
CILMOR ▷ AFRICAN TERROIR

CITRUSDAL

OLIFANTS RIVER MAP
CO-OP: Kelder St, Citrusdal
PO Box 41, Citrusdal 7340
TEL: 022 921-2233/5/6/7
FAX: 022 921-3937
citrusdalcellars@yebo.co.za
www.citrusdalcellars.co.za
GM: Christiaan De Witt

Lagrange
C'MASTER: Johan Delport
SALES: Mon–Fri 09:00–17:00, Sat 09:00–12:30
TASTINGS: no fee
TOURS: by appointment
PICNICS: welcome to bring
EST 1957. Area under vine: 1 200ha (75% white). Annual production: 6 500 tons, 5m litres
FLAGSHIP: Cardouw Res Cab

LISTED WINES

Cardouw
D WHITE: Chardonnay (W); Chenin (U)
RED: Cab (W); Res Cab (W); Pinotage (W)
Danckaert (export only)
D WHITE: Chardonnay (U); Chenin (U); S Blanc (U)
RED: Cab (W); Pinotage (W); Shiraz (W)

Goue Vallei
D SPARKLING: Brut (U, Chenin – export only); Pinot Gris Brut (U)
S SPARKLING: Vin Doux (U, White Muscadel)
D WHITE: Blanc de Blanc (U, Colombard); Chardonnay (W); Chenin (U); Classique Blanc (U, Chenin – export only); S Blanc (U)
OD PINK: Blanc de Noir (U, Tinta B & Cinsaut)
SS WHITE: Bukettraube (U); Late Vintage (U, Crouchen)
SS PINK: Classique Rosé (U, Pinotage – export only)
RED: Cab (W); Chianti (W, Cab & Cinsaut); Classic Rouge (U, Ruby Cab & Pinotage & Merlot & Tinta B); Pinotage (W); Shiraz (W)
FORTIFIED: Hanepoot Jerepiko (U); Red Jerepiko (U, Pinotage); White Muscadel (U)
PORT: Goue Vallei (W)
Impala (export only)
D WHITE: Chenin (U)
RED: Cab Merlot (U); Pinotage (W)

Ivory Creek
D WHITE: Chardonnay (U); Chenin (U); S Blanc (U); Sémillon (U)
RED: Cab (W); Pinotage (W); Royal Rouge (U, Ruby Cab & Pinotage & Merlot & Tinta B); Shiraz (W)

Masswine
Fortified for Roman Catholic Church: Masswine (U, Sherry & Colombar & White Muscadel)

Sonnigdal
OD WHITE: Grand Cru (U, Colombar)
S WHITE: Late Harvest (U, Chenin)
RED: Dry Red (U, Cinsaut)
Zaximus (export only)
D WHITE: Dry White (U, Colombar)
RED: Dry Red (U, Ruby Cab & Cinsaut)

HIGHLY RATED
Goue Vallei Hanepoot Jerepiko
NV – W3½ 02

CLAIRE ▷ CORDOBA
CLAIRE ▷ DOUGLAS GREEN
CLAIRVAUX ▷ APPELSDRIFT
CLARIDGE ▷ DE COMPAGNIE
CLEMENCE CREEK ▷ MOOIUITSIG

CLOOF

SWARTLAND MAP
P CELLAR: 4km past Darling Cellars, on gravel road
PO Box 125, Darling 7345
TEL: 022 492-2839
FAX: 022 492-3261
cloofwines@intekom.co.za
www.cloof.co.za
OWNER: Adv Johan van der Berg
C'MASTER: Christopher van Dieren
V'CULTURIST: Hardie van den Heever
SALES: Mon–Fri 10:00–16:00, Sat 10:00–14:00
TASTINGS: fee of R10 (refundable on purchase of wine)
FUTURES (EN PRIMEUR): yes
EST 1998. Area under vine: 166ha (69% red). Annual production: 550 tons, 350 000 litres
FLAGSHIPS: Cloof Cab, Cloof Pinotage

LISTED WINES

Cloof
D WHITE: Chenin (U)
RED: Bushvine (U, Pinotage & Cinsaut & Cab); Cab (W); Pinotage (W)

Groene Cloof
D WHITE: Chenin (U)
PINK: Rosé (U, Cinsaut)
RED: Pinotage (U)

HIGHLY RATED
Cloof Cabernet Sauvignon
2000 – W3½ 02
Cloof Pinotage
2000 – W3½ 02
1999 – W4½ PC 01

CLOS CABRIÈRE ▷ CABRIÈRE

CLOS DU CIEL

HELDERBERG MAP
P CELLAR: Off Eikendal Rd, Stellenbosch
PO Box 12830, Die Boord 7613
TEL: 021 855-2573
FAX: 021 887-6583
OWNER: Peter Aschke
C'MASTER: t.b.a.
VISITS: by appointment only
FLAGSHIP: Chardonnay
SINGLE VYD WINES: Chardonnay

LISTED WINES
D WHITE: Chardonnay (W)
RED: Pinot (W); Shiraz (W)

CLOS MALVERNE

STELLENBOSCH MAP
P CELLAR: Devon Valley Rd, Devon Valley, Stellenbosch
PO Box 187, Stellenbosch

7599
TEL: 021 865-2022
FAX: 021 865-2518
closma@mweb.co.za
www.closmalverne.co.za
OWNERS: Seymour & Sophia Pritchard
C'MASTER: IP Smit
SALES: Mon–Fri 10:00–16:30, Sat 10:00–13:00
TASTINGS: fee of R12
TOURS: by appointment
EST 1988. Area under vine: 25ha (90% red). Annual production: 370 tons, 259 000 litres
FLAGSHIP: Auret

LISTED WINES

D WHITE: S Blanc (U)
RED: Auret (W, Cab & Pinotage & Merlot); Cab (W); Cab Merlot (W); Cab Pinotage (W); Cab Shiraz (W); Pinotage (W); Res Pinotage (W); Shiraz (W)

HIGHLY RATED

Clos Malverne Auret
1999 – W4 02
1998 – JP 02
1997 – W4½ 00, DT 99, VD 99
Clos Malverne Cabernet Sauvignon
1999 – W3½ 01
Clos Malverne Cabernet Sauvignon Merlot
2001 – W3½ 02
Clos Malverne Cabernet Sauvignon Pinotage
1998 – W3½ 01
Clos Malverne Cabernet Sauvignon Shiraz
2000 – W4 02
1997 – W4 98
Clos Malverne Devonet
1997 – W4 00
Clos Malverne Pinotage
1998 – PT 99, W3½ 99
Clos Malverne Reserve Pinotage
2000 – W3½ 02
1999 – W3½ 01
1998 – JC 00, W3½ 00
1997 – W3½ 99
1996 – W3½ 98, PT 97

CLOVELLY

STELLENBOSCH MAP

P CELLAR: DevonVale, Stellenbosch – behind country club, off the M23 (Bottelary Rd)
Postnet Suite 215, Private Bag X5061, Stellenbosch 7599
TEL: 082 853-7190 (sales, cellar); 021 865-2511 (manor house)
FAX: 021 865-2511/2600
winesbyjf@hotmail.com
OWNER: York Partnership
DIRECTORS: Dean Toerien, Mineke Toerien Fourie, Jacques Fourie
C'MASTER: Jacques Fourie
VISITS: by appointment only
EST 1999. Area under vine: 3ha (70% red). Annual production: 50 tons, 27 500 litres

FLAGSHIPS: Cab, Chardonnay

LISTED WINES

D WHITE: Chardonnay (W)

RED: Cab (W); Triangle (W, Cinsaut & Cab & Pinotage)

HIGHLY RATED

Clovelly Cabernet Sauvignon
1999 – W3½ 01

COBBLERS HILL ▷ JORDAN

COLERAINE

PAARL MAP

P CELLAR: Pieter-Hugo St, S Paarl
PO Box 579, S Paarl 7624
TEL: 021 863-2073 (cellar); 021 863-3443 (sales)
FAX: 021 863-3443
info@coleraine.co.za
www.coleraine.co.za
OWNERS: Kerr family
DIRECTORS: Clive Bailey Kerr (MD), Hugh Keith Kerr
C'MASTER: Clive Bailey Kerr
VISITS: by appointment only
EST 1980 (producing wine since 1999). Area under vine: 30ha (100% red). Annual production: 100 tons, 90 000 litres
FLAGSHIP: Culraithin

LISTED WINES

Coleraine

RED: Cab Merlot (W); Shiraz (W)

Culraithin

RED: Cab (W)

Fire Engine

RED: Red (W, Merlot & Ruby Cab & Cinsaut)

HIGHLY RATED

Coleraine Cabernet Sauvignon Merlot
1999 – W3½ 01

Coleraine Shiraz
2000 – W3½ 02
1999 – W4½ 00

CONARI ▷ MONS RUBER
CONDÉ ▷ STARK-CONDÉ
CONSTANCE ▷ KLEIN CONSTANTIA
CONSTANCE ▷ SIGNAL HILL
CONSTANTIA ▷ G CONSTANTIA

CONSTANTIA UITSIG

CONSTANTIA MAP

P PRODUCER: Spaanschemat River Rd, Constantia
PO Box 402, Constantia 7848
TEL: 021 794-1810 (tastings, office)
FAX: 021 794-1812
wine@icon.co.za
www.constantiauitsig.co.za
OWNERS: David & Marlene McCay
GM: Tom Glover
C'MASTER: John Laubscher
V'CULTURIST: André Rossouw
SALES: Mon–Fri 09:00–17:00, Sat–Sun 10:00–16:00
ACCOM: Country Lodge, Tel 021 794-6500, Fax 021 794-7605
RESTAURANTS: Constantia

Uitsig & La Colombe, Mon–Sun 10:00–21:00, Tel 021 794-4480 (CU), Tel 021 794-2390 (LC)
EST 1990 (producing wine since 1993). Area under vine: 30ha (70% white). Annual production: 210 tons, 150 000 litres
FLAGSHIP: Sémillon Res

LISTED WINES

D WHITE: Chardonnay Res (W); Chardonnay Unwooded; S Blanc (U); Sémillon Res (W)
RED: Cab (W); Cab Merlot (W); Merlot (W)

HIGHLY RATED

Constantia Uitsig Reserve Chardonnay
2000 – W4 01
1997 – W4 98, VD 98
Constantia Uitsig Reserve Sémillon
1999 – W4½ 02
1998 – W4 00
1997 – W3½ 99

COPPOOLSE FINLAYSON

STELLENBOSCH MAP

P CELLAR: 21/23 Kenwil Drive, Okavango, Brackenfell
PO Box 4028, Old Oak 7537
TEL: 021 982-6175
FAX: 021 982-6296
wine@sentinel.co.za
www.cfwines.co.za
OWNERS: Rob Christiaan Coppoolse (MD), Walter Finlayson, Viv Grater
C'MASTER: Adele Dumbar
VISITS: by appointment only
EST 1994. No vineyards (négociant). Annual production: 800 tons, 500 000 litres
FLAGSHIPS: Sentinel wines

LISTED WINES

Kaya

D WHITE: Chenin (U)
RED: Shaka (U, Cinsaut & Pinotage & Cab)

Sentinel

D WHITE: Chardonnay (W); S Blanc (U)
RED: Cab (W); Merlot (W); Res Pinotage (W); Shiraz (W)
PORT: Cape Ruby (W, Tinta B & Touriga N)

HIGHLY RATED

Sentinel Chardonnay
1996 – W4 97
Sentinel Cabernet Sauvignon
1999 – W3½ 01
Sentinel Merlot
2000 – VD 01
1997 – PB 00
Sentinel Pinotage
2001 – W3½ 02
Sentinel Shiraz
2000 – PB 01
1998 – W4½ 00, PB 99, VD 99

CORDOBA

HELDERBERG MAP

P CELLAR: Strand Rd (R44), Somerset West

PO Box 5609, Helderberg 7135
TEL: 021 855-3744
FAX: 021 855-1690
mail@cordoba.co.za
www.cordoba.co.za
OWNER: Jannie Jooste
GM: Frans Badenhorst
C'MASTER: Christopher Keet
VISITS: by appointment only
EST 1982 (producing wine since 1994). Area under vine: 30ha (90% red). Annual production: 100 tons, 60 000 litres
FLAGSHIP: Cordoba Crescendo

LISTED WINES

Cordoba

D WHITE: Chardonnay (W)
RED: Crescendo (W, Cab Franc & Merlot); Merlot (W)

Mount Claire

RED: Mountain Red (W, Cab & Shiraz & Merlot)

HIGHLY RATED

Cordoba Cabernet Sauvignon
1997 – W3½ 01
Cordoba Crescendo
1999 – W4½ 02
1998 – W3½ 01
1997 – W4½ 00
1995 – JP 99, PB Trophy 98, W4 98
Cordoba Merlot Cabernet Sauvignon
1997 – W4 02
Cordoba Shiraz
1994 – W4½ 99

CORNERSTONE ▷ GRAHAM BECK
COUNTRY CELLARS ▷ SPAR

COWLIN

PAARL MAP

P CELLAR: Noble Hill Vineyards, 205 Main Rd, Simondium
PO Box 174, Simondium 7670
TEL: 083 628-5894
FAX: 021) 874-3855
cowlinwines@iafrica.com
www.cowlinwines.co.za
OWNERS: John Cowlin Family Trust, Bloemendal Trust, Rodney Cowlin
DIRECTORS: John Richard Cowlin (MD), Gideon Malherbe, Rodney Gerald Cowlin
GM: John Richard Cowlin
C'MASTER: to be appointed
V'CULTURIST: Gideon Malherbe
VISITS: by appointment
EST 1998 (producing wine since 2002). Area under vine: 15,3ha (83% red). Annual production: 10 tons, 6 800 litres
FLAGSHIP: Noble Hill

LISTED WINES

RED: Cab (W); Merlot (W); Noble Hill (W, Cab & Merlot)

CRAIGHALL ▷ AFRICAN WINES
CRANE ▷ GOEDVERWACHT
CRÈME DE TÊTE ▷ SIGNAL HILL
CRESCENDO ▷ CORDOBA

CRU MONRO ▷ VILLIERA
CULRAITHIN ▷ COLERAINE
CURRANT ABBEY ▷ OUDE WELLINGTON
CYRIL BACK ▷ CHARLES BACK

DALLA CIA ▷ MEERLUST
DANIE DE WET ▷ DE WETSHOF

DARLING CELLARS

SWARTLAND MAP

P CELLAR: R315 between Mamreweg & Darling
PO Box 114, Darling 7345
TEL: 022 492-2276
FAX: 022 492-2647
info@darlingcellars.co.za
www.darlingcellars.co.za
DIRECTORS: Wessel Rauch, Kristle Kirsten
GM: John Sheppard
C'MASTER: Abé Beukes
SALES: Mon–Thurs 08:00–17:00, Fri 08:00–16:00, Sat 09:00–12:00
TASTINGS: fee of R5
TOURS: by appointment
PICNICS: to purchase by arrangement; welcome to bring
EST 1949. Area under vine: 1 800ha (60% white). Annual production: 8 000 tons, 5m litres
FLAGSHIPS: Onyx wines

LISTED WINES

DC
D WHITE: Chardonnay (U); S Blanc (U)
RED: Black Granite Shiraz (W); Cab (W); Merlot (W); Pinotage (W); Shiraz (W)

Flamingo Bay
D WHITE: Chenin S Blanc (U)
PINK: Rosé (U, Cinsaut & Chenin)
RED: Cinsaut Cab (U)

Onyx
D WHITE: Chardonnay (W)
RED: Cab (W); Kroon (W, Shiraz & Pinotage); Pinotage (W); Shiraz (W)

Zantsi
DESSERT: Natural Sweet (U, Bukettraube & Chenin)

HIGHLY RATED

DC Sauvignon Blanc
2000 – W4½ 00
DC Cabernet Sauvignon
1999 – W3½ 00
1997 – W4 98
DC Pinotage
1999 – W3½ 01
DC Shiraz
1999 – W4 01
1997 – W3½ 00
DC Black Granite Shiraz
1999 – W3½ 02
GroeneKloof Pinotage
1998 – W4 00, SA Certificate 00, PT 99
GroeneKloof Shiraz
1999 – W4 02, JC 01, PB 00
1998 – W3½ 00
Onyx Kroon
2000 – W4 02
1999 – W3½ 01

DASCHBOSCH

SEE WORCESTER MAP

CO-OP: Louwshoek, Rawsonville
PO Box 174, Rawsonville 6845
TEL: 023 349-1110
FAX: 023 349-1980
cellar@daschbosch.co.za
www.daschbosch.co.za
DIRECTORS: Aling Stofberg (MD), D Viljoen, B du Plessis, A Stofberg, L Mollër
GM: Gerrit van Zyl
C'MASTER: Reino Kruger
V'CULTURIST: Schalk du Toit
SALES: Mon–Fri 08:00–17:00
TASTINGS: no fee
TOURS: by appointment
EST 1964 (trading until 2002 as Louwshoek-Voorsorg). Area under vine: 5 000ha (80% white). Annual production: 12 000 tons, 8m litres.
SINGLE VYD VINES: Chardonnay, S Blanc, Merlot, Pinotage

LISTED WINES

D WHITE: Chardonnay (W); S Blanc (U)
RED: Cab (W); Merlot (W); Pinotage (W); Ruby Cab (W)
FORTIFIED: Nectar de Provision (W, Colombar)

DASSIE'S ROOD ▷ BOTHA

DAVID FROST

PAARL MAP

P CELLAR: Voor Paardeberg Rd off the R44 to Wellington from N1
PO Box 7358, N Paarl 7623
TEL: 082 492-6721; 021 869-8339 (office)
FAX: 021 869-8732
frostms@global.co.za
www.frostwine.com
OWNERS: David & Michel Frost
GM: Michel Frost
C'MASTER: Jason Fisher
V'CULTURIST: Michel Frost
VISITS: by appointment only
EST 1994 (producing wine since 1998). Area under vine: 24ha (100% red). Annual production: 90 tons, 60 000 litres

LISTED WINES

RED: Gene Sarazen Cab (W, export only); Gene Sarazen Merlot (W, export only)

DC ▷ DARLING CELLARS
DEBUTANTE ▷ JACARANDA

DE COMPAGNIE

WELLINGTON MAP

P CELLAR: Bovlei Rd, off Bainskloof Pass Rd, Wellington
PO Box 395, Wellington 7654
TEL: 021 864-1241
FAX: 021 864-3620
orders@decompagnie.co.za
www.decompagnie.co.za
OWNER: De Compagnie Estate (Pty) Ltd
DIRECTORS: Johann Loubser (MD), Riana Scheepers

C'MASTER: Carl Allen
VISITS: by appointment only
N MONUMENT: manor house (1800)
EST 1972 (producing wine since 2002, previously under the name of Claridge). Area under vine: 13ha (72% red). Annual production: 80 tons, 48 000 litres

LISTED WINES
RED: Cab (W); Private Reserve (W, Merlot & Cab Franc & Cab)
BRANDY: Potstill (Chenin)

DE DOORNS

WORCESTER MAP
CO-OP: Kelder St, De Doorns – on N1 at entrance to the town
PO Box 129, De Doorns 6875
TEL: 023 356-2835 (tastings, sales); 023 356-2100 (cellar)
FAX: 023 356-2100
ddwk@freemail.absa.co.za
MD: Danie Koen
C'MASTER: Danie Koen
V'CULTURIST: Wilhelm le Roux
SALES: Mon–Fri 08:00–12:30 & 13:30–17:00, Sat 08:00–12:00
TASTINGS: no fee
TOURS: during harvest, by appointment
EST 1968. Area under vine: 100ha (70% white). Annual production: 1 800 tons, 1,3m litres
FLAGSHIP: Cabernet

LISTED WINES
D WHITE: Chardonnay (W)
OD WHITE: Colombar (U)
SS PINK: Late Harvest (U, Chenin & Colombar); Perlé Rose (U, Chenin & Pinotage)
S WHITE: Hanepoot (U); Medium Sjerrie (U); Red Muscadel (U); Sweet Sjerrie (U)
RED: Cab (W); Pinotage (W); Roodehof (W, Cab & Pinotage)

DEETLEFS

WORCESTER MAP
ESTATE: Porter St, Rawsonville
PO Box 36, Rawsonville 6845
TEL: 023 349-1260 (tastings, sales, cellar);
023 349-1377 (manor house)
FAX: 023 349-1951
info@deetlefs.com
www.deetlefs.com
OWNER: J de W Deetlefs
GM: Kobus Deetlefs
C'MASTER: Johan Lotz
V'CULTURIST: Theunis van Zyl
SALES: Mon–Thurs 08:00–17:00, Fri 08:00–16:00, Sat 10:00–13:00
TASTINGS: no fee
TOURS: by appointment
FUTURES (EN PRIMEUR): yes
EST 1822 (producing wine since 1900). Area under vine: 100ha (60% red). Annual

production: 1 500 tons, 1,2m litres
FLAGSHIP: Sémillon

LISTED WINES

Deetlefs
D WHITE: Chardonnay (W); Chenin (U); S Blanc (U); Sémillon (U)
OD WHITE: W Riesling (U)
RED: Merlot (W); Pinotage (W); Shiraz (W)

Stonecross
D WHITE: Chenin Chardonnay (W); S Blanc Sémillon (W)
RED: Merlot Pinotage (W); Pinotage (W)

HIGHLY RATED

Deetlefs Chenin Blanc 2000 – W3½ 02

DE FORELLEN ▷ LANZERAC

DE HEUVEL

TULBAGH MAP

ESTATE: R46, Tulbagh
PO Box 103, Tulbagh 6820
TEL: 023 231-0350 (cellar); 023 231-0939 (manor house)
FAX: 023 231-0938
bianco@lando.co.za
www.bianco.co.za
OWNER: Toni Bianco
C'MASTER: Toni Bianco
V'CULTURIST: Craig Bianco
SALES: Mon–Fri 08:30–12:30, Sat 09:00–13:00
TASTINGS: fee of R5
TOURS: by appointment
EST 1991 (producing wine since 1997). Area under vine: 15ha (100% red). Annual production: 80 tons

LISTED WINES

Bianco
RED: Cab (W); Pinotage (W); Shiraz (W)

HIGHLY RATED

Bianco Cabernet Sauvignon 1999 – W4 00

DE HOOPEN ▷ VINDIGO
DEKKERSVLEI ▷ MELLASAT
DE LA CAILLE ▷ MEERHOF

DELAIRE

STELLENBOSCH MAP

P CELLAR: Helshoogte Pass (R310), Stellenbosch
PO Box 3058, Stellenbosch 7602
TEL: 021 885-1756 (tastings, sales, cellar); 021 885-1694 (manor house)
FAX: 021 885-1270
delaire@iafrica.com
www.delairewinery.co.za
OWNER: Masaud Alikhani
GM: Jaco van der Westhuizen
C'MASTER: Bruwer Raats
V'CULTURIST: Paul Wallace
SALES: Mon–Fri 09:00–17:00, Sat 10:00–17:00, Sun 10:00–16:00
TASTINGS: fee of R12pp
TOURS: week-days by appoint-

ment at R10pp (no tours Jan–Apr)
ACCOM: Mountain Chalets (self-catering)
RESTAURANT: The Green Door, Mon–Sun 12:00–15:00, Tel 021 885-1149
PICNICS: available for purchase Sept–Apr
EST 1981 (producing wine since 1984). Area under vine: 24ha (60% red). Annual production: 150 tons, 120 000 litres
FLAGSHIPS: Merlot, Botmaskop

LISTED WINES

D WHITE: Chardonnay (U); S Blanc (U)
RED: Botmaskop (W, Cab); Cab Merlot (W); Merlot (W)

HIGHLY RATED

Delaire Chardonnay
1999 – W3½ 00
1997 – W4 98
Delaire Barrel Matured Sauvignon Blanc
1999 – W3½ 00
Delaire Sauvignon Blanc
1998 – W3½ 00
Delaire Botmaskop Cabernet Sauvignon
1998 – VD 01, W4 01
Delaire Cabernet Sauvignon Merlot
1999 – MG 01, W4 01
1998 – W3½ 00
Delaire Merlot
2000 – W3½ 02
1998 – W3½ 01
1997 – W3½ 00

DE LA TORTUE ▷ STOUMANN'S
DE LEUWEN JAGT ▷ SEIDELBERG

DELHEIM

STELLENBOSCH MAP

P CELLAR: Knorhoek Rd, Koelenhof
PO Box 10, Koelenhof 7605
TEL: 021 888-4600
FAX: 021 888-4601
delheim@delheim.com
www.delheim.com
OWNERS: Hans-Otto Hoheison, Michael Sperling
MD: M (Spatz) Sperling
GM: Victor Sperling
C'MASTER: Conrad Vlok
V'CULTURIST: Victor Sperling
SALES: Mon–Fri 09:00–17:00, Sat 09:00–15:30, Sun (Oct–Apr) 10:30–15:30
TASTINGS: fee of R15
TOURS: Mon–Fri 11:30 & 14:30, Sat 10:30 (by appointment for big groups)
RESTAURANT: Mon–Sat 12:00–15:00, Sun (Oct–Apr) 12:00–15:00, Tel 021 888-4607
EST 1941 (producing wine since 1951). Area under vine: 150ha (60% red). Annual production: 1 000 tons, 750 000 litres
FLAGSHIP: Grand Res
ESTATE WINE: Vera Cruz

SINGLE VYD WINE: Vera Cruz

LISTED WINES

Delheim

MCC SPARKLING: Brut (U, Pinot & Chardonnay & Pinot Meunier)

D WHITE: Chardonnay (W); Chardonnay Sur Lie (W); Heerenwijn (U, Chenin & Colombar & Chardonnay); S Blanc (U)

SS WHITE: Gewürz (U); Spatzendreck (U, Late Harvest)

OD PINK: Pinotage Rosé (U, Pinotage & Muscat de F)

RED: Cab (W); Dry Red (W); Grand Res (W, Cab); Merlot (W); Pinotage (W); Shiraz (W)

DESSERT: Edelspatz (U, Riesling, NLH)

Vera Cruz

RED: Merlot (W); Pinotage (W); Shiraz (W)

HIGHLY RATED

Delheim Cabernet Sauvignon
1999 – VD 01
1998 – PB 99, W4 99, VD 99

Delheim Grand Reserve
1998 – W3½ 01
1997 – W4 00
1995 – W4 98

Delheim Merlot
1996 – W4 99, VD 97

Delheim Pinotage
2000 – W4 02, PT 01
1999 – PT 00
1998 – W3½ 00, PT 99

Delheim Shiraz
1999 – W3½ 00
1998 – W4½ 99
1997 – W3½ 99

Vera Cruz Merlot
1998 – W3½ 00

Vera Cruz Pinotage
1998 – W3½ 00

Vera Cruz Shiraz
1999 – W3½ 02
1998 – JP 02, MG 01

DELLRUST

HELDERBERG MAP

P CELLAR: Old Firgrove Rd, Firgrove
PO Box 5666, Helderberg 7135
TEL: 021 842-2752/2457
FAX: 021 842-2456
dellrust@mweb.co.za
www.dellrust.co.za

OWNERS: A & E Bredell
C'MASTER: Albert Bredell
SALES: Mon–Fri 08:00–17:00, Sat 09:00–12:00
TASTINGS: no fee
TOURS: yes
EST 1988 (producing wine since 1998). Area under vine: 97ha (70% red). Annual production: 550 tons, 400 000 litres
FLAGSHIP: Three Vines

LISTED WINES

D WHITE: Chardonnay (U); Chenin (U); S Blanc (U)

RED: Merlot (W); Pinotage (W); Tinta Barocca Cinsaut (W); Three Vines (W, Merlot & Pinotage & Shiraz)
FORTIFIED: Jerepigo (U, Chenin)
PORT: Cape Vintage (W, Tinta B)

HIGHLY RATED

Dellrust Chardonnay
1999 – W4 00
Dellrust Merlot
2000 – W3½ 02
Dellrust Pinotage
1999 – W3½ 01
Dellrust Cape Vintage
2000 – W3½ 02

DE MEYE

PAARL MAP

P CELLAR: Cnr Old Paarl Rd & Muldersvlei Rd
PO Box 20, Muldersvlei 7607
TEL: 021 884-4131
FAX: 021 884-4154
demeye@cybertrade.co.za
OWNER: Jan Myburgh Family Trust
DIRECTORS: Jan Myburgh (MD), Philip Myburgh
C'MASTER: Marcus Milner
V'CULTURIST: Peter Meyer
SALES: Mon–Fri 09:00–17:00, Sat 09:00–13:00
TASTINGS: no fee
TOURS: by appointment
EST 1975 (producing wine since 1998). Area under vine: 65ha (90% red). Annual production: 200 tons, 120 000 litres
FLAGSHIP: Trutina

LISTED WINES

D WHITE: Chardonnay (W)
RED: Cab (W); Pinotage (W); Shiraz (W); Trutina (W, Cab & Shiraz)

HIGHLY RATED

De Meye Chardonnay
2000 – W3½ 01
De Meye Shiraz
2000 – W3½ 02
De Meye Trutina
2000 – W4 02

DERDEHEUVEL ▷ UITVLUCHT
DESIDERIUS ▷ JC LE ROUX

DE TOREN

STELLENBOSCH MAP

P CELLAR: Polkadraai Rd, Stellenbosch – M12 between Neethlingshof & Saxenburg
PO Box 48, Vlottenburg 7604
TEL: 021 881-3119
FAX: 021 881-3335
info@de-toren.com
www.de-toren.com
OWNER: Edenhall Trust
MD: Emil den Dulk
C'MASTER: Albie Koch
V'CULTURIST: Johan Joubert
VISITS: by appointment only
FUTURES (EN PRIMEUR): yes
EST 1991 (producing wine since 1999). Area under

vine: 21ha (80% red). Annual production: 100 tons, 70 000 litres

LISTED WINES

RED: Fusion V (W, Cab & Merlot & Cab Franc & Malbec & Petit Verdot)

HIGHLY RATED

De Toren Fusion V
2000 – W4½ 02
1999 – W4 01

DE TRAFFORD

STELLENBOSCH MAP

P CELLAR: Mont Fleur, Blaauwklip Rd, Stellenbosch
PO Box 495, Stellenbosch 7599
TEL: 021 880-1611
FAX: 021 880-1611
detrafford@adept.co.za
OWNER: David Trafford
C'MASTER: David Trafford
V'CULTURIST: Mavis Trafford
SALES: Fri & Sat 10:00–13:00
TOURS: by appointment
TASTINGS: no fee
EST 1983 (producing wine since 1992). Area under vine: 9ha (66% red). Annual production: 60 tons, 27 000 litres
FLAGSHIP: Elevation 393

LISTED WINES

D WHITE: Chenin (W); Keermont Chenin (W)
RED: Cab (W); Elevation 393 (W, Cab & Merlot & Shiraz); Keermont Cab (W, Cab & Merlot); Keermont Shiraz (W); Merlot (W); Pinot Noir (W); Shiraz (W)
DESSERT: Vin de Paille (W, Chenin)

HIGHLY RATED

De Trafford Chenin Blanc
1999 – W3½ CC 00
De Trafford Keermont Chenin Blanc
2001 – W4½ CC 02
1998 – W3½ 99
1997 – WINE 3½ 99
De Trafford Cabernet Franc
1999 – W3½ 00
De Trafford Cabernet Sauvignon
2000 – TT 02, W5 02
1998 – W4 01
1997 – PB 99
De Trafford Merlot
2000 – TT 02, W5 02
1999 – W4 01
De Trafford Pinot Noir
1999 – W4 01
De Trafford Shiraz
2000 – W4½ 02
1998 – W4½ 00
De Trafford Vin de Paille
2000 – W3½ 02
1999 – W4½ 01
1998 – PB 99

DE VILLIERS

PAARL MAP

P CELLAR: Estherdal, N Paarl
PO Box 714, N Paarl 7623
TEL: 021 864-2233
FAX: 021 864-2237

devwines@mweb.co.za
MD: Villiers de Villiers
C'MASTER: Hein Heesebeck
V'CULTURIST: Johan Theron
VISITS: by appointment only
EST 1695 (producing wine since 1698). Area under vine: 160ha (50% red). Annual production: 1 600 tons, 1,1m litres

LISTED WINES

RED: Cab (U); Cab (W); Merlot (U)

DEVONAIR

STELLENBOSCH MAP

P CELLAR: Devonvale Golf Course Rd, Stellenbosch
PO Box 148, Koelenhof 7605
TEL: 021 865-2353; Cell 083 463-4260
FAX: 021 865-2842
rumble@icon.co.za
OWNER: Devonair Farm (Pty) Ltd
MD: Bernie Rumble
W'MAKER: Danie Steytler (Kaapzicht)
V'CULTURIST: Manus Lotter
VISITS: by appointment only
EST 1998 (producing wine since 2000). Area under vine: 3,5ha (100% red). Annual production: 22 tons, 15 000 litres

LISTED WINES

RED: Cab (W)

HIGHLY RATED

Devonair Cabernet Sauvignon 2000 – VD 01

DEVON HILL

STELLENBOSCH MAP

P CELLAR: Blumberg Rd, off M23/Bottelary ~~Rd;~~ Devon Valley
PO Box 541, Stellenbosch 7599
TEL: 021 865-2453
FAX: 021 865-2444
devhill@mweb.co.za
www.devonhill.co.za
OWNERS: Jean Vogel, Jean Michel Conne (Switzerland)
MD: Jean Vogel
GM: Urs Allemann
C'MASTER: Theo Brink
VISITS: by appointment only
EST 1996. Area under vine: 50ha (80% red). Annual production: 270 tons, 170 000 litres
FLAGSHIPS: Devon Hill wines

LISTED WINES

Devon Hill

D WHITE: S Blanc (U)
RED: Bluebird (W, Cab & Pinotage & Merlot); Cab (W); Merlot (W); Pinotage (W)

Devon View

D WHITE: Chenin (U); S Blanc (U)
RED: Cab (W); Merlot (W); Pinotage (W); Shiraz (W)

HIGHLY RATED

Devon Hill Sauvignon Blanc 2001 – JC 02
Devon Hill Bluebird 2000 – VD 01

Devon Hill Cabernet Sauvignon
2000 – VD 01
Devon Hill Merlot
2000 – VD 01, JC 01
Devon Hill Noble Late Harvest
2000 – W3½ 01
Devon View Pinotage
1997 – W3½ 98

DE WAAL ▷ UITERWYK

DE WET

WORCESTER MAP
CO-OP: N1, north of Worcester
PO Box 16, De Wet 6853
TEL: 023 341-2710
FAX: 023 341-2762
dewetwynkelder@mweb.co.za
GM: Petrus Le Roux
C'MASTER: André van Dyk
SALES: Mon–Fri 18:00–17:00, Sat 09:00–12:00
TASTINGS: fee of 50c per wine
TOURS: by appointment
EST 1948. Area under vine: 980ha (90% white). Annual production: 16 000 tons, 128 000 litres
FLAGSHIPS: Pinotage, Pettilant Fronté

LISTED WINES

D WHITE: Chardonnay (W); Riesling (U); S Blanc (U)
OD WHITE: Bouquet Blanc (U, Colombar & Gewürz & White Muscadel)
SS PERLÉ: Pettilant Fronté (U, White Muscadel)
S WHITE: Special Late Harvest (U, Gewürz)
RED: Dry Red (W, Ruby Cab); Pinotage (W)

HIGHLY RATED

De Wet Port
1997 – PS Trophy 98
1996 – W3½ 99

DE WETSHOF

ROBERTSON MAP
ESTATE: R317 between Robertson & Bonnievale
PO Box 31, Robertson 6705
TEL: 023 615-1853/8/7
FAX: 023 615-1915
info@dewetshof.co.za
www.dewetshof.co.za
OWNER: Danie de Wet
C'MASTER: Danie de Wet
V'CULTURIST: George Thom
SALES: Mon–Fri 08:30–16:30, Sat 09:30–12:30
TASTINGS: no fee
TOURS: by appointment
EST 1949. Area under vine: 180ha (90% white). Annual production: 2 000 tons, 1,2m litres
FLAGSHIPS: De Wetshof Bateleur Chardonnay, De Wetshof Edeloes
SINGLE VYD WINES: Bateleur Chardonnay, D'Honneur Chardonnay, Edeloes, Muscadel

LISTED WINES

D WHITE: Bateleur

Chardonnay (W); Blanc De Wet (U, S Blanc); Blanc Fumé (W, S Blanc); Bon Vallon Chardonnay (U); D'Honneur Chardonnay (W); Finesse Chardonnay (W); S Blanc (U)
OD WHITE: R Riesling (U)
DESSERT: Edeloes (U, Riesling, NLH); Mine D'Or (U, Riesling, Natural Sweet)

HIGHLY RATED

Danie de Wet Limestone Hill Chardonnay
2000 – VD 00
De Wetshof Bon Vallon Chardonnay
1998 – VD 98
De Wetshof d'Honneur Chardonnay
1999 – VD 01
De Wetshof Rhine Riesling
1998 – W3½ 99
De Wetshof Edeloes
2000 – W3½ 02, JP 01
1998 – W4 99
De Wetshof Cape Muscadel
2000 – JC 02

DE ZOETE INVAL

PAARL MAP

ESTATE: end of Frater St, S Paarl
PO Box 591, S Paarl 7624
TEL: 021 863-2375
FAX: 021 863-2817
dezoeteinval@wine.co.za
www.dezoeteinval.co.za
OWNER: Adrian Frater Trust
C'MASTER: John Robert Frater
V'CULTURIST: Gerard Frater
SALES: Mon–Sat 09:00–17:00
PICNICS: welcome to bring
EST 1878 (producing wine since 1976). Area under vine: 30ha (50% red). Annual production: 100 tons
FLAGSHIPS: Cab, Port

LISTED WINES

D WHITE: Blanc de Blanc (U, S Blanc & Pinot Gris); Blush (U, S Blanc); Chardonnay (W)
SS WHITE: Late Harvest (U, S Blanc)
RED: Cab (U); Cab (W); Grand Rouge (W, Cab)
PORT: Vintage (W)

DGB

PRODUCING WHOLESALER: DGB (Pty) Ltd
PO Box 7896, Halfway House 1685
TEL: 011 653-1000
FAX: 011 653-1105
info@dgb.co.za
OWNERS: 40% personnel, 40% Brait, 20% Graham Beck
MD: Tim Hutchinson

LISTED WINES

See separate listings for Bellingham and Douglas Green

DIAMANT

PAARL MAP

P CELLAR: Suid-Agter Paarl Rd, Paarl
PO Box 509, S Paarl 7624
TEL: 021 863-3949 (sales); 021 863-4904 (cellar); 021 863-1508 (manor house)
FAX: 021 863-1508
pdeville@world-online.co.za
OWNER: Diamant Trust
MD: Daniel Malan
C'MASTER: Niel Malan
VISITS: by appointment only
EST 1693 (producing wine since 1992). Area under vine: 43ha (80% red). Annual production: 5 tons, 3 500 litres

LISTED WINES

PORT: Pinotage Port (W)

DIE KRANS

LITTLE KAROO MAP

P CELLAR: Station Rd, Calitzdorp
PO Box 28, Calitzdorp 6660
TEL: 044 213-3314
FAX: 044 213-3562
diekrans@mweb.co.za
OWNERS: Boets & Stroebel Nel
C'MASTERS: Boets & Stroebel Nel
SALES: Mon–Fri 08:00–17:00, Sat 09:00–16:00
TASTINGS: fee of R10pp for groups only
TOURS: by appointment
PICNICS: welcome to bring
EST 1890 (producing wine since 1964. Under vine: 40ha (50% red). Annual production: 550 tons, 400 000 litres
FLAGSHIPS: Vintage Res Port, Touriga Naçional

LISTED WINES

Die Krans

S SPARKLING: Spumante (U)
D WHITE: Chardonnay (U); Chenin (U)
SS WHITE: Golden Harvest (U)
D PINK: Rosé (U)
RED: Cab (W); Pinotage (U); Tinta Barocca (U); Touriga Naçional (W)
DESSERT: Heritage Collection (U, Chenin, Jerepigo); White Muscadel Jerepigo (U); White Muscadel Res (U)
PORT: Ruby (W, Tinta B & Touriga N & Souzão & Tinta R); Tawny (W, Tinta B & Touriga N & Cinsaut); Vintage (W, Tinta B & Touriga N & Souzão & Tinta R); Vintage Res (W, Tinta B & Touriga N & Souzão & Tinta R); White (W, Chenin)

Masters (made by Boplaas, Die Krans, Axe Hill)

PORT: Cape Vintage Res (W, Touriga N & Tinta B & Souzão)

HIGHLY RATED

Die Krans Touriga Naçional 2001 – W4 01
Die Krans Res White Muscadel 2001 – W4 02

1998 – W4 99
Die Krans White Muscadel Jerepigo
1997 – W4 98
Die Krans Cape Ruby
NV – DT 00
Die Krans Cape Tawny
W3½ 02
Die Krans Cape Vintage
1999 – W3½ 01
1996 – W3½ 98
Die Krans Vintage Reserve
1999 – W4 PS 02, JC 02
1997 – JP 00, VD 99, W4 PS Trophy 99
1995 – W4½ 98
Masters Cape Vintage Reserve
1998 – W4 00
1997 – W4 99

DIEMERSDAL

DURBANVILLE MAP

ESTATE: off Adderley Rd, Durbanville
PO Box 27, Durbanville 7551
TEL: 021 976-3361
FAX: 021 976-1810
joelouw@iafrica.co.za
OWNER: Tienie Louw
C'MASTER: T Louw
V'CULTURIST: Div van Niekerk
SALES: Mon–Fri 09:00–17:00, Sat 09:00–16:00
TOURS: by appointment
PICNICS: welcome to bring
EST 1700. Area under vine: 160ha (80% red). Annual production: 1 700 tons
FLAGSHIP: Private Collection

LISTED WINES

D WHITE: Chardonnay (W); S Blanc (U)
RED: Mathys (W, Cab & Merlot); Private Collection (W, Cab & Cab Franc & Merlot); Shiraz (W)

HIGHLY RATED

Diemersdal Pinotage
1998 – W4 00
Diemersdal Private Collection
1998 – W3½ 00
Diemersdal Shiraz
1998 – W3½ 01

DIEMERSFONTEIN

WELLINGTON MAP

P CELLAR: Jan van Riebeeck Drive (R301), Wellington
PO Box 41, Wellington 7654
TEL: 021 864-5050
FAX: 021 874-2095
wine@diemersfontein.co.za
www.diemersfontein.co.za
OWNERS: David & Susan Sonnenberg
GM: Anzell Adams
C'MASTER: Bertus Fourie
V'CULTURIST: Fred Steytler
SALES: Mon–Sun 10:00–16:30
TASTINGS: fee of R15
ACCOM: Country House, Tel 021 873-2671, Fax 021 873-4526; hopitality@diemers-fontein.co.za
PICNICS: available for purchase
EST 1965 (producing wine

since 2001). Area under vine: 49ha (80% red). Annual production: 180 tons, 115 000 litres
FLAGSHIPS: Carpé Diem wines
ESTATE WINES: all

LISTED WINES

RED: Cab (W); Carpé Diem Cab (W); Carpé Diem Merlot (W); Carpé Diem Pinotage (W); Merlot (W); Pinotage (W); Shiraz (W)

HIGHLY RATED

Diemersfontein Carpé Diem Merlot
2001 – W3½ 02
Diemersfontein Merlot
2001 – W3½ 02
Diemersfontein Carpé Diem Pinotage
2001 – TT 02, W5 02
Diemersfontein Pinotage
2001 – PT 01
Diemersfontein Shiraz
2001 – W4 02

DIE POORT

LITTLE KAROO MAP

P CELLAR: Herbertsdale, off N2 Highway, near Mossgas
PO Box 99, Herbertsdale 6505
TEL: 028 735-2406
FAX: 028 735-2347
OWNER: Jannie Jonker
C'MASTER: Jannie Jonker
SALES: usually Mon–Fri 08:00–17:00, Sat 08:00–13:00 – no wines produced for sale in 2002
TOURS: yes
EST 1959 (producing wine since 1961). Area under vine: 46ha (41% white). Annual production: 800 tons, 580 000 litres
FLAGSHIP: Hemelrood

LISTED WINES

S SPARKLING: Vin Doux
D WHITE: Blanc de Blanc; Cape St Blaize; Premier Grand Cru
SS WHITE: Camilla; Frohlich Stein; Late Harvest
RED: Cab; Hemelrood; Pinotage
DESSERT: Golden Jerepigo; Lissa Jonker Hanepoot; Raisin Jerepigo; Red Jerepigo; Selected Red Jerepigo; Selected White Jerepigo; Sweet Jerepigo; White Jerepigo
PORT: Port Pinotage; White Port

DIEU DONNÉ

FRANSCHHOEK MAP

P CELLAR: Uitkyk St, Franschhoek
PO Box 94, Franschhoek 7690
TEL: 021 876-2493
FAX: 021 876-2102
dieudonne@zsd.co.za
OWNER: Robert Maingard
MD: Stephan du Toit
C'MASTER: Stephan du Toit
V'CULTURIST: Hennie du Toit

SALES: Mon–Fri 08:30–16:00, Sat–Sun 10:00–15:00
TASTINGS: no fee
TOURS: by appointment
EST 1984. Area under vine: 40ha (60% red). Annual production: 10 tons, 125 000 litres
FLAGSHIP: Merlot
SINGLE VYD WINES: Cap Classique, Pinotage, S Blanc, Shiraz

LISTED WINES

MCC SPARKLING: Méthode Cap Classique (U, Chardonnay)
D WHITE: Chardonnay (U); Chardonnay (W)
OD WHITE: S Blanc (U)
S WHITE: Special Late Harvest (U, Chenin)
SS PINK: Rosé (U, Chardonnay & Cab)
RED: Cab (W); Cab Merlot (W); Merlot (W); Pinotage (W); Shiraz (W)

HIGHLY RATED

Dieu Donné Chardonnay
1999 – W3½ 01
1997 – W3½ 98
Dieu Donné Merlot
1999 – W4 01
1998 – W3½ 00
1997 – MG 99
Dieu Donné Noble Late Harvest
1997 – W3½ 01
Dieu Donné Special Late Harvest
1998 – W4 02

DISCOVERY ▷ FLAGSTONE

DISTELL

STELLENBOSCH MAP

PRODUCING WHOLESALER: Adam Tas Rd, Stellenbosch
PO Box 184, Stellenbosch 7599
TEL: 021 809-7000 (switchboard); 021 809-7569 (tastings, sales)
FAX: 021 883-2603; 021 887-2506 (vinoteque)
info@distell.co.za
www.distell.co.za
OWNERS: Rembrandt KWV Investments, SAB, Public
DIRECTORS: Jan Scanell (MD), SJ Genade, MJ Botha, HA Hëyl, TH Tar, AG Steyn, S Pittman, GJ Loubser, CJ Conrje, E Heyns
C'MASTER: Linley Schultz
V'CULTURIST: Ernst Le Roux
SALES: at Oude Libertas, Mon–Sat 08:30-17:00
TASTINGS: fee of R8,50
TOURS: by appointment, Mon–Fri 10:00 & 14:30
RESTAURANT: Oude Libertas, lunch Mon–Fri, dinner Mon–Sat, Tel 021 809-7495
PICNICS: available for purchase

LISTED WINES

See also separate listings for The Bergkelder, Drostdy,

Durbanville Hills, JC Le Roux, Lusan, Nederburg, Plaisir de Merle, Tukulu

Autumn Harvest
PERLÉ: Crackling

Capenheimer
WHITE: Perlé

Cellar Cask
D WHITE: Premier Grand Cru;

Johannisberger
SS WHITE: Premier Late Harvest; Premier Semi Sweet
ROSÉ: Johannisberger
RED: Premier Claret; Johannisberger Red

Chateau Libertas
RED: blend of varying varieties

5th Avenue
SPARKLING: from Chenin & Pinotage

Graça
OD WHITE: from S Blanc & Sémillon & Crouchen
PINK: Rosé (S Blanc & Sémillon & Crouchen & Pinotage)

Grand Mousseux
SPARKLING: Spumanté; Vin Doux; Vin Sec

Kellerprinz
WHITE: Grand Cru; Late Harvest

Monis
PORT: Very Old Tawny
SHERRY: Full Cream, Medium Cream, Dry
DESSERT: Collectors Moscato (Muscadel); Moscato (Red Muscadel); Vintage Muscadel (W)

Original Paarl Perlé
PERLÉ: blend

Overmeer
WHITE: Stein
RED: blend

Sedgwick's
SHERRY: Old Brown

Tassenberg
RED: blend

Taverna Rouge
RED: blend

Zonnebloem
D WHITE: Blanc de Blanc (U, Chenin & S Blanc); Chardonnay (W); Fine Art S Blanc (U); Grand Cru (U, Chenin & Colombar)' S Blanc (U); W Riesling (U)
RED: Cab (W); Fine Art Cab (W); Fine Art Merlot (W); Fine Art Pinot (W); Fine Art Shiraz (W); Laureat (W, Cab Franc & Cab & Merlot); Merlot (W); Pinotage (W); Shiraz (W)
DESSERT: Noble Late Harvest

Other:
NATURAL WINES: Lieberstein; Oom Tas; Royal King; Virginia; Witzenberg Late Harvest Edelgoud, Witzenberg Stein
FORTIFIED: Apex Cape Port, Apex White Malmsey; CWB White Port; Government House Port; Lombards Hanepoot; Multana Red, Multana White; Ship Sherry; Van der Stel Club Sherry

BRANDY: Chateau VO; Fish Eagle; Klipdrift Export; Klipdrift Premium; Kommando; Louis XIV; Martell; Mellowwood; Olof Bergh; Oude Meester Souverein 12 Year; Oude Meester Demant; Oude Meester VSOB; Paarl Rock; Paarl Rock 5 Year; Richelieu; Royal Oak; Van Ryn 10 Year; Van Ryn 12 Year; Van Ryn 15 Year; Van Ryn 20 Year; Van Ryn Collection Res; Viceroy 5

HIGHLY RATED

Monis Moscato
1992 – VD 00
Monis Muscadel
1992 – TT 02, W5 02, JC 02
Monis Very Old Tawny Port
1990 – JC 02
1989 – W4½ PS Trophy 02
NV – VD 01, W3½ 01, DT 00, W3½ 00, VD 99, W3½ 99, VD 97
Monis Medium Cream Sherry
VD 01, VD 00, VD 97
Monis Pale Dry Sherry
VD 01, VD 00
Monis VO Full Cream Sherry
VD 99
Zonnebloem Chardonnay
2000 – W3½ 01
Zonnebloem Fine Art Cabernet Sauvignon
1999 – W3½ 01
Zonnebloem Limited Edition Cabernet Sauvignon
1995 – W3½ 98
Zonnebloem Merlot
1999 – W3½ 02
Zonnebloem Noble Late Harvest
1996 – W3½ 02

DOMEIN DOORNKRAAL

LITTLE KAROO MAP

P CELLAR: Doornkraal, De Rust
PO Box 14, De Rust 6650
TEL: 044 251-6715 (tastings, sales); 044 241-2156 (cellar); 044 241-2556 (manor house)
FAX: 044 241-2548
doornkraal@xsinet.co.za
OWNER: Swepie le Roux
MD: Piet le Roux
C'MASTER: Maria le Roux
SALES: Mon–Fri 09:00–17:00, Sat 08:00–13:00
TASTINGS: fee of R5pp for groups of 5 or more
TOURS: by appointment
RESTAURANT: Jemima's, Tues–Sat 12:00–14:00 & 18:30–22:00, Sun 18:30–22:00, Tel 044 272-0808
jemima@yahoo.com
PICNICS: available for purchase
EST 1890. Area under vine: 35ha (50% red). Annual production: 200 tons, 14 000 litres
FLAGSHIPS: Tickled Pink, Pinta
WINES WITHOUT SULPHUR: Generaal Jerepigo

LISTED WINES

Doornkraal

S SPARKLING: Tickled Pink (U, Pinotage & Chenin & Muscat de F)
OD WHITE: Kuierwyn (U, Chenin & Colombard & Muscat de F)
RED: Kannaland (U, Pinotage & Merlot & Cab); Merlot (W); Merlot Pinotage (W)
FORTIFIED: Diep Tanige Port (W, Touriga N & Tinta B, Tawny Port); Generaal Jerepigo (U, Muscat d A); Jerepigo (U, Muscat de F); Pinta (U, Pinotage & Tinta B); White Port (W, Chenin)

HIGHLY RATED

Doornkraal Pinta
NV – VD 01

DOMINION

NÉGOCIANT: Shiraz House, Brandwacht Estate, Stellenbosch
Postnet Suite 280, Private Bag X29, Somerset West 7129
FAX: 021 883-8782
info@domionwines.co.za
DIRECTORS: Johann Laubser (MD), C Collard, T Hacker, J Kwast
SALES: at retail stores only
FLAGSHIP: Milestone Reserve

LISTED WINES

D WHITE: Milestone Res (W, Chardonnay); Vlaeberg (U, S Blanc)

DOMINIQUE ▷ LOUISVALE

DOOLHOF

WELLINGTON MAP

P CELLAR: on R301 between Wellington & Ceres
PO Box 157, Wellington 7654
TEL: 021 864-1154
FAX: 021 864-1984
doolhof@ezinet.co.za
www.maizevalleywines.co.za
OWNER: DN Holt
DIRECTORS: Martin Latimer (MD), Derick Holt, June Holt
GM: Hendrik Loubscher
C'MASTER: Bertus Fourie
V'CULTURIST: Hendrik Loubscher
SALES: Mon–Fri 08:00–17:00, Sat 09:00–13:00
TASTINGS: fee of R10
N MONUMENT: entire farm
EST 1997 (producing wine since 2001). Area under vine: 35ha (80% red). Annual production: 40 tons, 28 000 litres

LISTED WINES

D WHITE: Chardonnay (U); S Blanc (U)
RED: Cab (W); Merlot (W); Pinotage (W); Shiraz (W)

DOORNROODT ▷ AAN DE DOORNS

DORNIER

STELLENBOSCH MAP

P CELLAR: Blaauwklippen Rd,

Paradyskloof
PO Box 8518, Stellenbosch 7599
TEL: 021 880-0557
FAX: 021 880-1499
info@dornierwines.co.za
www.dornierwines.co.za
OWNER: Christoph Dornier
C'MASTER: Ian Naudé
V'CULTURIST: Lukas de Kock
VISITS: due to open to the public in 2003
EST 1995 (producing wine since 2002). Area under vine: 55ha (80% red)

LISTED WINES

To be announced

DOUBLE CABERNET ▷ NORDALE
DOUELLE ▷ WELGEMEEND

DOUGLAS

NORTHERN CAPE

CO-OP: De Villiers St, Douglas
PO Box 47, Douglas 8730
TEL: 053 298-8314
FAX: 053 298-1845
wynkelder@gwk.co.za
OWNER: GWK Ltd
MD: J van Dyk
C'MASTER: P Le Roux
V'CULTURIST: D Kershoff
SALES: Mon–Fri 08:00–17:00
TASTINGS: no fee
TOURS: by appointment
ACCOM: René's Guesthouse, Tel 082 266-9038
RESTAURANT: Die Geelhuis, Mon–Sun 10:00–22:00, Tel 053 298-2579
EST 1968. Area under vine: 400ha (85% white). Annual production: 7 000 tons, 4m litres

LISTED WINES

Avoca

D WHITE: Chardonnay (U)
SS WHITE: Gewürz (W)
RED: Cab (U); Classic Red (W, Merlot & Ruby Cab); Merlot (U)
FORTIFIED: Ruby Port (U); Red Muscadel (W)

DOUGLAS GREEN

NÉGOCIANT based in Wellington – not open to the public
PO Box 79, Groot Drakenstein 7680
TEL: 021 874-1011 (office); 021 873-1001 (production)
FAX: 021 874-1690
douglasgreen@dgb.co.za
www.douglasgreen.co.za
OWNER: DGB (Pty) Ltd
MD: Tim Hutchinson
W'MAKER: Gerhard Carstens
SALES: at retail stores only
EST 1983. Annual production: 3,1m litres
FLAGSHIP: St Augustine

LISTED WINES

D WHITE: Chardonnay (W); Saint Vincent (U, S Blanc & Chenin); S Blanc (U)
SS WHITE: Saint Morand (U,

Chenin & Muscat d A)
S WHITE: Saint Anna (U, Chenin & Gewürz, Natural Sweet)
S ROSÉ: Saint Claire (U, Chenin & Gewürz & Pinotage, Natural Sweet)
SS RED: Saint Celine (U, Cinsaut)
RED: Cab (W); Pinotage (W); Saint Raphael (U, Cinsaut); St Augustine (W, Cab & Merlot & Cab Franc)
SHERRY: Dry Fino (W); Full Cream (W); Medium Cream (W)
PORT: Ruby (W); Springbok History (W, Tinta F & Tinta B & Souzão & Pontac & Malvasia Rey & Tempranillo)

HIGHLY RATED

Douglas Green Cabernet Sauvignon Shiraz 1995 – VD 97
Douglas Green St Augustine 1996 – VD 99

DOWN TO EARTH ▷ VILLIERA
DRAGON ▷ ASHANTI
DRAGON TREE ▷ FLAGSTONE

DRAKENSIG

PAARL MAP

P CELLAR: Klapmuts/Franschhoek Rd
PO Box 22, Simondium 7670
TEL: 021 874-3881; 082 825-2503
FAX: 021 874-3882
drakensig@mweb.co.za
OWNER: Marais Viljoen
VISITS: by appointment only
EST 1999. Area under vine: 13ha (100% red). Annual production: 50 tons, 37 500 litres

LISTED WINES

D WHITE: S Blanc (U)
RED: Cab (W); Pinotage (W); Shiraz (W)

DROSTDY

TULBAGH MAP

WINERY: Winterhoek Rd (cellar), Drostdy Musem (sales), Tulbagh
PO Box 9, Tulbagh 6820
TEL: 023 230-1086; 023 230-0203 (Drostdy Museum)
FAX: 023 230-0510
mskein@distell.co.za
www.drostdywines.co.za
OWNER: Distell(Pty) Ltd
MD: Jan Scanell
C'MASTER: Frans du Toit
SALES: at Drostdy Museum, Mon–Fri 08:30–17:00, Sat 09:00–17:00, Sun 14:30–17:00
TASTINGS: fee of R10
N MOMUMENT: tasting room/museum (1804) built as a magistrate's court, and houses collecion of early Cape furniture
PICNICS: welcome to bring
EST 1973. Annual production:

15m litres
FLAGSHIPS: Drostdy-Hof Adelpracht, Drostdy-Hof Chardonnay, Drostdy-Hof Extra Light, Drostdy-Hof Merlot

LISTED WINES

Drostdy-Hof

D WHITE: Chardonnay (W); Premier Grand Cru (U, Chenin & Colombar & Riesling & S Blanc); S Blanc (U); Steen (U, Chenin)
OD WHITE: Extra Light (U, Chenin & White Muscadel)
SS WHITE: Late Harvest (U, Chenin); Stein (U, Chenin & Palomino & Sémillon & Colombar)
S WHITE: Adelpracht (U, Chenin, SLH)
RED: Cab (W); Claret (U, Tinta B & Pinotage & Cab); Merlot (W); Pinotage (W); Ruby Cab (W) Ryk Tulbagh
SHERRY: Medium Cream (W, Chenin); Pale Dry (W, Chenin)

DUET ▷ NEDERBURG
DUKE ▷ WAMAKERSVALLEI
DUNCAN'S CREEK ▷ RICKETY BRIDGE
DU PLESSIS > HAVANA HILLS

DU PREEZ

WORCESTER MAP

ESTATE: Excelsior Farm, off Old N1, Rawsonville
PO Box 12, Rawsonville 6845
TEL: 023 349-1642
FAX: 023 349-1642
dupreezestate@intekom.co.za
OWNER: HL du Preez (Snr)
MD: Hendrik Lodewyk du Preez (Snr)
C'MASTER: Hennie du Preez (Jnr)
SALES: Mon–Fri 09:00–16:00
TASTINGS: fee of R10
TOURS: by appointment
PICNICS: welcome to bring
EST 1919 (producing wine since 1926). Area under vine: 150h (65% white). Annual production: 2 000 tons, 1,4m litres
FLAGSHIPS: Merlot, Polla's Red, S Blanc, Shiraz

LISTED WINES

D WHITE: Chardonnay (W); S Blanc (U)
RED: Polla's Red (W, Ruby Cab & Pinotage & Shiraz & Petit Verdot); Merlot (W); Pinotage (W); Res Merlot (W); Shiraz (W)
DESSERT: Hanepoot (U)

HIGHLY RATED

Du Preez Polla's Red 2000 – TT 02, W5 02

DURBANVILLE HILLS

DURBANVILLE MAP

P CELLAR: Tygervalley Rd/M13 – joint venture between Distell and local producers
PO Box 3276, Durbanville 7511

TEL: 021 558-1300
FAX: 021 559-8169
Dhills@distell.co.za
www.durbanvillehills.co.za
OWNERS: Distell (50%), seven growers (45%), Workers Trust (5%)
DIRECTORS: Hennie Heÿl (chairman), MJ Botha, KG Green, HN De Villiers, B Louw
MGR: Martin Moore
C'MASTER: Martin Moore
V'CULTURIST: Johan Pienaar (consulting)
SALES: Mon–Fri 09:00–16:00, Sat 09:30–14:30, Sun 11:00–15:00
TASTINGS: fee of R15
TOURS: Mon–Fri 11:00 & 15:00
N MONUMENTS: old wooden basket press (1870), original muzzle-loading cannon (1750) from Kanonkop in Durbanville
RESTAURANT: @ the Hills, Tues–Sun 12:00–15:00
EST 1999. Area under vine: see various Durbanville cellar listings. Annual production: 4 200 tons, 2,9m litres (80% sold off in bulk)
FLAGSHIPS: single-vineyard wines
SINGLE VYD WINES: Biesjes Craal S Blanc, Caapmans Cab Merlot, Luipaardsberg Merlot

LISTED WINES

D WHITE: Biesjes Craal S Blanc (U); Chardonnay (W); Rhinofields Chardonnay (W); Rhinofields S Blanc (U); S Blanc (U)
RED: Cab (W); Caapmans Cab Merlot (W); Luipaardsberg Merlot (W); Merlot (W); Pinotage (W); Rhinofields Merlot (W); Shiraz (W)

HIGHLY RATED

Durbanville Hills Sauvignon Blanc
1999 – VD 00
Durbanville Hills Rhinofields Sauvignon Blanc
2001 – W4 02
Durbanville Hills Cabernet Sauvignon
1999 – W3½ 00
Durbanville Hills Shiraz
2000 – W4 02

DU TOITSKLOOF

WORCESTER MAP

CO-OP: R101, Rawsonville – off N1
PO Box 55, Rawsonville 6845
TEL: 023 349-1601
FAX: 023 349-1581
info@dutoitskloof.co.za
www.dutoitskloof.com
DIRECTORS: Johan De Wet, Francois Botha, Lammie Smith, Chrisjan Smith, Jacobus Stofberg
C'MASTER: Philip Jordaan
V'CULTURIST: Schalk du Toit

SALES: Mon–Fri 08:00–17:00, Sat 08:30–12:00 – wine shop on N1 open Mon–Fri 09:00–17:30, Sat 10:00–17:00, Sun 11:00–16:00
TASTINGS: fee of R5
TOURS: by appointment for fee of R10
EST 1962. Area under vine: 750ha (70% white). Annual production: 11 000 tons, 7,5m litres
SINGLE VYD WINES: Bukettraube, Hanepoot Jerepigo

LISTED WINES

D SPARKLING: Brut (U, Chardonnay & S Blanc)
D WHITE: Blanc de Blanc (U, Chenin & Riesling); Chardonnay (W); Chenin (U); Riesling (U); S Blanc (U)
SS WHITE: Bukettraube (U); Late Vintage (U, Chenin & Muscat d A)
S WHITE: Special Late Harvest (U, Chenin & Muscat d A, Natural Sweet)
OD PINK: Blanc de Noir (U, Pinotage)
RED: Cab (W); Merlot (W); Pinotage (W); Pinotage Ruby Cab (W); Shiraz (W)
DESSERT: Noble Late Harvest (U, Muscat d A)
FORTIFIED: Hanepoot Jerepigo (U); Red Muscadel (U)
PORT: Cape Ruby (W, Pinotage)

HIGHLY RATED

Du Toitskloof Merlot
2001 – W3½ 02
Du Toitskloof Shiraz
2001 – W3½ 02
Du Toitskloof Hanepoot Jerepigo
1999 – VD 00, W4 00
1997 – W4 00
1995 – W4 00
Du Toitskloof Red Muscadel
1996 – W3½ 98

DVB ▷ LIEVLAND
D'VINE ▷ SWARTLAND

EAGLEVLEI

STELLENBOSCH MAP
P CELLAR: Old Paarl Rd, Muldersvlei
PO Box 969, Stellenbosch 7600
TEL: 021 880-1846
FAX: 021 880-1846
avanhels@adept.co.za
www.eaglevlei.co.za
OWNERS: Steve & Jean Weir, André van Helsdingen
PROPRIETOR: A van Helsdingen
C'MASTER: Etienne le Riche
V'CULTURISTS: Paul Wallace, Henry Fisk
VISITS: not open to the public
EST 1998. Area under vine: 11ha (100% red). Annual production: 25 tons, 15 000 *l*

FLAGSHIP: Cabernet

LISTED WINES

RED: Cab (W); Cab Merlot (W); Pinotage (W)

HIGHLY RATED

EagleVlei Pinotage 2000 – W4 02

EARTH ▷ HARTSWATER
EBONY ▷ ASARA
ECLIPSE ▷ AVONDALE
EDELKEUR ▷ NEDERBURG
EDELOES ▷ DE WETSHOF
EDELSPATZ ▷ DELHEIM
EDELROOD ▷ NEDERBURG
EDMOND ▷ RUPERT & ROTHSCHILD
ED'S RESERVE ▷ THELEMA
EERSTERIVIER ▷ STELLENBOSCH VYDS

EIKEHOF

FRANSCHHOEK MAP

P CELLAR: La Motte, Franschhoek

PO Box 222, Franschhoek 7690

TEL: 021 876-2469

FAX: 021 876-2469

eikehof@mweb.co.za

OWNER: Francois Malherbe

C'MASTER: F Malherbe

VISITS: by appointment only

FUTURES (EN PRIMEUR): yes

EST 1902 (producing wine since 1992). Area under vine: 46 ha (60% red). Annual production: 70 tons, 50 000 litres

FLAGSHIP: Bush Vine Sémillon

SINGLE VYD WINE: Bush Vine Sémillon

LISTED WINES

D WHITE: Bush Vine Sémillon (W); Chardonnay (W); Sémillon (W)

RED: Cab (W); Merlot (W); Shiraz

HIGHLY RATED

Eikehof Bush Vine Sémillon 2000 – W4 01

EIKENDAL

HELDERBERG MAP

P CELLAR: R44 between Stellenbosch & Somerset West

PO Box 2261, Dennesig 7601

TEL: 021 855-1422/3

FAX: 021 855-1027

vineyards@eikendal.com

www.eikendal.com

OWNER: Substantia (AG)

DIRECTORS: Prof R Saager (MD), HJ Saager, S Saager, CH Saager

GM: Hennie Pool

C'MASTER: Lizelle Gerber (J Krammer consulting)

V'CULTURIST: J Pienaar (consultant)

SALES: Mon–Fri 09:00–17:00, Sat 09:00–16:00, Sun 10:30–16:00

TASTINGS: fee of R10pp

TOURS: Dec–Feb, 11:30 & 14:30

ACCOM: Lodge, Tel 021 855-

3617, Fax 021 855-2862
lodge@eikendal.com
RESTAURANT: Nov–Apr, Sun–Fri 12:00–14:00
PICNICS: available to purchase
EST 1982. Area under vine: 62ha (60% red). Annual production: 500 tons, 350 000 litres
FLAGSHIPS: Chardonnay, Classique

LISTED WINES

D WHITE: Blanc de Blanc (U, S Blanc & Chenin & Pinot Blanc); Blanc Fumé (W, S Blanc); Chardonnay (W); Chardonnay Unwooded; S Blanc (U); Res Chardonnay (W); Verdi (U, S Blanc & Chenin & Sémillon)
OD WHITE: Chenin Blanc (W)
SS WHITE: Special Late Harvest (U, Chenin)
D PINK: Rosé (U, Merlot – export only)
RED: Cab (W); Classique (W, Cab & Merlot & Cab Franc); Merlot (W); Pinot (W); Rossini (W, Merlot & Cab Franc & Cab & Shiraz); Rouge (W, Cinsaut & Cab & Pinot & Cab Franc & Merlot); Shiraz (W)

HIGHLY RATED

Eikendal Chardonnay
2000 – JC 01
1998 – W4 99
1997 – VD 97
1996 – W4 97
Eikendal Reserve Chardonnay
1999 – W3½ 01
Eikendal Classique
1999 – W3½ 02
1998 – W3½ 00
Eikendal Reserve Cabernet Sauvignon
1999 – W4½ 01
Eikendal Merlot
1999 – W3½ 01

ELEGANCE ▷ AVONDVREDE
ELEGANCE ▷ NEDERBURG
ELEMENTS ▷ HARTSWATER

ELEPHANT PASS

FRANSCHHOEK MAP

P CELLAR: Verdun Rd, Franschhoek
PO Box 415, Franschhoek 7690
TEL: 021 876-3666
FAX: 021 876-2219
OWNERS: Ann & Peter Wrighton
C'MASTER: Jean Daneel
V'CULTURIST: Dave Maseke
VISITS: by appointment only
ACCOM: Owl Cottage, Tel 021 876-3280
EST 1994. Area under vine: 5ha (50% red). Annual production: 9 000 litres
FLAGSHIP: Celebration
SINGLE VYD WINES: all

LISTED WINES

D WHITE: Celebration Blanc

Fumé (W, S Blanc); Chardonnay (W)
RED: Merlot (W, Merlot & Cab)

HIGHLY RATED

Elephant Pass Merlot
1998 – W3½ 00

ELEVATION 393 ▷ DE TRAFFORD
ELROY'S PAWT ▷ KLOOFZICHT
ELSJES COREAL ▷ D'VILLE HILLS
ELUSIVO ▷ MONS RUBER
EMINENCE ▷ NEDERBURG
EMPEREUR ▷ SIGNAL HILL
ENGELTJIPIPI ▷ TJG
ENIGMA ▷ PORTERVILLE
EQUUS ▷ ZANDVLIET
ETHAN ASHER ▷ KLOOFZICHT

ETIENNE LE RICHE

STELLENBOSCH MAP

P CELLAR: Leef-op-Hoop, Jonkershoek, Stellenbosch
PO Box 6295, Stellenbosch 7612
TEL: 021 887-0789
FAX: 021 887-0789
lerichewines@adept.co.za
OWNER: Etienne le Riche
DIRECTORS: Etienne le Riche (MD), Marcelle le Riche
C'MASTER: Etienne le Riche
SALES: Mon–Fri 08:30–16:30
TASTINGS: no fee
ACCOM: Kleine Hoop, Tel 021 887-8958
EST 1997. Annual production: 50 tons, 32 000 litres
FLAGSHIP: Le Riche Cab Res

LISTED WINES

Leef op Hoop
RED: Cab (W); Cab Merlot (W)

Le Riche
RED: Cab (W); Cab Merlot (W); Cab Res (W)

HIGHLY RATED

Le Riche CWG Auction Reserve Cabernet Sauvignon
1999 – W3½ 02

Le Riche Reserve Cabernet Sauvignon
2000 – W4 02, JC 02
1999 – JP 01
1998 – W4½ 00
1997 – W4 00, JP 99, PB 98, VD 98

EVENTIDE ▷ MISCHA
EVITA'S ▷ SWARTLAND

EXCELSIOR

ROBERTSON MAP

ESTATE: PO Box 17, Ashton 6715
TEL: 023 615-1980/2050
FAX: 023 615-2019
info@excelsior.co.za
www.excelsior.co.za
OWNERS: Stephen & Freddie de Wet
C'MASTER: Jaco Marais
V'CULTURISTS: S & F De Wet
VISITS: not open to the public
EST 1914. Area under vine: 300ha (60% red). Annual production: 3 000 tons, 2,1m litres

FLAGSHIPS: S Blanc, Special Res
SINGLE VYD WINE: Cab

LISTED WINES

D WHITE: Chardonnay (W); S Blanc (U)
RED: Cab (U); Merlot (W); Shiraz (W); Special Res (W, Cab)

HIGHLY RATED

Excelsior Sauvignon Blanc
2000 – W3½ 00
Excelsior Merlot
2001 – W4 02
Excelsior Vlakteplaas Muscadel
2000 – VD 01

FAIRHILLS ▷ ASHWOOD

FAIRSEAT

NÉGOCIANT: Wynberg, Cape Town
PO Box 53058, Kenilworth 7745
TEL: 021 797-1951
fairseat@mweb.co.za
OWNER: Dick Davidson
SALES: at retail stores only

LISTED WINES

D WHITE: Chardonnay (W)
RED: Ruby Cab (W)

FAIR VALLEY

Wines made at Charles Back's Fairview cellar
PO Box 583, S Paarl 7624
TEL: 021 863–2450
FAX: 021 863–2591
fairwine@mweb.co.za
OWNER: Fair Valley Workers Association
MD: Tommie Fortuin
C'MASTER: Awie Adolf
SALES: at Fairview
EST 1999

LISTED WINES

D WHITE: Chenin (U)
RED: Pinotage (W)

FAIRVIEW

PAARL MAP

P CELLAR: Suid-Agter Paarl Rd, S Paarl – see separate listing for wines of Charles Back
PO Box 583, S Paarl 7642
TEL: 021 863-2450
FAX: 021 863-2591
fairwine@mweb.co.za
www.fairview.co.za
OWNER: Charles Back – see separate listing for Charles Back
C'MASTER: Anthony de Jager
V'CULTURIST: Johan Botha
SALES: Mon–Fri 08:30–17:00, Sat 08:30–13:00
TASTINGS: fee of R10pp
EST 1699. Area under vine: 160ha (90% red).
FLAGSHIPS: Fairview Akkerbos Chardonnay, Fairview Beacon Shiraz, Fairview Oom PagEl Sémillon, Fairview Primo Pinotage, Fairview Solitude Shiraz

🍷 **LISTED WINES**
See separate listing under Charles Back

FAITHFUL HOUND ▷ MULDERBOSCH
FARRAGO ▷ BWC
FAT BASTARD ▷ ROBERTSON
FELICITÈ ▷ CAPE BAY
FEMINITE ▷ LEMBERG
FERENT ▷ ZANDVLIET
5TH AVENUE ▷ DISTELL
FINE ART ▷ DISTELL
FINESSE ▷ DE WETSHOF
FIRE ▷ HARTSWATER
FIRE ENGINE ▷ COLERAINE
FISH EAGLE ▷ LE GRAND CHASSEUR
FIVE HEIRS ▷ LANZERAC
FIVE SOLDIERS ▷ RUSTENBERG
FLAGSHIP ▷ CHARLES BACK

FLAGSTONE

CAPE TOWN

P CELLAR: Dock House, Portswood Square, V&A Waterfront, Cape Town
PO Box 51564, V&A Waterfront 8002
TEL: 021 425-3766 (sales); 021 418-5974 (cellar)
FAX: 021 425-3764
terraceroad@icon.co.za
www.flagstonewines.com
OWNER: Bruce Stanley Jack
C'MASTER: Bruce Jack
VISITS: by appointment only
EST 1998. Area under vine: 7ha (66% red). Annual production: 720 tons, 288 000 litres
FLAGSHIPS: The Music Room Cab, The Poetry Collection Pinot, Two Roads Chardonnay, Writer's Block Pinotage
SINGLE VYD WINES: BK5, Elim Vineyards, Heywood House, Music Room, Strata Series, Two Roads, Writer's Block

🍷 **LISTED WINES**
D WHITE: The Berrio (U, S Blanc); Free Run (U, S Blanc); Heywood House Barrel Fermented (S Blanc); Noon Gun (W, Chenin & S Blanc & Sémillon); Resolute 166 (W, Chardonnay); Semaphore (W, Pinot Gris); Strata Series Chenin (W); Strata Series The Rumour Mill (W, Viognier); Two Roads (W, Chardonnay)
RED: BK5 (W, Pinot); Dragon Tree (W, Cab & Pinotage); Longitude (W, Merlot & Cab & Pinot & Shiraz & Pinot Gris & Pinotage & Cab Franc); Strata Series George's Blend (W, Pinot & Merlot); Strata Series Rockwood Cab (W); Strata Series Shindig Shiraz (W); Strata Series The Discovery (W, Pinot Meunier); The Music Room (W, Cab); The Poetry Collection (W, Pinot); Writer's Block (W, Pinotage)
PORT: The Whole Bangshoot (W, Barbera &

Carignan & Mataro & Sangiovese & Tinta F & Souzão & Cab & Mourvèdre)

HIGHLY RATED

Flagstone Heywood House Barrel Fermented Sauvignon Blanc
2001 – W3½ 02

Flagstone Two Roads Chardonnay
1999 – W3½ 00

Flagstone Longitude
2000 – VD 01

Flagstone The Music Room Cabernet Sauvignon
2000 – W4 02
1999 – W4 01

Flagstone The Poetry Collection Pinot Noir
2000 – W3½ 02

Flagstone Strata George's Blend Pinot Noir Merlot
2000 – VD 01

FLAMINGO ▷ ROOIBERG
FLAMINGO BAY ▷ DARLING CELLARS
FLEERMUISKLIP ▷ LUTZVILLE
FLEUR DU CAP ▷ BERGKELDER
FLOREAL ▷ ONDERKLOOF
FOREGON ▷ OUDEKLOOF
FLORENCE ▷ VERGELEGEN
FORGE MILL ▷ FRANSCHHOEK VYDS
FORRESTER ▷ KEN FORRESTER

FORT SIMON

STELLENBOSCH MAP

ESTATE: Fischers Rd, off Bottelary Rd, Kuils River

PO Box 43, Sanlamhof 7532

TEL: 021 906-0304 (tastings, sales); 021 903-8034 (cellar); 021 906-2549 (manor house)

FAX: 021 906-0304

fortsim@iafrica.com

OWNER: Renier Uys

DIRECTORS: Renier Uys (MD), Piet Uys, Marinus Bredell

C'MASTER: Marinus Bredell

V'CULTURIST: Renier Uys

SALES: Mon–Fri 09:30–17:00, Sat 10:00–14:00

TASTINGS: fee of R1

TOURS: by appointment

EST 1965 (producing wine since 1998). Area under vine: 62ha (55% red). Annual production: 250 tons, 190 000 litres

FLAGSHIP: Chardonnay

LISTED WINES

D WHITE: Chardonnay (W); Chenin Barrel Fermented; Res Chenin Barrel Fermented; Res S Blanc Barrel Fermented; S Blanc (U)

RED: Anna Simon (W, Merlot & Pinotage); Merlot (W); Pinotage (W); Shiraz (W)

HIGHLY RATED

Fort Simon Chardonnay
1998 – W4 99

Fort Simon Reserve Barrel Fermented Sauvignon Blanc
2000 – W4 02

Fort Simon Anna Simon
1998 – DT 99

Fort Simon Merlot
1998 – W3½ 01
Fort Simon Merlot Pinotage
1998 – VD 99

FOUR COUSINS ▷ VAN LOVEREN
FOUR SISTERS ▷ RUITERSVLEI
IV SPEARS ▷ SPIER
1482 ▷ MONT DESTIN

FRAAI UITZICHT

ROBERTSON MAP

P CELLAR: Klaasvoogds East, between Robertson & Ashton/ Montagu on R60 (R62)
PO Box 97, Robertson 6705
TEL: 023 626-6156
FAX: 023 626-6156
info@fraaiuitzicht.co.za
www.fraaiuitzicht.com
OWNER: Axel Spanholtz (MD), Mario Motti
VISITS: by appointment only
ACCOM: Fraai Uitzicht 1798
RESTAURANT: Fraai Uitzicht 1798, Wed–Sun 11:00–22:00
EST 18th Century (producing wine since 2000. Area under vine: 12ha (80% red). Annual production: 7 tons, 5 000 litres

LISTED WINES

RED: Merlot (W)

FRANCISKANER ▷ SIMONSIG

FRANSCHHOEK VYDS

FRANSCHHOEK MAP

CO-OP: Main Rd, Franschhoek
PO Box 52, Franschhoek 7690
TEL: 021 876-2086
FAX: 021 876-3440
info@franschhoekvineyards.com
www.franschhoekvineyards.com
OWNERS: various producers...
MGR: Daan Coetzee
C'MASTER: Deon Truter
SALES: Mon–Fri 09:30–17:00, Sat 10:00–16:00, Sun 11:00–15:00
TASTINGS: fee of R10
RESTAURANT: La Cotte, lunch Tues–Sun, Tel 021 876-2521
EST 1945. Annual production 2 300 tons, 1,3m litres
FLAGSHIPS: Franschhoek Cellars wines

LISTED WINES

Forge Mill

D WHITE: Chenin Blanc (U); S Blanc (U); S Blanc Sémillon (U)
RED: Cinsaut Cab (U); Pinotage (U); Shiraz (W)

Franschhoek Cellars

D WHITE: Chenin (U); S Blanc (U)
RED: Merlot(W); Pinotage (W)

La Cotte

D WHITE: Blanc de Blanc (U, Chenin & Sémillon & Colombard); S Blanc (U)
SS WHITE: Semi-Sweet (U, Chenin & Hanepoot)
RED: Grand rouge (U, Cinsaut & Pinotage)

GILGA

NÉGOCIANT: wines made at Overgaauw
PO Box 28746, Danhof 9310
TEL: 082 579 6282
FAX: 051 446-9029
PROPRIETORS: Chris Joubert, Eugene Viljoen
W'MAKER: Chris Joubert
SALES: by arrangement or at retail stores
FLAGSHIP: Shiraz
SINGLE VYD WINE: Shiraz

LISTED WINES
RED: Shiraz (W)

HIGHLY RATED
Gilga Shiraz
2000 – JP 02
1998 – W4½ 00

GLACIERE ▷ SIGNAL HILL

GLEN CARLOU

PAARL MAP

P CELLAR: Simondium Rd, Klapmuts
PO Box 23, Klapmuts 7625
TEL: 021 875-5528
FAX: 021 875-5314
glencarl@mweb.co.za
www.glencarlou.co.za
OWNERS: Hess Group (Berne), Finlayson family (SA)
DIRECTORS: Walter Finlayson (MD), DB Finlayson, Dr M Lienhard
C'MASTER: David Finlayson
V'CULTURIST: Marius Cloete
SALES: Mon–Sun 09:00–16:45
TASTINGS: fee of R5
TOURS: by appointment
FUTURES (EN PRIMEUR): yes
EST 1981 (producing wine since 1989); Area under vine: 65ha (50% red). Annual production: 450 tons, 300 000 litres
FLAGSHIPS: Chardonnay, Grand Classique

LISTED WINES
D WHITE: Chardonnay (W);

Verdelho (W)
RED: Grand Classique (W, Cab & Merlot & Cab Franc & Malbec & Petit Verdot); Pinot (W); Shiraz (W); Tortoise Hill (W, Cab & Zin & Shiraz & Touriga N)
DESSERT: Natural Sweet (W, Chardonnay)
PORT: Cape Vintage (W, Touriga N & Tinta B)

HIGHLY RATED

Glen Carlou Chardonnay
2001 – W4 02
1999 – W3½ 00
1998 – W4½ 99
1997 – W4 98
Glen Carlou Reserve Chardonnay
1999 – PB 00
Glen Carlou Devereux Chenin Blanc
1997 – W4 99
Glen Carlou Grand Classique
2000 – W4 02
1999 – W3½ 01
1998 – W4 00
1997 – PB 99
1996 – W4 99, VD 98
1995 – W4 98
Glen Carlou Shiraz
2001 – W4 02
Glen Carlou Cape Vintage
1997 – JC 00
1996 – W4 99

GLENHURST ▷ QUOIN ROCK

GLENVIEW

NÉGOCIANT: wines made at Rostberg, Stellenbosch
PO Box 32234, Camps Bay 8040
TEL: 021 438-1080
FAX: 021 438-3167
bayexport@kingsley.co.za
OWNER: Robin Marks
W'MAKER: Robin Marks
SALES: by arrangement only, or at retail stores – business not open to the public
EST 1998. Annual production: 60 000 litres (60% red)
FLAGSHIP: Dry Red

LISTED WINES

D WHITE: Dry White (U, Chenin)
RED: Dry Red (U, Cinsaut & Cab & Merlot & Shiraz)

GLENWOOD

FRANSCHHOEK MAP
P CELLAR: Robertsvlei, Franschhoek
PO Box 204, Franschhoek 7690
TEL: 021 876-2044 (sales, cellar); 021 876-2257 (tastings)
FAX: 021 876-3338
glenwood@adept.co.za
www.glenwoodvineyards.co.za
OWNER: Alistair Gordon Wood
GM: Dawid Petrus Burger
C'MASTER: DP Burger
SALES: Mon–Fri 09:00–13:00 (office hours),

otherwise by appointment
TASTINGS: fee of R10
TOURS: by appointment if outside office hours
PICNICS: welcome to bring
EST 1989 (producing wine since 2000). Area under vine: 22ha (75% white). Annual production: up to 50 tons, up to 32 500 litres
FLAGSHIPS: Chardonnay, Shiraz
SINGLE VYD WINES: Chardonnay, Sémillon

LISTED WINES

D WHITE: Chardonnay Barrel-Fermented & Matured; Chardonnay Wooded; Sémillon Wooded; S Blanc (U); Sémillon (W); Sémillon Barrel-Fermented & Matured
RED: Shiraz (W)

HIGHLY RATED

Glenwood Barrel Fermented Chardonnay
2001 – W3½ 02

GLORIEUX ▷ L'ÉMIGRÉ
GLUWEIN ▷ WOOLWORTHS
GOAT ROTI ▷ CHARLES BACK
GOATS DO ROAM ▷ CHARLES BACK

GOEDE HOOP

STELLENBOSCH MAP
ESTATE: Fischers Rd (off Bottelary Rd), Koelenhof
PO Box 25, Kuils River 7579
TEL: 021 903-6286
FAX: 021 906-1553
goede@adept.co.za
www.goedehoop.co.za
OWNER: Pieter Bestbier
MD: Petrus Johannes Bestbier
W'MAKERS: Pieter Bestbier, Willie Malherbe
V'CULTURIST: Johan De Beer
VISITS: by appointment only
N MONUMENT: house built in 1800s
EST early 1900s (producing wine since 1935). Area under vine: 100ha (60% red). Annual production: 300 tons, 200 000 litres
FLAGSHIP: Cabernet, Vintage Rouge

LISTED WINES

D WHITE: Chardonnay (W); S Blanc (U)
RED: Cab (W); Pinotage (W); Shiraz (W); Vintage Rouge (W, Merlot & Cab)

HIGHLY RATED

Goede Hoop Cabernet Sauvignon
1999 – W3½ 02
1997 – W4 00
Goede Hoop Vintage Rouge
1999 – W3½ 02
1998 – W3½ 00

GOEDVERTROUW

WALKER BAY MAP
ESTATE: ±10km from Bot River on Van der Stel pass Rd

PO Box 37, Bot River 7185
TEL: 028 284-9769
FAX: 028 284-9443
OWNERS: Arthur & Elreda Pillman
C'MASTER: Arthur Pillman
VISITS: by appointment only
EST 1988. Area under vine: 8ha (70% red). Annual production: 20 tons, 11 000 litres
FLAGSHIP: Pinot Noir

LISTED WINES

D WHITE: Chardonnay (U); S Blanc (U)
RED: Cab (W); Pinot (W)

GOEDVERWACHT

ROBERTSON MAP

ESTATE: between Robertson & Bonnievale
PO Box 128, Bonnievale 6730
TEL: 023 616-3430 (tastings, sales, cellar); 023 616-2845 (manor house)
FAX: 023 616-2073
goedverwachtestate@lando.co.za
www.goedverwacht.co.za
OWNER: Jan du Toit
C'MASTER: Jan du Toit
V'CULTURIST: Danie Visser
SALES: Mon–Fri 09:00–16:30, Sat 10:00–13:00
TASTINGS: no fee
TOURS: by appointment
PICNICS: welcome to bring
EST 1960. Area under vine: 100ha (78% white). Annual production: 1 600 tons, 100 000 litres
FLAGSHIP: Crane White Colombar

LISTED WINES

D WHITE: Crane White Colombar (U); Great Expectations Chardonnay (W); The Good Earth S Blanc (U)
PINK: Shiraz Rosé (U)
RED: Cab Merlot (W); Crane Red Merlot (W)

GÔIYA GAAN ▷ WEST CORP
GÔIYA KGEISJE ▷ WEST CORP
GOLDEN ALIBAMA ▷ DISTELL
GOLDEN TRIANGLE ▷ STELLENZICHT
GOOD EARTH ▷ GOEDVERWACHT
GOOD EVERYDAY ▷ V JOHNSON'S

GOUDINI

WORCESTER MAP

CO-OP: Goudyn, Rawsonville
PO Box 132, Rawsonville 6845
TEL: 023 349-1090
FAX: 023 349-1095
winesales@goudiniwine.co.za
www.goudiniwines.co.za
DIRECTORS: Hennie Hugo (MD), DJ Richter, AA le Roux, TC Botha, NM Dippenaar, DPH Jordaan
GM: Hennie Hugo
C'MASTER: Hennie Hugo
V'CULTURIST: Hendrik Myburgh
SALES: Mon–Fri 08:00–17:00,

Sat 09:30–12:30
TASTINGS: no fee
TOURS: by appointment
PICNICS: available by arrangement
EST 1948. Area under vine: 900ha (80% white). Annual production: 18 000 tons, 15m litres
FLAGSHIPS: Chenin, Res Ruby Cab, Res Sémillon, Ruby Cab Merlot

LISTED WINES

D SPARKLING: Brut (U, S Blanc)
OD SPARKLING: Demi Sec (U, S Blanc)
D WHITE: Blanc de Blanc (U, Clairette Blanche); Chardonnay (W); Riesling (U); S Blanc (U); Sémillon Res (W); Umfiki (U, Sémillon & Chardonnay)
OD WHITE: Chenin (U); Colombar (U)
SS WHITE: Special Late Harvest (U, Chenin)
SS PINK: Blanc de Noir (U, Cinsaut); Rosé (U, Pinotage)
RED: Cab (W); Pinotage (W); Rodini (U, Pinotage & Cinsaut); Res Ruby Cab (W); Ruby Cab Merlot (W); Shiraz (W)
DESSERT: Hanepoot (W)
PORT: Port (U, Cinsaut)

HIGHLY RATED

Goudini Ruby Cabernet
1998 – MG 00
Goudini Reserve Ruby Cabernet
1999 – VD 01
Goudini Ruby Cabernet Merlot
1999 – W3½ 00, DT 99
Goudini Hanepoot
1998 – W3½ 00

GOUDVELD

FREE STATE MAP

ESTATE: between Welkom & Theunissen
PO Box 1091, Welkom 9460
TEL: 057 352-8650
FAX: 057 352-8650
OWNER: Michael van der Merwe
C'MASTER: Merkil Alers
SALES: Mon–Sat 08:00–17:00
TASTINGS: no fee
TOURS: yes
PICNICS: welcome to bring
EST 1978 (producing wine since 1984). Area under vine: 27ha (80% red). Annual production: 285 tons, 200 000 litres
FLAGSHIPS: Port, Goue Nektar

LISTED WINES

D WHITE: Grand Cru (U, Chardonnay)
S WHITE: Goue Nektar (W, Hanepoot); Keldertrots (W, Colombar); Late Harvest (U, Chenin & Colombar)
PINK: Semi Rosé (U, Chenin & Pinotage)

RED: Cab (W); Pinot (W); Pinotage (W); Ruby Cab (W)
PORT: Port (W, Muscadel & Pinotage)

GOUE VALLEI ▷ CITRUSDAL
GOUTTE D'OR ▷ BEAUMONT
GOUVERNEURS ▷ G CONSTANTIA
GOVERNMENT HOUSE ▷ DISTELL
GRAÇA ▷ DISTELL

GRACELAND

STELLENBOSCH MAP

P CELLAR: Stellenrust Rd
PO Box 7066, Stellenbosch 7599
TEL: 021 881-3121
FAX: 021 881-3341
pmcn@iafrica.co.za
www.gracelandvineyards.com
OWNERS: Paul & Susan McNaughton
C'MASTER: Susan McNaughton
V'CULTURIST: Susan McNaughton
VISITS: by appointment only
FUTURES (EN PRIMEUR): yes
ACCOM: Graceland guest house
EST 1990 (producing wine since 1998). Area under vine: 9ha (100% red). Annual production: 45 tons, 30 000 litres
SINGLE VYD WINE: Cab

LISTED WINES

RED: Cab (W); Merlot (W); Shiraz (W)

HIGHLY RATED

Graceland Cabernet Sauvignon
2000 – W3½ 02
Graceland Merlot
2000 – W3½ 02
1999 – W3½ 01

GRAHAM BECK

R'SON & F'HOEK MAPS

P CELLARS: Robertson Cellar on R60 between Robertson & Worcester; Coastal Cellar on R45 outside Franschhoek
PO Box 724, Robertson 6705
TEL: 023 626-1214 (HO, Robertson Cellar); 021 874-1258 (Coastal Cellar)
FAX: 023 626-5164
cellar@grahambeckwines.co.za
www.grahambeckwines.co.za
OWNER: Graham Beck
GM: Gary Baumgarten
C'MASTERS: Pieter Ferreira (Robertson), Charles Hopkins (Coastal)
V'CULTURIST: Leon Dippenaar
SALES: at Robertson Cellar, Mon–Fri 09:00–17:00, Sat 10:00–15:00 – Coastal Cellar not open to the public
TOURS: by appointment
EST 1985 (producing wine since 1991). Area under vine: 210ha (55% white). Annual production: 3 000 tons, 900 000 litres
FLAGSHIPS: Brut Blanc de Blancs, Old Road Pinotage, The Ridge Syrah

SINGLE VYD WINES: Old Road Pinotage; The Ridge Syrah

LISTED WINES

Graham Beck

D WHITE: Sangiovese (W)

Graham Beck Coastal

D WHITE: S Blanc (U); Viognier (W)
RED: Cab (W); Merlot (W); Pinotage (W); Pinno (U, Pinotage); Shiraz (W)

Graham Beck Robertson

MCC SPARKLING: Brut (W, Chardonnay & Pinot); Brut Blanc de Blancs (W, Chardonnay); Sparkling Pinotage (W)
D WHITE: Chardonnay (W); Waterside White (W, Chardonnay & Colombard)
RED: Railroad Red (W, Cab & Shiraz); The Ridge Syrah
FORTIFIED: Rhona Muscadel (U, Muscat de F)

HIGHLY RATED

Graham Beck Brut
NV – W4 00, SA Trophy 00
Graham Beck Brut Blanc de Blancs
1996 – SA Trophy 02, W4 01
1992 – W4 97
Graham Beck Chardonnay (Robertson)
2001 – W3½ 02
2000 – W3½ 02
Graham Beck Sauvignon Blanc (Coastal)
2001 – W3½ 01
Graham Beck Cabernet Sauvignon (Coastal)
1999 – W3½ 01
Graham Beck Cornerstone Cabernet Sauvignon (Coastal)
1998 – W3½ 01
Graham Beck Merlot (Robertson)
1998 – W4 00
Graham Beck Pinotage (Coastal)
1999 – JC 00
1998 – W4½ 99, PT 99
Graham Beck Railroad Red (Robertson)
2001 – W3½ 02
Graham Beck Shiraz (Coastal)
2000 – W4 02, SA Certificate 02
1999 – W4½ 00
1998 – VD 99, W3½ 02
Graham Beck The Ridge Shiraz (Robertson)
2000 – W3½ 02
1999 – W3½ 01
1998 – W4½ 00
1997 – W3½ 99, DT 98
Graham Beck Rhona Muscadel (Robertson)
1996 – W4 98

GRAND CLASSIQUE ▷ GLEN CARLOU
GRAND MOUSSEUX ▷ DISTELL

GRANGEHURST

HELDERBERG MAP

P CELLAR: Eikendal Rd (off R44), Stellenbosch
PO Box 206, Stellenbosch 7599

TEL: 021 855-3625
FAX: 021 855-2143
winery@grangehurst.co.za
OWNER: Jeremy Walker
C'MASTER: Jeremy Walker
V'CULTURIST: Thys Greeff (consultant)
SALES: Mon–Fri 09:00–16:00
TASTINGS: no fee
TOURS: by appointment
EST 1992. Area under vine: 12ha (100% red). Annual production: 52 tons, 30 600 litres
FLAGSHIP: Nikela, CWG wines

LISTED WINES

RED: Cab Merlot (W); CWG Auction Blend (W, Cab & Pinotage & Merlot); CWG Cab (W); Nikela (W, Cab & Pinotage & Merlot); Pinotage (W)

HIGHLY RATED

Grangehurst Cabernet Sauvignon Merlot
1998 – W4 01
1995 – W4 98
Grangehurst CWG Auction Reserve Cabernet Sauvignon
1998 – W4 01
1996 – W3½ 99
1995 – JP 99, W5 98
Grangehurst Merlot
1994 – W4 97
Grangehurst Nikela
1997 – W3½ 01, DT 99
Grangehurst Pinotage
1997 – W4 99
1996 – W3½ 98
1995 – PT 97

GRANITE ▷ DARLING CELLARS
GRATITUDE ▷ DISTELL
GRAVEL HILL ▷ HARTENBERG
GREAT EXPECTATIONS ▷ GOEDVERWACHT
GREAT WHITE ▷ V JOHNSON'S
GREEN WINE ▷ LADISMITH
GROENE CLOOF ▷ CLOOF

GROOT CONSTANTIA

CONSTANTIA MAP

ESTATE: Groot Constantia Rd Private Bag X1, Constantia 7848
TEL: 021 794-5128 (tastings, sales, cellar); 021 794-5067 (manor house)
FAX: 021 794-1999
cellar@grootconstantia.co.za
www.grootconstantia.co.za
OWNER: Groot Constantia Trust (company 21)
DIRECTORS: De La Bat, J van Niekerk, H Bailey, ZA Bukani, Dr MA Cluver, HMJ du Preez, JM Robert, IB van der Vyver, AJ van Velden
GM: Jean Naude
C'MASTER: Boela Gerber
V'CULTURIST: Kallie Bröcker
SALES: Dec–Apr Mon–Sun 09:00–18:00, May–Nov Mon–Sun 10:00–17:00
TASTINGS: fee of R14
TOURS: by appointment – 7 tours a day

Dec–Apr (10:00–16:00); 3 tours a day May–Nov (10:00, 11:00, 15:00)
N MONUMENTS: Manor House & Jonkershuis (1685), Cloete Cellar (1791)
RESTAURANT: Jonkershuis, Mon 10:00–16:30, Tues–Sat 09:–22:00, Sun 09:00–17:00, Tel 021 794-6255; Tavern, Apr–Oct Mon–Sun 10:00–17:00, Nov–Mar Mon–Sun 10:00–18:00, Tel 021 794-1144
PICNICS: available for purchase; welcome to bring
EST 1685. Area under vine: 100ha (60% red). Annual production: 600 tons, 400 000 litres
FLAGSHIPS: Gouverneurs Res wines

LISTED WINES

MCC SPARKLING: Cap Classique (U, Chardonnay)
D WHITE: Chardonnay (W); Constantia Blanc (U, S Blanc & Sémillon); Gouverneurs Chardonnay Res (W); S Blanc (U)
OD WHITE: Bouquet Blanc (U, Morio Muscat & S Blanc); W Riesling (U)
RED: Cab (W); Constantia Rood (W, Merlot & Pinotage); Gouverneurs Res (W, Cab & Merlot & Cab Franc); Merlot (W); Merlot Shiraz (W); Pinotage (W); Shiraz (W)
PORT: Cape Ruby (W, Tinta B & Touriga N); Port (W, Tinta B & Touriga N)

HIGHLY RATED

Groot Constantia Chardonnay
2000 – W3½ 01
1999 – W3½ 00
Groot Constantia Reserve Chardonnay
1997 – W3½ 98
Groot Constantia Weisser Riesling
2001 – W3½ 02
1999 – JP 00
Groot Constantia Gouverneurs Reserve
2000 – W3½ 02
1996 – W4 98
Groot Constantia Merlot
2000 – W3½ 02
Groot Constantia Pinotage
1999 – W4 00, PT 00
Groot Constantia Shiraz
1999 – W4½ 00

GROOT EILAND

WORCESTER MAP

CO-OP: Goudini Rd, Rawsonville
PO Box 93, Rawsonville 6845
TEL: 023 349-1140/1941
FAX: 023 349-1801
grooteiland@lando.co.za
OWNERS: Johan F Bock, Stephen Bock, WS Rix, C De Villiers, Johan PJ Stofberg, WJ du Toit Snr, Jan D Hugo
DIRECTORS: J du Toit

Stofberg (MD), WJ du Toit, TCB Jordaan, PD du Toit Jnr
GM: FH Schlünz
C'MASTER: FH Schlünz
SALES: Mon–Fri 08:30–12:30 & 13:30–17:00
TASTINGS: fee of R5pp for large groups
TOURS: by appointment
EST 1962. Area under vine: 600ha (90% white). Annual production: 10 000 tons, 7,5m litres
FLAGSHIP: Chardonnay
SINGLE VYD WINE: Shiraz

LISTED WINES

D WHITE: Chardonnay (W); Chenin (U)
OD WHITE: Meander (U, Chenin & S Blanc)
SS WHITE: Honigtraube (U, Hanepoot & Chenin)
OD PINK: Rosé (U, Ruby Cab & Colombar)
RED: Cab (W); Merlot (W); Pinotage (W); Shiraz (W)

GROOTE POST

SWARTLAND MAP

P CELLAR: Darling Hills Rd, off R27
Private Bag X1, Bloubergrant 7443
TEL: 022 492-2825 (tastings, cellar); 021 557-0606 (sales); 022 492-2822 (manor house)
FAX: 022 492-2693 (cellar); 021 557-8280 (sales)
gpwines@iafrica.com
www.grootepost.com
OWNERS: Peter & Nicolas Pentz
GM: Robert Anderson
C'MASTER: Lukas Wentzel
V'CULTURIST: Jannie de Clerk
SALES: Mon–Fri 08:30–16:30, Sat 08:30–13:00
TASTINGS: no fee
TOURS: by appointment at R5pp
N MONUMENTS: Klawer Valley house (1780), Groote Post house (1706)
PICNICS: available for purchase; welcome to bring
EST 1993 (producing wine since 1999). Area under vine: 100ha (50% red). Annual production: 340 tons, 195 000 litres
FLAGSHIP: S Blanc

LISTED WINES

D WHITE: Chardonnay Wooded; Chenin (U); S Blanc (U); Unwooded Chardonnay
RED: Merlot (W); Pinot (W); Shiraz (W)

HIGHLY RATED

Groote Post Chardonnay 1999 – W4 00
Groote Post Chenin Blanc 2000 – JC 01
Groote Post Sauvignon Blanc 1999 – W4 99

GRÜNBERGER ▷ BERGKELDER

GRUNDHEIM

LITTLE KAROO MAP

P CELLAR: Oudtshoorn
PO Box 400, Oudtshoorn 6620
TEL: 044 272-6927
FAX: 044 272-2627
grundheim@absamail.co.za
OWNER: Danie Grundling
C'MASTER: Dys Grundling
SALES: Mon–Sun 08:00–17:00
TASTINGS: no fee
EST 1920 (producing wine since 1940). Area under vine: 25ha (50% red); Annual production: 260 tons, 182 000 litres
FLAGSHIP: Red Muskadel

LISTED WINES

SWEET: Red Muskadel (U); Red Muskadel (W); White Muskadel
PORT: Ruby (W, Cab & Cinsaut); Vintage Port (W, Tinta B & Cinsaut & Cab)

GUARDIAN PEAK ▷ RUST EN VREDE

GUSTO

NÉGOCIANT: PO Box 6045, Uniedal, Stellenbosch 7612
TEL: 082 807-4447
FAX: 021 881-3095
gustowwines@hotmail.com
OWNERS: Nicolette de Kock, PG Slabbert
W'MAKERS: PG Slabbert, Nicolette de Kock
VISITS: not open to the public
EST 2002
SINGLE VYD WINE: S Blanc

LISTED WINE

D WHITE: S Blanc (U)

GWENDOLYN ▷ SAXENBURG

HAGELSBERG ▷ MIDDELVLEI

HAMILTON RUSSELL

WALKER BAY MAP

ESTATE: Hemel-en-Aarde Valley, R320, outside Hermanus
PO Box 158, Hermanus 7200
TEL: 028 312-3595 (sales, cellar); 028 312-1791 (tastings)
FAX: 028 312-1797
hrv@hermanus.co.za
OWNER: Anthony Hamilton-Russell
C'MASTER: Kevin Grant
SALES: Mon–Fri 09:00–17:00, Sat 09:00–13:00
TASTINGS: no fee
TOURS: by appointment
PICNICS: welcome to bring
EST 1981. Area under vine: 60ha (60% red). Annual production: 101 tons, 58 300 litres
FLAGSHIPS: Chardonnay, Pinot

LISTED WINES

D WHITE: Chardonnay (W)
RED: Pinot (W)

HIGHLY RATED

Hamilton Russell Chardonnay

2001 – W4½ 02
1997 – W3½ 98
1996 – W4 97
Hamilton Russell Ashbourne Chardonnay
1996 – W4 97
Hamilton Russell Pinot Noir
2001 – W3½ 02
2000 – VD 01, JC 01
1999 – W3½ 01, PB 00, VD 00
1998 – W4 99
1997 – JP 99
1996 – W4 98
Hamilton Russell Ashbourne Pinot Noir
1997 – W4 99

HARTENBERG

STELLENBOSCH MAP

ESTATE: Bottelary Rd (M23), Koelenhof
PO Box 69, Koelenhof 7605
TEL: 021 865-2541
FAX: 021 865-2153
info@hartenbergestate.com
www.hartenbergestate.com
OWNER: Hartenberg Holdings (Pty) Ltd
DIRECTORS: Paddy Bomford (MD), James Browne, Tanya Browne, David Geerdts, Japie Krige, Heleen Rabe, Fiona MacKenzie, Ken MacKenzie, Carl Schultz, Rod Tyson
C'MASTER: Carl Schultz
V'CULTURIST: Frans Snyman
SALES: Mon–Sat 09:00–17:00
TASTINGS: no fee except for large groups
RESTAURANT: vintners platter served in summer, soup in winter; 12:00–14:00
EST 1690s (producing wine under current ownership since 1987). Area under vine: 85ha (68% red). Annual production: 500 tons, 350 000 litres
FLAGSHIPS: Res Chardonnay, Gravel Hill Shiraz
SINGLE VYD WINE: Gravel Hill

LISTED WINES

D WHITE: Chardonnay (W); Chatillon (U, Sémillon & Chenin); S Blanc (W)
SS WHITE: R Riesling (U)
D PINK: Occasional Rosé (U, Zin & Merlot & Pinotage)
RED: Cab (W); Cab Shiraz (W); Gravel Hill Shiraz (W); Merlot (W); Pinotage (W); Pontac (W); Shiraz (W); Zinfandel (W)

HIGHLY RATED

Hartenberg Barrel Fermented Chenin Blanc
1997 – W4 98
Hartenberg Chardonnay
1998 – W3½ 99
1997 – W3½ 98
Hartenberg Reserve Chardonnay
1998 – W3½ 99
Hartenberg Rhine Riesling
2001 – W3½ 02

Hartenberg Cabernet Sauvignon
1997 – W4 02
1995 – VD 98
Hartenberg Cabernet Sauvignon Merlot
1995 – W3½ 98
Hartenberg Cabernet Sauvignon Shiraz
2000 – W4 02
Hartenberg Merlot
1996 – W4½ 99, VD 00
1995 – PB 98, W3½ 98, VD 97
Hartenberg Shiraz
1996 – DT 98
1995 – W3½ 98, VD 97
Hartenberg L'Estreux
1998 – W4 99
1997 – W3½ 99

HARTSWATER

NORTHERN CAPE

COMPANY: Hartswater, Vryburg Main Rd
PO Box 2335, Hartswater 8570
TEL: 053 474-0700
FAX: 053 474-0975
wynkelder@senwes.co.za
OWNER: SENWES
DIRECTORS: PR Maree (MD), F Strydom, B Klint
C'MASTER: PR Maree
V'CULTURIST: Ryan Puttick
SALES: Mon–Fri 08:00–13:00 & 14:00–17:00
TASTINGS: no fee
TOURS: by appointment
RESTAURANT: Die Wynhuis, Mon–Fri 09:00–17:00, Tel 053 474-0907
PICNICS: welcome to bring

LISTED WINES

Elements

OD WHITE: Wind (U, Chenin & Colombar)
ROSÉ: Fire (U, Ruby Cab & Chenin, Natural Sweet)
RED: Earth (U, Pinotage & Ruby Cab)

Hinterland

D WHITE: Chardonnay (U)
OD WHITE: Chenin (U)
RED: Ruby Cab (U)
FORTIFIED: Jerepigo (U, Colombar & Fernão Pires)

HAUTE CABRIÈRE ▷ CABRIÈRE

HAVANA HILLS

DURBANVILLE MAP

P CELLAR: Van Schoorsdrif Rd, off N7, Philadelphia, West Coast
PO Box 46241, Kernkrag 7440
TEL: 021 972-1106
FAX: 021 972-1105
rambo@telkom.sa.net
OWNER: JPH du Plessis
C'MASTER: Nico Vermeulen
V'CULTURIST: Rudi Benn
VISITS: by appointment only
EST 1999. Area under vine: 50ha (90% red). Annual production: 150 tons, 100 000 litres
FLAGSHIPS: Bizweni wines, Du Plessis Res wines

LISTED WINES

Bizweni

RED: Cab (W); Merlot (W)

Havana Hills

D WHITE: Chardonnay (W); S Blanc (U)

RED: Cab (W); Cab Merlot (W); Merlot (W); Du Plessis Res Cab (W); Du Plessis Res Cab Merlot (W); Du Plessis Res Merlot (W); Du Plessis Res Shiraz (W); Shiraz (W)

HIGHLY RATED

Havana Hills Chardonnay
2000 – W3½ 02

Havana Hills Sauvignon Blanc
2001 – W4 02

Havana Hills Merlot
2000 – W3½ 02

Havana Hills Shiraz
2000 – W4 02

Havana Hills Du Plessis Reserve Cabernet Sauvignon
2000 – W4 02
1999 – W4½ 01, VD 00

Havana Hills Du Plessis Reserve Cabernet Sauvignon Merlot
2000 – W4 02, JP 02

Havana Hills Du Plessis Reserve Shiraz
2000 – W4 02, VD 01
1999 – VD 01, W4 01

HAWEQUAS ▷ MONT DU TOIT

HAZENDAL

STELLENBOSCH MAP

ESTATE: Bottelary Rd (M23), Stellenbosch
PO Box 336, Stellenbosch 7599
TEL: 021 903-5112
FAX: 021 903-0057
info@hazendal.co.za
OWNER: Dr Mark Voloshin
MD: Del Terwin
C'MASTER: Ronell Wiid
V'CULTURIST: Schalk du Toit (Vinpro)
SALES: Mon–Fri 08:30–16:30, Sat–Sun 10:00–15:00
TASTINGS: no fee
TOURS: weekdays 11:00 & 15:00 at a fee of R10
N MONUMENTS: gabled cellar (1781), H-plan homestead (1790) – whole farm declared a national monument
RESTAURANT: Hazendal Hermitage Restaurant, Mon–Sun 12:00–14:30
PICNICS: available for purchase by arrangement
EST 1699 (producing wine under current ownership since 1996).
Area under vine: 60ha (70% white).
Annual production: 400 tons, 180 000 litres
FLAGSHIPS: Hazendal Cab, Hazendal Shiraz

LISTED WINES

Hazendal
D WHITE: Chardonnay (W); Chenin Blanc Wooded; S Blanc (U)
RED: Cab Shiraz (W); Merlot (W)

Kleine Hazen
D WHITE: Bushvine Chenin (U)
SS WHITE: Konynwijn (U, Chenin)
OD PINK: Blanc de Noir (U, Pinot)
RED: Res Red (U, Cab & Shiraz & Merlot)

HIGHLY RATED
Hazendal Chardonnay 1998 – W3½ 99
Hazendal Wooded Chenin Blanc 2000 – W4 02
Hazendal Shiraz Cabernet Sauvignon 1998 – DW 99

HEERENWIJN ▷ DELHEIM

HELDERBERG

HELDERBERG MAP

COMPANY: Winery Rd, Firgrove (Stellenbosch Vineyards)
PO Box 465, Stellenbosch 7599
TEL: 021 842-2371/2
FAX: 021 842-2373
E-MAIL, WEBSITE: see Stellenbosch Vineyards
OWNER: Stellenbosch Vineyards
C'MASTER: Chris Keilly
V'CULTURIST: Francois de Villiers
SALES: Mon–Fri 09:00–17:30, Sat 09:00–17:00
TASTINGS: fee varies according to the size of the group
RESTAURANT: Mon–Sat 12:00–14:30, breakfast on booking, Tel 021 842-2012

LISTED WINES
D WHITE: Chardonnay (W); Chenin S Blanc (U); S Blanc (U)
OD WHITE: Chenin (U)
RED: Cab (W); Merlot Cab (W); Pinotage (W); Shiraz (W)

HELDERKRUIN

STELLENBOSCH MAP

P CELLAR: Annandale Rd (off R44), Stellenbosch
PO Box 91, Stellenbosch 7599
TEL: 021 881-3926 (tastings); 021 881-3899 (cellar)
FAX: 021 881-3898
helkruin@iafrica.com
OWNER: Niel du Toit
C'MASTER: Koos Bosman
V'CULTURIST: Pietie Goosen
SALES: Mon–Sat 10:00–17:00
TASTINGS: no fee
TOURS: by appointment
EST 1969 (producing wine since 1998). Area under vine: 100ha (70% red). Annual production: 200 tons, 40 000 *l*
FLAGSHIP: Shiraz

LISTED WINES

Helderkruin

MCC SPARKLING: Phyllis Hands (U, Chardonnay)
D WHITE: Chardonnay (W); S Blanc (U)
RED: Cab (W); Merlot (W); Pinotage (W); Shiraz (W)
PORT: Workers Vintage (W, Merlot & Cab & Shiraz)

Yellow Cellar

D WHITE: Chenin (U)
RED: Dry Red (U)

HIGHLY RATED

Helderkruin Phyllis Hands
1998 – W3½ 01

HEMELROOD ▷ DIE POORT
HENRI ROSELT ▷ ROOIBERG
HERCULES PARAGON ▷ SIMONSVLEI
HERE XVII ▷ DISTELL
HERFSGOUD ▷ KANGO
HERITAGE ▷ DIE KRANS
HEYWOOD HOUSE ▷ FLAGSTONE

HIDDEN VALLEY

P PRODUCER: wines made at Grangehurst from grapes grown in Devon Valley & Helderberg
PO Box 12577, Die Boord 7613
TEL: 021 855-0296 (sales)
FAX: 021 855-0297
info@hiddenvalleywines.com
www.hiddenvalleywines.com
OWNER: David C Hidden
GM: Trish Harris
W'MAKER: Jeremy Walker
V'CULTURIST: Johan (Grobbie) Grobbelaar
SALES: at Grangehurst
EST 1960s (Devon Valley farm). Producing wine since 1996. Area under vine: 21ha (100% red). Annual production: 54 tons, 31 500 litres
FLAGSHIP: Pinotage
SINGLE VYD WINE: Pinotage

LISTED WINES

RED: Cab (W); Pinotage (W)

HIGHLY RATED

Hidden Valley Cabernet Sauvignon
1999 – W3½ 02
Hidden Valley Pinotage
1998 – PT 00
1997 – PT 99, W5 PC 99
1996 – W4 98, PT 98

HIGH CONSTANTIA

CONSTANTIA MAP

P CELLAR: Groot Constantia Rd – 100m before bottom bottom gates of Groot Constantia
PO High Constantia, Groot Constantia Rd, Constantia 7800
TEL: 021 794-7171 (cellar); 083 300-2064 (D van Niekerk)
FAX: 021 794-7999
david@highconstantia.co.za
OWNERS: David van Niekerk, Bob de Villiers
C'MASTER: Roger Arendse

V'CULTURIST: David van Niekerk
VISITS: by appointment only
FUTURES (EN PRIMEUR): yes
EST 1992 (producing wine since 1999). Area under vine: 14ha (88% red). Annual production: 33 tons, 19 800 litres
FLAGSHIP: Sebastiaan
SINGLE VYD WINES: all

LISTED WINES

MCC SPARKLING: Cap Classique (U, Pinot & Chardonnay)
D WHITE: Chardonnay (W)
RED: Cab (W); Cab Franc (W); Sebastiaan (W, Cab Franc & Cab)

HILDENBRAND

WELLINGTON MAP

ESTATE: Horseshoe Rd, 4,5km outside Wellington
PO Box 270, Wellington 7655
TEL: 021 873-4115; 082 656-6007
FAX: 021 873-4115
info@wine-estate-hildebrand.co.za
www.wine-estate-hildebrand.co.za
OWNER: Reni Hildenbrand
C'MASTER: Reni Hildenbrand
SALES: Mon–Sun 10:00–16:00
TASTINGS: fee of R5
TOURS: at a fee of R10pp
N MONUMENTS: main house (1796), cellar (1853)
ACCOM: Klein Rhebokskloof
RESTAURANT: Hildenbrand's Table, lunch 12:00–15:00
EST 1991 (producing wine since 1999). Area under vine: 18ha (50% red). Annual production: 80 tons (50 tons to Bovlei), 17 400 litres
FLAGSHIPS: Cab Barrique, Sémillon
SINGLE VYD WINES: Cab, Chardonnay, Chenin, Sémillon

LISTED WINES

D WHITE: Chardonnay (U); Chardonnay Barrique; Chenin (U); Chenin Chardonnay (U); Sémillon (W)
RED: Cab (U); Cab Barrique; Shiraz (W)

HILL & DALE ▷ STELLENZICHT
HILLTOP ▷ DISTELL
HINTERLAND ▷ HARTSWATER
HIPPO CREEK ▷ PICARDI REBEL
HONEY BLOSSOM ▷ DISTELL
HONIGTRAUBE ▷ GROOT EILAND

HOOPENBURG

STELLENBOSCH MAP

P CELLAR: R101, Muldersvlei – off the Stellenbosch/Klipheuwel Rd
PO Box 1233, Stellenbosch 7599
TEL: 021 884-4221
FAX: 021 884-4904
info@hoopenburgwines.co.za
www.hoopenburgwines.co.za
OWNERS: Ernst Gouws, Dr

Heinz Schmitz
C'MASTER: Ernst Gouws
SALES: Mon–Fri 08:30–17:00, Sat 09:30–12:30
TASTINGS: fee of R10
EST 1992. Area under vine: 36ha (70% red). Annual production: 350 tons, 200 000 litres
FLAGSHIP: Chardonnay
SINGLE VYD WINES: Chardonnay, Merlot, Pinot

LISTED WINES

D WHITE: Chardonnay (W); S Blanc (U); Winemakers Selection Chardonnay (W)
RED: Cab (W); Pinot (W); Merlot (W); Pinotage (W); Shiraz (W); Winemakers Selection Cab (W); Winemakers Selection Merlot (W)

HIGHLY RATED

Hoopenburg Chardonnay 2000 – W3½ 01
Hoopenburg Pinotage 2000 – W3½ 01
Hoopenburg Pinot Noir 2000 – JC 02
1997 – W3½ 99

HOPE MARGUERITE ▷ BEAUMONT

HUGUENOT

WELLINGTON MAP

PRODUCING WHOLESALER:
Verseilles St, Wellington
PO Box 275, Wellington 7654
TEL: 021 864-1277
FAX: 021 873-2075
OWNER: Kosie Botha
DIRECTORS: Gert Brynard (MD), JC Botha, GM Botha, GM Brynard
GM: Gert Brynard
C'MASTER: Bill Matthee
VISITS: not open to the public

LISTED WINES

Huguenot

OD WHITE: Premier Grand Cru (U, Chenin); Stein (U, Chenin)
SS WHITE: Late Harvest (U, Chenin & Clairette Blanche)
S WHITE: Hanepoot (U)
RED: Cab (U); Pinotage (U); Smooth Red (U, Cinsaut)
DESSERT: Red Jerepico (U, Red Muscadel); White Jerepico (U, White Muscadel)
PORT: Ruby Red (U); Invalid (U); Tawny (U)

Valentine

S SPARKLING: Vin Doux (U, Chenin & Clairette Blanche)

HUMBRO ▷ SIMONSVLEI
HUNTING FAMILY ▷ SLALEY

ICEBERG ▷ RIJK'S
IMOYA ▷ KWV
IMPALA ▷ CITRUSDAL
IMPERIAL ▷ BOVLEI
INANDA ▷ YONDER HILL
INDABA ▷ CAPE CLASSICS
INDALO ▷ SWARTLAND

INFINITI ▷ STELLENBOSCH VYDS

INGWE

HELDERBERG MAP

P CELLAR: off De Beers Rd, Strand
PO Box 563, S Paarl 7624
TEL: 021 858-1063, 021 852-7346
FAX: 021 858-1063
ingwewine@eject.co.za
OWNER: Ingwe Wine Est (Pty) Ltd
DIRECTORS: AJM Moueix (MD), G Knox, F Baard
GM: F Baard
C'MASTER: PJ Geyer
V'CULTURIST: F Baard
VISITS: by appointment only
EST 1998. Area under vine: 25ha (88% white). Annual production: 100 tons, 60 000 litres

LISTED WINES

D WHITE: Chardonnay (W); S Blanc (U)
RED: Shiraz Cab Merlot (W)

INSENSATA ▷ AVONTUUR
INSPIRATION ▷ VILLIERA

INT WINE SERVICES

NÉGOCIANT: International Wine Services – wines made at Stellenbosch Vineyards' cellars
POSTAL ADDRESS: 3 Eldorado, Military Rd, Tamboerskloof 8001
TEL: 083 321-6884
FAX: 021 424-8683
Iwsinsa2@new.co.za
www.wine-info.co.uk
OWNER: Waverley Wines, Scottish & Newcastle
MD: Allan Westray
W'MAKER: Kym Milne MW
SALES: by arrangement
FLAGSHIPS: Apostles Falls wines

LISTED WINES

Apostles Falls
D WHITE: Chardonnay (W)
RED: Cab (W); Merlot (W)

Rylands Grove
D WHITE: Barrel Fermented Chenin; Chardonnay (W); Chenin Colombard (U); Dry Muscat (U, Muscat d A)
RED: Cab Merlot (W); Cinsaut Tinta Barocca (W)

HIGHLY RATED

Rylands Grove Barrel Fermented Chenin Blanc 2001 – W4 02

IONA

ELGIN MAP

P PRODUCER: Highlands, Elgin
PO Box 527, Grabouw 7160
TEL: 028 284-9678
FAX: 028 284-9078
gunn@corpdial.co.za
OWNER: 1st David Rd (Pty) Ltd
MD: Andrew Gunn
W'MAKER: Miles Mossop
V'CULTURIST: Mathew van

Heerden
SALES: "any reasonable hour"
TASTINGS: no fee
N MONUMENT: Herbert Baker homestead (±1910)
ACCOM: guest house
PICNICS: welcome to bring
EST 1998 (producing wine since 2001). Area under vine: 12ha (60% white). Annual production: 30 tons, 20 000 litres

LISTED WINES
D WHITE: S Blanc (U)

ISABELLA ▷ MURATIE
IVORY ▷ ASARA
IVORY CREEK ▷ CITRUSDAL

JACARANDA

WELLINGTON MAP
ESTATE: 3km north of Wellington, Old Hermon Rd
PO Box 121, Wellington 7654
TEL: 021 864-1235
jacaranda@iafrica.com
OWNERS: Jan & Trish Tromp
C'MASTER: Jan Tromp
SALES: Mon–Fri 10:00–17:00, Sat 10:00–14:00
TASTINGS: no fee
TOURS: no appointment necessary
ACCOM: Jacaranda Cottage (self-catering & B&B)
PICNICS: welcome to bring
EST 1976 (producing wine since 1994). Area under vine: 2,8ha (82% white). Annual production: 5 tons, 2 000 litres
FLAGSHIP: Debutante
SINGLE VYD WINES: all

LISTED WINES
D WHITE: Chenin (U)
RED: Cab (W); Debutante (W, Cab & Merlot); Dry Red (W, Cab & Merlot)
DESSERT: Jerepigo (U, Chenin)

HIGHLY RATED
Jacaranda Merlot 1999 – W4 01

JA CLIFT

P CELLAR: Main St, S Paarl
PO Box 515, S Paarl 7624 (D Clift)
PO Box 3231, Paarl 7620 (L Loots)
PO Box 6020, Paarl 7622 (E Jordaan)
TEL: 021 872-9779 (sales); 083 228-7855 (D Clift); 082 554-6357 (W Everson); 083 270-5706 (L Loots); 021 863-1529 (E Jordaan); 083 630-8183 (A Smook)
FAX: 021 871-1619
dehoop@mweb.co.za (D Clift)
we@intekom.co.za (W Everson)
ljl@iafrica.com (L Loots)
asmook@mweb.co.za
OWNER: JA Clift (Pty) Ltd

W'MAKERS: Derek Clift (Mason's Hill), Lieb Loots (Blyde), Dr Edwin Jordaan (Bergheim), Anthony Smook, William Everson
VISITS: by appointment only
N MONUMENT: cellar building (1800s)
EST 2000

LISTED WINES

Anthony Smook
RED: Cab (W); Cab Merlot (W); Pinotage (W); Shiraz (W)

Bergheim
RED: Pinotage (W)

Blyde
RED: Merlot (W); Pinotage (W); Shiraz Cab (W)

Mason's Hill
RED: Shiraz (W)

William Everson
D WHITE: Chardonnay (W)
RED: Pinotage (W); Shiraz (W)

JACOBSDAL

STELLENBOSCH MAP

ESTATE: R306, Kuils River
PO Box 11, Kuils River 7579
TEL: 021 905-1360
FAX: 021 905-1360
dumas@iafrica.com
www.jacobsdal.co.za
OWNERS: Dumas Ondernemings (Pty) Ltd; Dumas family
C'MASTER: Cornelius Dumas
V'CULTURIST: Hannes Dumas
VISITS: by appointment only
EST 1910 (producing estate wine since 1974). Area under vine: 100ha (50% red). Annual production: 200 tons, 150 000 litres

LISTED WINES

RED: Pinotage (W)

HIGHLY RATED

Jacobsdal Pinotage
1995 – PT 98, VD 98
1994 – PT 97

JAKO'S ▷ STELLENDRIFT
JACQUES BRUÉRE ▷ BON COURAGE
JACQUES PINARD ▷ DISTELL
JANE'S ▷ BEAUMONT

JANÉZA

ROBERTSON MAP

P CELLAR: Langverwacht, Bonnievale
PO Box 306, Bonnievale 6730
TEL: 023 616-2848
FAX: 023 616-2848
jan.eza@lando.co.za
OWNERS: Jan & Eza Wentzel
C'MASTER: Jan Wentzel
VISITS: by appointment only
EST 1991 (producing wine since 2001). Area under vine: 7ha (80% red). Annual production: 35 tons, 19 000 litres

LISTED WINES

D WHITE: Chardonnay (W); S Blanc (U)
RED: Tresuva (W, Cab & Shiraz & Merlot)

JC LE ROUX

STELLENBOSCH MAP

P CELLAR: Devon Valley Rd, Devon Valley
PO Box 184, Stellenbosch 7599
TEL: 021 865-2590
FAX: 021 865-2585
jclr@dist.co.za
www.jcleroux.co.za
OWNER: Distell
MD: Jan Scanell
C'MASTER: Melanie van der Merwe
V'CULTURIST: Willem Laubscher
SALES: Mon–Fri 08:30–16:30, Sat 10:00–16:00, Sun (Sept–Apr) 10:00–15:00
TASTINGS: fee of R15
TOURS: Mon–Fri 10:00, 11:30, 15:00; Sat 10:00, 12:00; Sun (Sept–Apr) 11:00, 12:00; Mid-Dec to 12 Jan 10:00, 11:00, 12:00, 14:00, 15:00 – for a fee of R15 (includes tasting)
RESTAURANT: Tues–Sat 10:00–15:00 (breakfast & lunch); Sun (Sept–Apr) 10:00–13:00 (brunch)
EST 1996 (House of JC le Roux). Wines produced since 1984. Area under vine: 28ha (99% white). Annual production: 6 000 tons, 5,4m litres
FLAGSHIPS: JC le Roux Chardonnay, JC le Roux Pinot, Pongrácz

LISTED WINES

Desiderius
MCC SPARKLING: from Chardonnay & Pinot (U)

JC le Roux
MCC SPARKLING: Chardonnay (U); La Vallée (U, Pinot); Pinot (U); Pinot Rosé (U, Merlot & Cab & Pinot)
D SPARKLING: S Blanc (U)
S SPARKLING: La Chanson (U, Pinotage & Shiraz); Le Domaine (U, S Blanc & Muscadel)

Pongrácz
MCC SPARKLING: Pongrácz (U, Pinot & Chardonnay)

HIGHLY RATED

JC le Roux Chardonnay
1996 – W4 02, JC 02
1990 – W4 98
JC le Roux Pinot Noir
1989 – VD 99, W4 98
Pongrácz
NV – W4 00, W3½ 99, W4 97

JD ▷ JEAN DANEEL

JEAN DANEEL

FRANSCHHOEK MAP

P CELLAR: Verdun Rd, Guldenheuwel, Franschhoek
PO Box 518 Franschhoek 7690
TEL: 021 876-3638 (sales); 082 553-2101 (cellar)

FAX: 021 886-2753
jdwines@worldonline.co.za
OWNERS: J Daneel, Gert Grobe
C'MASTER: Jean Daneel
VISITS: by appointment only
EST 1997. Annual production: 40 tons, 26 000 litres
FLAGSHIPS: Signature

LISTED WINES

JD

D WHITE: Chenin (W); Chenin S Blanc (U)
RED: Cab Merlot (W); Signature (W, Cab & Merlot)

JEAN LE LONG ▷ BOSCHENDAL
JEAN LE RICHE ▷ SIMONSIG
JOHN MARTIN ▷ BACKSBERG
JOHN X MERRIMAN ▷ RUSTENBERG

JONKHEER

ROBERTSON MAP

P CELLAR: Main St, Bonnievale
PO Box 13, Bonnievale 6730
TEL: 023 616-2137
FAX: 023 616-3146
info@jonkheer.co.za
OWNERS: Andries & Dirk Jonker
DIRECTORS: PK Albertyn (MD), PK Albertyn Snr, NC Jonker
GM: PK Albertyn
C'MASTER: Erhard Roothman
V'CULTURIST: Dirk Jonker
VISITS: by appointment only
EST 1912 (producing wine since 1930). Area under vine: 15ha (70% white). Annual production: 2 000 tons, 1,5m *l*
FLAGSHIPS: Jonkheer Family Reserve wines, Jonkheer Muscatheer

LISTED WINES

Bakenskop

D WHITE: S Blanc (U)
RED: Cab (W); Merlot (W)
FORTIFIED: Red Muscadel (U, Muscat de F); White Muscadel (U, Muscat de F)

Jonkheer

D WHITE: Chardonnay (U); Family Res Chardonnay (W)
RED: Cab Merlot (W); Family Res Cab (W); Pinotage (W)
FORTIFIED: Muscatheer (U, Muscat de F)

HIGHLY RATED

Bakenskop Rooi Volsoet Muskadel
2001 – W3½ 02
1999 – VD 01
Bakenskop Wit Volsoet Muskadel
1999 – TG 02, W5 02
1995 – VD 01
1991 – W4 98
Jonkheer Pinotage
2000 – W3½ 02

JOOSTENBERG

PAARL MAP

P CELLAR: Joostenberg Divisional Rd, Muldersvlei (north of N1/R304 junction)
PO Box 82, Elsenburg 7607
TEL: 021 884-4932

FAX: 021 884-4052
joostenberg@mweb.co.za
www.joostenberg.co.za
OWNER: Myburgh Winery (Pty) Ltd
C'MASTER: Tyrrel Myburgh
V'CULTURIST: Anette Myburgh
SALES: Mon–Fri 09:00–17:00
TASTINGS: no fee
N MONUMENT: manor house (1752)
EST 1800s (producing wine until 1947, and then again from 1999). Area under vine: 38ha (56% white). Annual production: 200 tons, 130 000 litres
FLAGSHIPS: Noble Late Harvest, Natural Sweet

LISTED WINES

D WHITE: Barrel-Fermented Chardonnay; Chenin (W); Riesling (U); Unwooded Chardonnay
RED: Merlot (W, Cab & Merlot); Shiraz (W)
DESSERT: Chenin Natural Sweet (W, Chenin)

JORDAN

STELLENBOSCH MAP

P CELLAR: Stellenboschkloof Rd, Vlottenburg
PO Box 12592, Die Boord, Stellenbosch 7613
TEL: 021 881-3441
FAX: 021 881-3426
info@jordanwines.com
www.jordanwines.com
OWNERS: Jordan family
DIRECTORS: Gary Jordan (MD), Ted Jordan, Kathy Jordan
W'MAKERS: Gary & Kathy Jordan, Johan Kruger
V'CULTURISTS: Gary Jordan
SALES: Nov–Apr Mon–Fri 10:00–16:30 Sat 09:30–14:30; May–Oct Mon–Fri 10:00–16:30 Sat 09:30–12:30
TASTINGS: fee of R10, refundable with purchases
TOURS: by appointment, at a fee of R10
PICNICS: welcome to bring
EST 1982 (producing wine since 1993). Area under vine: 96ha (50% red). Annual production: 700 tons, 500 000 litres
FLAGSHIP: Cobblers Hill
SINGLE VYD WINE: Cobblers Hill

LISTED WINES

Bradgate

D WHITE: Chenin S Blanc (U)
RED: Cab Merlot (W)

Jordan

D WHITE: Blanc Fumé (W, S Blanc); Chameleon White (U, S Blanc & Chardonnay); Chardonnay (W); Chenin (W); S Blanc (U)
OD WHITE: R Riesling (U)
RED: Cab (W); Chameleon Red (W, Cab & Merlot); Merlot (W); Cobblers Hill (W,

Cab & Merlot & Cab Franc)

HIGHLY RATED

Bradgate Chenin Blanc
Sauvignon Blanc
2001 – W3½ 02
Jordan Chardonnay
2001 – TG 02, W5 02, JC 02
2000 – W4 01, JC 01
1998 – W3½ 99
1996 – W4 97
Jordan Chenin Blanc
2001 – TG 02, W5 02, DT 01
Jordan Rhine Riesling
2001 – W3½ 02
Jordan Sauvignon Blanc
2002 – JC 02
1999 – W4 99
Jordan Blanc Fumé
2001 – TG 02, W5 02
1999 – W3½ 00
1997 – W4 97
1996 – W4 97
Jordan Cabernet Sauvignon
1998 – JC 01
1997 – W3½ 00
1996 – W3½ 99
Jordan Chameleon Cab Merlot
2000 – W3½ 02
1998 – VD 00
Jordan Cobblers Hill Reserve
1999 – W4½ 02
1998 – W4½ 01
1997 – W3½ 00
Jordan Merlot
2000 – W4 02
1999 – W3½ 01, VD 00, JC 00
1996 – W4 99
1995 – VD 97

JOSEPH'S HAT ▷ ASHANTI

JOUBERT-TRADAUW

LITTLE KAROO MAP

P CELLAR: R62, 12km west of Barrydale
PO Box 15, Barrydale 6750
TEL: 028 572-1619 (tastings, sales, cellar); 028 572-1315 (manor house)
FAX: 028 572-1315
joubert.r62@lando.co.za
www.joubert-tradauw.co.za
OWNER: J A Joubert
C'MASTER: Meyer Joubert
SALES: Mon–Fri 10:00–16:00
TASTINGS: fee of R5
TOURS: by appointment
ACCOM: Lentelus B&B
RESTAURANT: R62 Deli
PICNICS: available for purchase, by arrangement
EST 1950 (producing wine since 1999). Area under vine: 30ha (80% red).
Production: 300 tons, 25 000 *l*
FLAGSHIPS: R62 Cab Merlot, Chardonnay
SINGLE VYD WINES: Chardonnay

LISTED WINES

D WHITE: Chardonnay (W)
RED: R62 Cab Merlot (W); R62 Merlot Cab (W)

HIGHLY RATED

Joubert-Tradauw Chardonnay
2001 – W3½ 02

JOURDAN ▷ CABRIÈRE

JP BREDELL

HELDERBERG MAP

P CELLAR: Helderzicht, Winery Rd, Firgrove, Somerset West
PO Box 5266, Helderberg 7135
TEL: 021 842-2478
FAX: 021 842-3124
jpbredellwyne@new.co.za
OWNER: Anton Bredell
DIRECTORS: Donald Keys (MD), Anton Bredell
C'MASTER: Anton Bredell
SALES: Mon–Fri 08:–17:00
TASTINGS: fee of R5
TOURS: by appointment
EST 1933 (producing wine since 1965). Area under vine: 86ha (95% red). Production: 850 tons, 800 000 *l*
FLAGSHIPS: Cape Vintage Res Port, Shiraz

LISTED WINES

Bredell's

RED: Cab (W); Cab Merlot (W); Merlot (W); Pinotage (W); Shiraz (W)
PORT: Cape Vintage Reserve (W, Tinta B & Touriga N & Touriga Francessa & Souzão); LBV (W, Tinta B & Souzão)

HIGHLY RATED

Bredell's Merlot
1998 – W4 02
Bredell's Pinotage
1998 – W4 00
1996 – W4 98
Bredell's Shiraz
1999 – W4 01
1997 – W4 99
1996 – W4 98
Bredell's Cape LBV
1996 – W4 PS 01, VD 99
Bredell's Cape Vintage Reserve
1998 – VD 01, W4 01, DT 00, JC 00, JP 01
1997 – VD 00, JC 00, W4 00, JP 00
1995 – JP 99, W4½ 98
Bredell & Nel Cape Vintage Reserve
1997 – W4 01, SA Trophy 01

JUDEX ▷ BRAHMS
JUICY RED ▷ WOOLWORTHS
JULIA ▷ AVONDALE
JULIETTE ▷ STARKE

KAAIMANSGAT ▷ B FINLAYSON
KAAPSE VONKEL ▷ SIMONSIG

KAAPZICHT

STELLENBOSCH MAP

ESTATE: Bottelary Rd (M23)
PO Box 5, Sanlamhof 7532
TEL: 021 906-1620
FAX: 021 906-1622
kaapzicht@mweb.co.za
www.kaapzicht-wines.com
OWNER: Steytdal Farm (Pty) Ltd
DIRECTORS: Danie Louis Steytler (MD), George Louis Steytler, Zelda MW Steytler
C'MASTER: Danie Steytler

V'CULTURIST: George Steytler
SALES: Mon–Thurs 09:00–16:30, Fri 09:00–16:00, Sat 09:00–12:00
TASTINGS: fees for large groups only
TOURS: by appointment
ACCOM: The Vineyard Cottage, Tel 082 737-8329
PICNICS: welcome to bring
EST 1946 (producing wine since 1984). Area under vine: 146ha (70% red). Annual production: 700 tons, 500 000 litres
FLAGSHIPS: Steytler wines

LISTED WINES

Cape View
RED: Dry Red (U, Cinsaut & Cab & Merlot); Rouge De Kaap (U, Cab & Shiraz)

Friesland
D WHITE: Chenin (U)
RED: Dry Red (U, Cinsaut & Cab & Merlot)

Kaapzicht
D WHITE: Chenin (U); S Blanc (U)
RED: Bin 3 (W, Cab & Merlot); Cab (W); Classic Red (U, Cinsaut & Cab & Merlot); Estate Red (U, Cab & Shiraz); Merlot (W); Pinotage (W); Shiraz (W)

Steytler
RED: Pinotage (W); Vision (W, Cab & Pinotage & Merlot)

HIGHLY RATED

Cape View Cabernet Sauvignon
1996 – VD 97
Friesland Bin 3
1998 – W3½ 00
Kaapzicht Sauvignon Blanc
1999 – W3½ 00
1997 – W4 97
Kaapzicht Cabernet Sauvignon
1995 – PB 98
Kaapzicht Reserve Cabernet Sauvignon
1996 – W4 99
Kaapzicht Bin 3
2001 – W4 02
Kaapzicht Pinotage
2000 – JC 01
1998 – W4 99, PT 99
1996 – VD 97, PT 97
Kaapzicht Reserve Pinotage
1997 – W4½ 99, PT 98, VD 98
1996 – W4 98
Kaapzicht Shiraz
2000 – W4 02
1997 – W3½ 00
Kaapzicht Hanepoot Jerepigo
1998 – VD 99
Steytler Pinotage
1999 – W4 01, PB 00, PT 00, MG Trophy 00
1998 – W4½ PC 00
Steytler Vision
2000 – W4 02, JC 02, VD 01
Woolworths Terroir Stellenbosch Kaapzicht Cab Merlot
1997 – W3½ 99

KADETTE ▷ KANONKOP
KALKVELD ▷ ZANDVLIET
KALLISTA ▷ PARADYSKLOOF

KANGO

LITTLE KAROO MAP

CO-OP: Van der Riet St, Oudtshoorn
PO Box 46, Oudtshoorn 6620
TEL: 044 272-6065
FAX: 044 279-1038 (office); 044 279-2339 (cellar)
antonwyn@mweb.co.za
www.kangowines.com
DIRECTORS: Jannie Maree (MD), Stefan Maree, Vicci Schoeman, Theuns Mulder, Guilleaume Swiegers
GM: Johan Brits
C'MASTER: Anton Nel
V'CULTURIST: Willem Botha
SALES: Mon–Fri 08:00–16:30, Sat 09:00–13:00
TASTINGS: no fee
TOURS: by appointment
EST 1975. Area under vine: 295ha (82% white). Annual production: 3 000 tons, 2,2m litres
FLAGSHIPS: Mont Noir wines

LISTED WINES

Garob

SS WHITE: Grand Cru (U, Colombard & Chenin); Late Harvest (U, Colombard & Chenin)
RED: Selected Red (U, Ruby Cab & Cinsaut)
FORTIFIED: Red Muscadel (U)

Kango

BRANDY: Hanepoot; VO
WITBLITS: Vuurvoël
LIQUEUR: Pepperment; Van der Hum

Mont Noir

D WHITE: Grand Cru (U, Chenin & S Blanc & Chardonnay); S Blanc (U)
SS WHITE: Johannisberger (U, Colombard & White Muscadel)
RED: Dry Red (U, Ruby Cab & Merlot); Merlot (U); Pinotage (W)

Rijkshof

S SPARKLING: Vin Doux (U, Chenin & Hanepoot)
SS WHITE: Herfsgoud (U, Gewürz & Ferdinand de Lesseps)
RED: Claret (U, Ruby Cab & Cinsaut)
FORTIFIED: Golden Jerepigo (U, White Muscadel); Hanepoot (U); Red Jerepigo (U, Red Muscadel); Red Muscadel (U); White Muscadel (U)
PORT: Ruby (W, Tinta B)

Vlakteplaas

FORTIFIED: Red Muscadel (U); White Muscadel (U)

KANNALAND ▷ D DOORNKRAAL

KANONKOP

STELLENBOSCH MAP

ESTATE: R44, between Stellenbosch & Klapmuts
PO Box 19, Elsenburg 7607
TEL: 021 884-4656
FAX: 021 884-4719
wine@kanonkop.co.za
OWNERS: Johann & Paul Krige
C'MASTER: Beyers Truter
SALES: Mon–Fri 08:30–17:00, Sat 09:00–12:30
TASTINGS: no fee
EST 1910 (producing wine since 1931). Area under vine: 100ha (100% red). Annual production: 400 tons, 225 000 litres
FLAGSHIPS: Cab, Paul Sauer, Pinotage

LISTED WINES

RED: Cab (W); Kadette (W, various varieties); Paul Sauer (W, Cab & Merlot & Cab Franc); Pinotage (W)

HIGHLY RATED

Kanonkop Cabernet Sauvignon
1998 – TG 02, W5 02
1995 – W4 99
1994 – W4 97
1993 – W3½ 99
1992 – W4 99
1991 – W4 99
Kanonkop Kadette
1998 – DT 99
Kanonkop Pinotage
2000 – PT 01
1999 – W4½ 01, PT 00
1998 – W4 00, JP 00, MG 99, VD 99, PT 99
1997 – W4 99, PT 98, VD 98
1995 – PT 97
Kanonkop CWG Auction Reserve Pinotage
1999 – W4 01
Kanonkop Paul Sauer
1998 – TT 02, W5 02, JP 02, VD 01
1997 – VD 00, W3½ 00
1996 – JP 00, VD 99, W4 99
1995 – JP 99, W4½ 98, VD 98

KANU

STELLENBOSCH MAP

P CELLAR: Polkadraai Rd (R306), Stellenbosch
PO Box 548, Stellenbosch 7599
TEL: 021 881-3808
FAX: 021 881-3514
info@kanu.co.za
www.kanu.co.za
OWNER: Hydro Holdings
MGR: Ben Truter
C'MASTER: Teddy Hall
V'CULTURIST: Johann Schloms
SALES: Nov–Feb Mon–Fri 10:00–17:00 Sat 09:00–13:00; Mar–Oct Mon–Fri 10:00–16:30
TASTINGS: fee of R2 per glass
TOURS: by appointment
EST 1998. Area under vine: 70ha (50% red). Annual production: 600 tons
FLAGSHIPS: Cab, Chenin Wooded
SINGLE VYD WINES: Chardonnay,

Chenin Wooded, Shiraz

LISTED WINES

D WHITE: Chardonnay (W); Chenin (U); Chenin Wooded; S Blanc (U)

RED: Cab (W); Keystone (W, Cab & Merlot); Merlot (W); Red (U, Ruby Cab & Cinsaut); Shiraz (W)

DESSERT: Kia-Ora Noble Late Harvest (W, Chenin)

HIGHLY RATED

Kanu Chardonnay
2000 – W4½ 01
1998 – SA Certificate 00, W4 99
Kanu Limited Release Wooded Chenin Blanc
2001 – W4 02, PB 01, DT 01
2000 – TG 02, W5 02, CC 02, DT 01
1999 – W4 CC 01, PB 00
Kanu Chenin Blanc
2001 – W4 02, DW 01
Kanu Bulkamp Sauvignon Blanc
2001 – W3½ 01
2000 – W3½ 00
Kanu Cabernet Sauvignon
1999 – W3½ 01
1998 – W3½ 00, VD 01, JC 00
Kanu Keystone
1999 – W4½ 02, VD 01
1998 – W3½ 01
Kanu Merlot
1999 – VD 01
1998 – PB 99, W3½ 99
Kanu Red
1998 – W4 00
Kanu Shiraz
1998 – W4 00
Kanu Noble Late Harvest
1999 – W4 00

KAP SEKT ▷ NEDERBURG

KATBAKKIES

NÉGOCIANT: PO Box 21675, Kloof St 8008
TEL: 021 424-6883
FAX: 021 426-1967
avdwalt@inds.co.za
OWNER: Andries van der Walt
W'MAKER: A van der Walt
SALES: by appointment only
FUTURES (EN PRIMEUR): yes
EST 1999. Annual production: 2,5 tons, 1 600 litres
SINGLE VYD WINES: all

LISTED WINES

D WHITE: Arché Chenin in Hout (W); Skilpadvlei Chenin (W)

KAYA ▷ COPPOOLSE FINLAYSON
KEERMONT ▷ DE TRAFFORD
KEIZER'S CREEK ▷ ROODEZANDT
KELDERTROTS ▷ GOUDVELD
KELKIE ▷ BONNIEVALE
KELLERPRINZ ▷ DISTELL

KEN FORRESTER

HELDERBERG MAP

P CELLAR: Scholtzenhof farm, Forrester Vineyards, Winery Rd (off R44), between

Stellenbosch & Somerset West
PO Box 1253, Stellenbosch 7599
TEL: 021 855-2374
FAX: 021 855-2373
marketing@kenforrester-wines.com
www.kenforresterwines.com
OWNERS: Ken & Teresa Forrester
W'MAKER: Martin Meinert – wines made at Meinert cellar
V'CULTURIST: Shawn Smit, Aiden Morton (consultant)
VISITS: by appointment only
TASTINGS: at 96 Winery Rd Restaurant
ACCOM: Goose Cottage
RESTAURANT: 96 Winery Road, lunch Mon–Sun, dinner Mon–Sat, Tel 021 842-2020
EST 1993. Area under vine: 33ha (50% red). Annual production: 300 tons, 195 000 litres
FLAGSHIP: Forrester Family Res Grand Chenin

LISTED WINES

Forrester Family Reserve
D WHITE: Chenin (W)
RED: Grenache Syrah Res (W); Shiraz (W)

Forrester's
D WHITE: Petit Chenin (U)

Ken Forrester
D WHITE: Chenin (W); S Blanc
RED: Grenache Syrah (W)

T
SWEET: Noble Late Harvest (W, Chenin)

HIGHLY RATED

Ken Forrester Chenin Blanc
2001 – DT 01
2000 – W3½ 02
1995 – W4 98
Ken Forrester Barrel Reserve Chenin Blanc
2000 – DT 01
Ken Forrester Show Reserve Chenin Blanc
2000 – W3½ 01
T Noble Late Harvest
2000 – TG 02, W5 02
1999 – W3½ 00
Woolworths Ken Forrester Barrel Chenin Blanc
2001 – W3½ 02
Woolworths Barrel Reserve Chenin Noble Late Harvest (Ken Forrester)
1999 – W3½ 00

KEVIN ARNOLD ▷ WATERFORD
KEYSTONE ▷ KANU

KHANYA

P PRODUCER: wines of Nico Vermeulen, winemaker at Havana Hills
Van Schoorsdrift Rd, Philadelphia
PO Box 46241, Kernkrag 7440
TEL: 021 972-1106
FAX: 021 972-1105
rambo@telkomsa.net

OWNER: Nico Vermeulen
W'MAKER: N Vermeulen
VISITS: by appointment only
EST 1999. Annual production: 3 600 litres

LISTED WINES

D WHITE: S Blanc (U)
RED: Cab Merlot (W); Shiraz Cab (W)

KIA-ORA ▷ KANU
KIDDUSH ▷ ZANDWIJK

KLAWER

OLIFANTS RIVER MAP

CO-OP: next to N7, 3km from Klawer/Vredendal turn-off
PO Box 8, Klawer 8145
TEL: 027 216-1530
FAX: 027 216-1561
klawerwyn@kingsley.co.za
CHAIRMAN: Willem Botha
C'MASTER: Bob de Villiers
V'CULTURIST: Klaas Coetzee
SALES: Mon–Fri 08:00–17:00, Sat 09:00–12:00
TASTINGS: fee of R5pp for groups of 15 & more
TOURS: by appointment
PICNICS: available for purchase by arrangement; welcome to bring
EST 1953. Area under vine: 1253ha (87% white). Annual production: 24 000 tons, 18m litres

LISTED WINES

Birdfield

D WHITE: Chardonnay (W); Chenin (U); S Blanc (U)
RED: Merlot (U); Merlot Shiraz (W); Pinotage (U)

HIGHLY RATED

Klawer Red Muscadel 2001 – VD 01
Klawer White Muscadel 1997 – W3½ 98

KLAWERVLEI

STELLENBOSCH MAP

ESTATE: Koelenhof Rd (R304), Stellenbosch
PO Box 144, Koelenhof 7605
TEL: 021 865-2746
FAX: 021 865-2415
OWNER: Hermann Feichtenschlager
C'MASTER: Hein Feichtenschlager
SALES: Mon–Fri 10:00–17:00, Sat 10:00–13:00
TASTINGS: fee of R5
EST 1960s (producing wine since 1994). Area under vine: 35ha (75% red). Annual production: 175 tons
FLAGSHIP: Merlot
SINGLE VYD WINES: all
ORGANIC WINES: all

LISTED WINES

D WHITE: Chenin (U)
SS WHITE: Chenin (U)
RED: Cab (W); Merlot (W); Pinotage (W)

KLEIN BABYLONSTOREN ▷ B'BERG

KLEINBOSCH ▷ CAPE WINE CELLARS

KLEIN CONSTANTIA

CONSTANTIA MAP

ESTATE: Klein Constantia Rd, Constantia
PO Box 375, Constantia 7848
TEL: 021 794-5188
FAX: 021 794-2464
info@kleinconstantia.com
www.kleinconstantia.com
OWNERS: Duggie & Lowell Jooste
MD: Lowell Jooste
C'MASTER: Ross Gower
V'CULTURIST: Kobus Jordaan
SALES: Mon–Fri 08:30–17:00, Sat 09:00–13:00
TASTINGS: fee for tour groups of R15pp
TOURS: by appointment
N MONUMENTS: homestead, old wine cellar (±1820)
EST 1982 (producing wine since 1986). Area under vine: 81ha (68% white). Annual production: 450 tons, 15 000 *l*
FLAGSHIP: Vin de Constance

LISTED WINES

D WHITE: Chardonnay (W); S Blanc (U, S Blanc & Sémillon)
OD WHITE: R Riesling (U)
RED: Cab (W); Marlbrook (W, Cab & Merlot & Cab Franc); Pinot (W); Shiraz (W)
DESSERT: Vin de Constance (W, Muscat de F)

HIGHLY RATED

Klein Constantia Rhine Riesling
1998 – JC 00
1995 – W4 02
Klein Constantia Sauvignon Blanc
1997 – VD 97
1996 – W4½ 98
Klein Constantia Sémillon
1997 – W4 02
Klein Constantia Cabernet Sauvignon
1997 – W4½ 00
1995 – W3½ 99
1994 – W4 99
1992 – W4 99
1991 – W4 99
Klein Constantia CWG Cabernet Sauvignon
1995 – W3½ 98
Klein Constantia Reserve Cabernet Sauvignon
1997 – TG 02, W5 02
1993 – W4 99
Klein Constantia Marlbrook
1998 – W3½ 02
1995 – W4 99
Klein Constantia Pinot Noir
1998 – JC 00
Klein Constantia Shiraz
1999 – W4 02
1998 – W4 01
1997 – W4½ 99, VD 99, PB 98, DT 98
Klein Constantia Vin de Constance
1997 – W4½ 02, PB 01

1996 – W4½ 01, JP 01
1995 – W4 00, JP 00
1994 – W4½ 99, JP 99
1993 – W4 98

Klein Constantia Sauvignon Blanc Botrytis
1998 – VD 01, W4 01, MG 00, PB 99
1996 – PB 98

KLEINE DRAKEN ▷ ZANDWIJK
KLEINE HAZEN ▷ HAZENDAL

KLEINE ZALZE

STELLENBOSCH MAP

P CELLAR: Strand Rd (R44), Technopark turn-off
PO Box 12837, Die Boord 7613
TEL: 021 880-0717
FAX: 021 880-0716
quality@kleinezalze.co.za
www.kleinezalze.com
OWNERS: Kobus (JW) Basson, & Jan (JG) Malan
MD: Kobus Basson
C'MASTER: Willem (WF) Loots
V'CULTURIST: Schalk du Toit
SALES: Mon–Sat 09:00–17:00, Sun 11:00–16:00
TASTINGS: fee of R10pp, refundable with wine purchases exceeding R50
TOURS: by appointment, at a fee of R15pp
ACCOM: guest cottages
RESTAURANT: Tues–Sun 11:30–14:30
PICNICS: available for purchase
EST 1683 (producing wine since 1695). Area under vine: 54ha (85% red). Annual production: 1 500 tons, 1m litres
FLAGSHIPS: Vineyard Selection wines
SINGLE VYD WINES: Vineyard Selection

LISTED WINES

D WHITE: Chardonnay (U); Chenin Bush Vines (U); S Blanc (U); Vyd Sel Chardonnay Barrel Fermented; Vyd Sel Chenin Barrel Fermented
RED: Cab (W); Gamay Noir (U); Merlot (W); Pinotage (W); Red Zin (U, Cinsaut & Ruby Cab & Cab); Vyd Sel Cab Barrel Matured; Vyd Sel Merlot Barrel Matured; Vyd Sel Pinotage Barrel Matured; Vyd Sel Shiraz Barrel Matured

HIGHLY RATED

Kleine Zalze Barrel Fermented Chenin Blanc
201 – JC 02
Kleine Zalze Sauvignon Blanc
2001 – W3½ 01, MG 01
Kleine Zalze Cabernet Sauvignon
1999 – VD 01
Kleine Zalze Shiraz
1998 – MG 00

KLEIN GUSTROUW

STELLENBOSCH MAP

ESTATE: Jonkershoek, Stellenbosch
PO Box 6064, Stellenbosch 7612
TEL: 021 887-4556
FAX: 021 887-4556
OWNERS: Chris & Athalie McDonald
C'MASTER: Chris McDonald
VISITS: by appointment only
N MONUMENT: Cape Dutch homestead (1817)
EST 1981 (producing wine since 1993). Area under vine: 16ha (78% white). Annual production: 21 tons, 11 000 litres

LISTED WINES

RED: Cab Merlot (W)

HIGHLY RATED

Klein Gustrouw Cabernet Sauvignon Merlot
1999 – W3½ 02
1998 – W4 01
1997 – W4 00
1995 – PB 99

KLEIN SIMONSVLEI ▷ NIEL JOUBERT

KLEINVALLEI

PAARL MAP

P CELLAR: off Waterfall Rd, Daljosafat Valley, Paarl
PO Box 9060, Klein Drakenstein 7628
TEL: 021 868-8662
FAX: 021 868-3130
kleinvallei@mweb.co.za
OWNER: Dr PJ van Schaik
DIRECTORS: Pieter Jacobus van Schaik (MD), S Fischer
VISITS: by appointment only
EST 2000. Area under vine: 10ha (90% red). Annual production: 10 tons, 5 500 litres
FLAGSHIP: Cabernet

LISTED WINES

D WHITE: Chardonnay (W)
RED: Cab (W); Pinotage (W)

KLOMPZICHT

PAARL MAP

P CELLAR: in Paarl; private
PO Box 6353, Uniedal 7612
TEL: 021 867-0085; 083 229-5031
FAX: 021 867-0085
kcwc@iafrica.com
OWNERS: FJ & AD Klomp
MD: FJ Klomp
W'MAKERS: Hempies du Toit, M Carmichael Green
V'CULTURIST: Fred Duckett
VISITS: not open to the public
EST 1998. Area under vine: 24ha (100% red). Annual production: 140 tons, 14 000 litres
SINGLE VYD WINES: Merlot, Shiraz

LISTED WINES

RED: Merlot (W); Shiraz (W); Shiraz Cab (W)

KLOOFZICHT

TULBAGH MAP

P CELLAR: Straatskerk Rd, Tulbagh
PO Box 364, Tulbagh 6820
TEL: 023 230-0658
FAX: 023 230-0774
bevparr@icon.co.za
OWNERS: John & Beverley Parr
MD: John Hilliary Parr
C'MASTER: JH Parr
SALES: Mon–Sun 09:00–16:00
TASTINGS: no fee
TOURS: yes
PICNICS: welcome to bring
EST 1989. Area under vine: 2ha (100% red). Annual production: 12 tons, 8 400 litres
FLAGSHIP: Alter Ego
SINGLE VYD WINE: Alter Ego

LISTED WINES

D WHITE: Ethan Asher Sémillon (W)
RED: Asher Pinotage (W); Alter Ego (W, Cab & Merlot); Ethan Asher Cab (W)
FORTIFIED: Elroy's Pawt (W, Cab & Merlot)

KLOOVENBURG

SWARTLAND MAP

P CELLAR: Botmaskloof Pass, Riebeek-Kasteel Rd (R46)
PO Box 2, Riebeek-Kasteel 7307
TEL: 022 448-1635
FAX: 022 448-1035
kloovenburg@wacess.new.co.za
www.kloovenburg.co.za
OWNER: Pieter du Toit
C'MASTER: Pieter du Toit
V'CULTURIST: Kobus van Graan
SALES: Mon–Fri 09:00–16:30, Sat 09:00–14:00
TASTINGS: no fee
TOURS: by appointment
PICNICS: welcome to bring
EST 1860 (producing wine since 1860). Area under vine: 130ha (70% red). Annual production 200 tons, 120 000 litres
FLAGSHIPS: Chardonnay, Shiraz
SINGLE VYD WINES: Cab, Chardonnay, S Blanc, Merlot

LISTED WINES

D WHITE: Chardonnay (W)
RED: Cab (W); Merlot (W); Pinotage (W); Shiraz (W)

HIGHLY RATED

Kloovenburg Chardonnay 2000 – W3½ 01
Kloovenburg Shiraz 2000 – PB Trophy 01

KNIGHT ▷ TRAVINO

KNORHOEK

STELLENBOSCH MAP

P CELLAR: Knorhoek Rd, off R44, Koelenhof
PO Box 2, Koelenhof 7605
TEL: 021 865-2627 (sales); 021 865-2114 (tastings, manor

house)
FAX: 021 865-2627
office@knorhoek.co.za
www.knorhoek.co.za
OWNERS: Hansie & James van Niekerk
C'MASTER: Willem Loots
V'CULTURIST: Corius Visser
SALES: Mon–Sat 10:00–17:00
TASTINGS: no fee
TOURS: by appointment
ACCOM: guest house, Tel 021 865-2114, guesthouse@knorhoek.co.za
EST ±1900 (producing wine since 1997). Area under vine: 100ha (80% red). Annual production: 120 tons, 90 000 litres
FLAGSHIP: Cabernet

LISTED WINES

D WHITE: S Blanc (U)
RED: Cab (W); Pinotage (W)

HIGHLY RATED

Knorhoek Cabernet Sauvignon
1997 – W4½ 00, VD 99
Knorhoek Pinotage
1998 – W3½ 00
1997 – W4 00

KOELENHOF

STELLENBOSCH MAP

COMPANY: R304, 1km from the N1 en route to Stellenbosch
PO Box 1, Koelenhof 7605
TEL: 021 865-2020
FAX: 021 865-2796
koelwyn@mweb.co.za
OWNERS: 75 shareholders
DIRECTORS: JJN Cloete, FNH Wirth, PJ Johnson, JN Gerber, K Bellingan
GM: WCL de Vries
C'MASTERS: Wilhelm de Vries, Andrew de Vries
V'CULTURIST: Herman du Preez
SALES: Mon–Thurs 09:30–17:00, Fri 09:30–16:30, Sat 09:00–13:00
TASTINGS: no fee
PICNICS: welcome to bring
EST 1941. Area under vine 60% white. Annual production: 10 000 tons, 7,2m litres
FLAGSHIP: Pinotage Rosé

LISTED WINES

SPARKLING: Sparkling (U, S Blanc & Chenin)
D WHITE: S Blanc (U)
OD WHITE: Koelenhoffer (U, S Blanc & Chenin)
SS WHITE: Koelnektar (U, Gewürz)
SS PINK: Pinotage Rosé (U)
RED: Cab (W); Koelenberg (W, Cab & Merlot); Merlot (W); Pinotage (W)
PORT: Port (W, Pinotage)

HIGHLY RATED

Koelenhof Merlot
1999 – W3½ 02, VD 01
1998 – W4 99

KOELNEKTAR ▷ KOELENHOF
KONINGSHOF ▷ AVONDVREDE

KONYNWIJN ▷ HAZENDAL
KRANS ▷ DIE KRANS

KRANSKOP

ROBERTSON MAP

P CELLAR: Klaasvoogds East, 4km along gravel road off Robertson/Ashton Rd
PO Box 18, Klaasvoogds 6707
TEL: 023 626-3200
FAX: 023 626-3200
OWNER: TD Smit
C'MASTER: TD Smit
VISITS: by appointment only
EST 1980 (producing wine since 2001). Area under vine: 40ha (60% red). Annual production: 40 tons, 25 000 litres

LISTED WINES

D WHITE: Chardonnay (W)
RED: Cab (W); Merlot (W); Shiraz (W)

KRONE BOREALIS ▷ TJG
KRONENDAL ▷ CAPE CLASSICS
KUIERWYN ▷ DOMEIN DOORNKRAAL
KUMALA ▷ WESTERN WINES
KUMKANI ▷ STELLENBOSCH VYDS
KUPFERBERGER ▷ BERGKELDER
KWARTET ▷ NIETVOORBIJ

KWV

PAARL MAP

COMPANY: La Concorde, 57 Main St, S Paarl
PO Box 528, S Paarl 7624
TEL: 021 807-3900 (HO), 021 807-3008 (KWV wine shop)
FAX: 021 807-3349
customer@kwv.co.za
www.kwv-international.com
OWNERS: KWV Group Ltd
DIRECTORS: Vernon Davis (MD), DP van Staden, P Momberg, GL De Kock
C'MASTER: Sterik de Wet
SALES: at Wine Emporium, Kohler St, S Paarl, Mon–Sat 09:00–16:30, Tel 021 807-3007 – also at Laborie
EST 1918. Annual production: 9 000 tons, 12,6m litres bought in (6m litres under own labels)
FLAGSHIPS: Abraham Perold, Cathedral Cellar
SINGLE VYD WINE: Abraham Perold Shiraz

LISTED WINES – see also separate listing for Laborie

Abraham Perold

RED: Shiraz (W)

Cathedral Cellar

D WHITE: Chardonnay (W); S Blanc (U)
RED: Cab (W); Merlot (W); Pinotage (W); Shiraz (W); Triptych (W, Cab & Merlot & Shiraz)

KWV

D WHITE: Chardonnay (W); Chêne (W, Chenin); Chenin (U); S Blanc (U); Steen (U, Chenin)
RED: Cab (W); Merlot (W);

Pinotage (W); Roodeberg (W, Cab & Merlot & Shiraz); Shiraz (W)
FORTIFIED: Vintage Red Muscadel (W)
SHERRY: Medium Cream Very Old (W)
PORT: Full Ruby (W, Tinta B & Souzão); Full Tawny (W, Tinta B & Souzão); Millennium (W, Cinsaut & Tinta); Port (W, Tinta B & Souzão)
LIQUEUR: Van der Hum
BRANDY: 3 Star (Chenin & Colombard, 43% Vol); 10 Year Old (Chenin & Colombard & Cinsaut, 38% Vol); Diamond Jubilee (Chenin & Colombard & Cinsaut, 38% Vol); 20 Year Old (Chenin & Colombard & Cinsaut, 38% Vol); XO (Chenin & Colombard & Cinsaut, 40% Vol); Very Special Cape Brandy (Chenin & Colombard, 36% Vol); 5 Star VO (Chenin & Colombard, 43% Vol); VSOP (Chenin & Colombard & Cinsaut, 40% Vol); Imoya VSOP (Chenin & Colombard & Cinsaut, 40% Vol)

Pearly Bay
S SPARKLING: Celebration (U, Muscat d A)
D WHITE: Cape White (U, Chenin)
RED: Cape Red (W, Cinsaut & Ruby Cab & Pinotage)

Robert's Rock
D WHITE: Chardonnay Sémillon (W); Chenin Chardonnay (W)
RED: Cab Merlot (W); Cinsaut Ruby Cab (W); Merlot Cinsaut (W); Pinotage Pinot Noir (W); Shiraz Malbec (W)

HIGHLY RATED

Cathedral Cellar Chardonnay
2001 – W4½ 02
2000 – VD 01
Cathedral Cellar Cabernet Franc
1996 – SA Trophy 99
Cathedral Cellar Cabernet Sauvignon
1998 – W3½ 02
1995 – VD 97
Cathedral Cellar Merlot
1995 – PB 98
Cathedral Cellar Pinotage
1996 – PT 98, VD 98
Cathedral Cellar Shiraz
1997 – VD 99
1996 – VD 99
Cathedral Cellar Triptych
1998 – W3½ 02, MG 00
1996 – MG 98, VD 98
1995 – VD 98
KWV Pinotage
2000 – W3½ 02
KWV Roodeberg
1995 – VD 97
KWV Port
1999 – W4 02
KWV Full Tawny
W4 99, VD 98

KWV Light Tawny
1980 – W3½ 02
KWV Vintage Port
1997 – DT 00, W3½ 00, SA Trophy 00
KWV 1820 Full Cream Sherry
NV – VD 01
KWV VO Sherry
VD 99

LA BERYL ▷ CHARLES BACK

LABORIE

PAARL MAP

ESTATE: Taillefert St, Off Main Rd (R45), Paarl
PO Box 528, S Paarl 7624
TEL: 021 807-3390 (tastings, sales); 021 807-3196 (cellar)
FAX: 021 863-1955
therongi@kwv.co.za
www.kwv-international.com
OWNER: KWV International
MD: Vernon Davis
GM: Herman Le Roux
C'MASTER: Gideon Theron
V'CULTURIST: H van Reenen
SALES: in summer Mon–Sun 09:00–17:00; in winter Mon–Sat 09:00–17:00
TASTINGS: fee of R7
TOURS: by appointment, at a fee of R15
N MONUMENTS: manor house (±1800), cellar (1691)
RESTAURANT: Icampanelli, Tues–Sun 12:30–16:00 (kitchen closes 14:30), Tues–Sat 18:30–late (kitchen closes 21:30), Tel 021 807-3095
PICNICS: available for purchase
EST 1691 (producing wine since 1974). Area under vine: 39ha (70% red). Annual production: 500 tons, 339 000 litres
FLAGSHIP: Cabernet

LISTED WINES

MCC SPARKLING: Cap Classique (U, Chardonnay & Pinot)
D WHITE: Chardonnay (W); S Blanc (U)
RED: Cab (W); Merlot (W); Merlot Cab (W); Pinotage (W)
DESSERT: Pineau De Laborie (W, Pinotage)
BRANDY: Alambic (W, Chardonnay)

HIGHLY RATED

Laborie Brut
1996 – W4½ 01
1993 – W4 99
1992 – W4 98
Laborie Chardonnay
1997 – W3½ 98
Laborie Cabernet Sauvignon
2000 – VD 01
1998 – W3½ 00
1997 – W4 99
1996 – W3½ 98
Laborie Merlot Cabernet Sauvignon
1998 – W3½ 00

LA BRI

FRANSCHHOEK MAP

P CELLAR: Franschhoek
PO Box 180, Franschhoek 7690
TEL: 021 876-2086 (tastings, sales); 021 876-2593 (cellar)
FAX: 021 876-3197
info@la-bri.co.za
OWNER: RD Hamilton
MGR: J Haasbroek
C'MASTER: t.b.a. (busy changing)
V'CULTURIST: Jean Daneel (consultant)
VISITS: by appointment only
EST 1694 (producing wine since 1984). Area under vine: 18ha (90% red). Annual production: 150 tons, 90 000 litres

LISTED WINES

D WHITE: Ltd Release Chardonnay (W); Ltd Release Sémillon (W); Res Chardonnay (W); Sauvage (U, S Blanc)
RED: Ltd Release Cab (W); Res Cab Merlot (W)

LA CAVE ▷ WAMAKERSVALLEI
LA CHANSON ▷ JC LE ROUX
LA COTTE ▷ FRANSCHHOEK VYDS

LA COURONNE

FRANSCHHOEK MAP

ESTATE: Robertsvlei Rd, Franschhoek
PO Box 448, Franschhoek 7690
TEL: 021 876-2110/3939
FAX: 021 876-4168
lacouronne@mweb.co.za
OWNER: Ernst Pichler
DIRECTORS: Robert John McKee (MD), Werner Priklepil
C'MASTER: Dominic Burke
V'CULTURIST: Andrew Teubes
SALES: Mon–Fri 09:00–16:00, weekends by appointment
TASTINGS: fee of R25
ACCOM: hotel, Tel 021 876-2770
RESTAURANT: breakfast, lunch & dinner Mon–Sun, 07:00–21:00, Tel 021 876-2770
EST 1992. Area under vine: 12ha (80% red). Annual production: 60 000 litres
FLAGSHIP: Cabernet

LISTED WINES

D WHITE: Chardonnay (W); S Blanc (U)
RED: Cab (W); Ménage á Trois (W, Cab & Cab Franc & Merlot); Merlot (W); Rogues Rouge (W, Cab & Cab Franc & Merlot)

HIGHLY RATED

La Couronne Menage a Trois
2000 – W4 02
1999 – W3½ 01

LADISMITH

LITTLE KAROO MAP

CO-OP: 1 Nywerheid St, Ladismith
PO Box 56, Ladismith 6655

TEL: 028 551-1042
FAX: 028 551-1930
lkws@telkomsa.net
GM: André Simonis
C'MASTER: André Simonis
V'CULTURIST: Willem Botha
SALES: Mon–Fri 08:00–17:00
TASTINGS: no fee
TOURS: by appointment
EST 1939 (producing wine since 1948). Annual production: 8 000 tons
ORGANIC WINE: Green Wine S Blanc

LISTED WINES

African Sunset
D WHITE: Blanc de Blanc (U)
SS WHITE: Late Harvest (U)
PINK: Rosé (U)
RED: Ruby Red (U)

Green Wine
D WHITE: S Blanc (U)

Towerkop
D WHITE: Chardonnay (U); Chardonnay (W); Chenin (U); Colombar (U); Riesling (U)
OD WHITE: Stein (U, Chenin & Colombar)
RED: Ruby Cab (U)

LA FEMINITE ▷ LEMBERG
LA GRATITUDE ▷ DISTELL

LAIBACH

STELLENBOSCH MAP
P CELLAR: R44 between Stellenbosch & Klapmuts
PO Box 7109, Stellenbosch 7599
TEL: 021 884-4511
FAX: 021 884-4848
info@laibach.co.za
www.laibach.co.za
OWNERS: Dr Rudulf Kuehnen, Dr Petra Laibach-Kuehnen
GM: Michael Malherbe
C'MASTER: Stefan Dorst
V'CULTURIST: M Malherbe
SALES: Mon–Fri 09:00–17:00, Sat 09:00–13:00
TASTINGS: fee of R5
TOURS: by appointment
EST 1994 (producing wine since 1997). Area under vine: 39ha (80% red). Annual production: 300 tons, 210 000 litres
FLAGSHIP: Friedrich Laibach

LISTED WINES
D WHITE: Chardonnay (W)
OD WHITE: Chenin (U)
SS WHITE: Wooded Chenin
RED: Cab (W); Cab Merlot (W); Classic Cape Red (U, Cab & Cinsaut & Pinotage); Friedrich Laibach (W, Cab); Merlot (W); Pinotage (W)

HIGHLY RATED
Laibach Chardonnay
1997 – W3½ 98
Laibach Chenin Blanc
1997 – W4 98
Laibach Cabernet Sauvignon Merlot
1997 – W3½ 00, PB 98
Laibach Merlot

1998 – W3½ 00
Laibach Pinotage
1997 – W4 99
Laibach Natural Sweet
1997 – W3½ 98

L'AMI SIMON ▷ L'AVENIR

LAMMERSHOEK

SWARTLAND MAP

P CELLAR: April's Kloof, Malmesbury
PO Box 597, Malmesbury 7300
TEL: 022 482-2835 (tastings, sales); 022 482-3044 (cellar)
FAX: 022 482-2835
kretzelp@intekom.co.za
www.lammershoekwinery.co.za
OWNERS: P Kretzel, A Kretzel, G Stephan, S Stephan
MD: Paul Kretzel
C'MASTER: Paul Kretzel
V'CULTURIST: Anna Kretzel
SALES: Mon–Sun 09:00–17:00
TASTINGS: fee of R10
TOURS: by appointment
PICNICS: welcome to bring
EST 1971 (producing wine since 1999). Area under vine: 110ha (60% white). Annual production: 150 tons, 100 000 litres

LISTED WINES

D WHITE: Chardonnay (W)
OD WHITE: Chenin (W); Viognier (W)
RED: Cab Merlot (W); Pinotage (W); Shiraz Carignan (W, Shiraz & Carignan & Grenache & Viognier); Tinta Barocca (W); Zinfandel (W)
DESSERT: Hârslevelü Noble Late Harvest (W)
PORT: Cape Ruby (W, Zin)

HIGHLY RATED

Lammershoek Barrique Chenin Blanc
2000 – W4 02
Lammershoek Cabernet Sauvignon Merlot Barrique
2000 – W3½ 02

LA MOTTE

FRANSCHHOEK MAP

P CELLAR: Main Rd (R45)
PO Box 94, Paarl 7622
TEL: 021 876-3119
FAX: 021 876-3446
cellar@la-motte.co.za
www.la-motte.com
OWNER: Hanneli Koegelenberg
DIRECTORS: Hanneli & Hein Koegelenberg
C'MASTER: Jacques Borman
V'CULTURIST: Pietie le Roux
SALES: Mon–Fri 09:00–16:30, Sat 10:00–15:00
TASTINGS: fee of R5
TOURS: by appointment, at a fee of R10
N MONUMENTS: manor house & jonkershuis (1751), cellar (1872), water mill (1793)
RESTAURANT: cheese platters Nov–Mar

EST 1752 (producing wine since 1970). Area under vine: 108ha (65% red). Annual production: 800 tons, 480 000 litres
FLAGSHIP: Shiraz

LISTED WINES

La Motte
D WHITE: Chardonnay (W); Ltd Release Chardonnay (W); S Blanc (U)
RED: Cab (W); Millennium (W, Cab & Merlot & Cab Franc); Shiraz (W)

Schoone Gevel (export only)
D WHITE: Chardonnay (W)
RED: Merlot (W)

HIGHLY RATED

La Motte Cabernet Sauvignon
1998 – W4½ 02
1993 – W3½ 98
La Motte Shiraz
1999 – PB 01
1995 – W3½ 00
La Motte Millennium
1997 – W3½ 01

LANDAU ▷ BLAAUWKLIPPEN

LANDAU DU VAL

FRANSCHHOEK MAP
P CELLAR: La Brie, Franschhoek
PO Box 104, Franschhoek 7690
TEL: 082 410-1130 (tastings, sales); 021 876-2317 (manor house)
FAX: 021 876-3369
OWNER: Basil Landau
C'MASTER: Jean Daneel
V'CULTURIST: Jaco Schwenke
VISITS: by appointment only
N MONUMENT: manor house (1787)
EST ±1700 (producing wine since 1995). Area under vine: 15ha (100% white). Annual production: 1 000 litres
SINGLE VYD WINES: S Blanc, Sémillon
WINES WITHOUT SULPHUR: S Blanc, Sémillon

LISTED WINES

D WHITE: S Blanc (U); Sémillon (W)

HIGHLY RATED

Landau du Val Sémillon
1998 – W4 99

LANDSKROON

PAARL MAP
P CELLAR: Suid-Agter-Paarl Rd, S Paarl (off R44)
PO Box 519, S Paarl 7624
TEL: 021 863-1039
FAX: 021 863-2810
landskroon@mweb.co.za
www.landskroonwines.com
OWNERS: Paul & Hugo de Villiers' Family Trusts
C'MASTER: Paul de Villiers (Jnr)
V'CULTURIST: Hugo de Villiers (Jnr)
SALES: Mon–Fri 09:00–17:00, Sat 09:00–13:00
TASTINGS: no fee

TOURS: by appointment
ACCOM: cottage, self-catering
PICNICS: welcome to bring
EST 1881 (producing wine since 1900s). Area under vine: 270ha (90% red). Annual production: 1 300 tons, 1m litres
FLAGSHIPS: Paul de Villiers wines

LISTED WINES

D WHITE: Chardonnay (W); Chenin (U); S Blanc (U)
OD PINK: Blanc de Noir (U, Pinot & Pinotage)
RED: Cab (W); Cab Franc Merlot (W); Cinsaut (W); Cinsaut Shiraz (W); Merlot (W); Paul de Villiers Cab (W); Paul de Villiers Shiraz (W); Pinotage (W); Shiraz (W)
DESSERT: Morio Muscat Jerepico (U)
VINTAGE PORT: Port (W, Tinta B & Tinta R & Souzão)

HIGHLY RATED

Landskroon Cabernet Sauvignon
1999 – W3½ 01
Landskroon Shiraz
1998 – W4 00
Landskroon Port
1997 – SA Trophy 02, W4½ PS Trophy 01, DW 00, PB 00
1996 – W3½ 00, SA Certificate 00, VD 99
1995 – W4 99, SA Trophy 99, VD 98
1994 – W3½ 98

LANDZICHT

FREE STATE

CO-OP: Voortrekker St, Jacobsdal
PO Box 94, Jacobsdal 8710
TEL: 053 591-0164
FAX: 053 591-0145
landzicht@inext.co.za
OWNER: GWK Ltd
MD: J van Dyk
GM: Ian D Sieg
C'MASTER: Chrisna Botha
V'CULTURIST: Dirk Malan
SALES: Mon–Fri 08:00–13:00 & 14:00–17:00, Sat 09:00–12:00
TASTINGS: no fee
TOURS: by appointment
N MONUMENT: Anglo Boer War graves
ACCOM: Veenwouden
PICNICS: welcome to bring
EST 1976. Area under vine: 380ha (70% white). Annual production: 4 000 tons, 3m litres
FLAGSHIPS: Cab, White Muscadel

LISTED WINES

D WHITE: Blanc de Blanc (U, Colombar & Chenin & S Blanc
S WHITE: Gewürztraminer (U, Natural Sweet)
RED: Landzicht (W, Cab); Merlot (W); Pinotage (W);

Shiraz (W)
FORTIFIED: Hanepoot (U); Muscadel (U, White Muscadel); Red Muscadel (U); Sherry (W)

HIGHLY RATED

Landzicht Hanepoot
2001 – W3½ 02
Landzicht Wit Muskadel
2001 – W4 02

LANGVERWACHT

ROBERTSON MAP

CO-OP: Robertson/Stormsvlei Rd, Bonnievale
PO Box 87, Bonnievale 6730
TEL: 023 616-2815
FAX: 023 616-3059
GM: CJ Gerber
C'MASTER: CJ Gerber
V'CULTURIST: B Stipp
SALES: Mon–Fri 09:00–12:30 & 13:30–17:00
TASTINGS: no fee
TOURS: by appointment
EST 1956. Area under vine: 600ha (90% white). Annual production: 10 000 tons
FLAGSHIPS: Colombar, Ruby Cab

LISTED WINES

WHITE: Chardonnay (U); Colombar (U); Colombar Chardonnay (U)
RED: Ruby Cab (U); Shiraz (U)

LANOY ▷ BOSCHENDAL

LANZERAC

STELLENBOSCH MAP

P CELLAR: Jonkershoek Rd, Uniedal, Stellenbosch
PO Box 6233, Uniedal, Stellenbosch 7612
TEL: 021 886-5641 (tastings, sales, cellar); 021 887-1132 (manor house)
FAX: 021 887-6998
wine@lanzerac.co.za
www.lanzeracwines.co.za
OWNER: Dr Christo Wiese
GM: Wynand Hamman
C'MASTER: Wynand Hamman
SALES: Mon–Fri 09:00–16:30, Sat 10:00–14:00
TASTINGS: fee of R15
TOURS: Mon–Fri 11:00 & 15:00
ACCOM: hotel, Lanzerac Manor, Tel 021 887-1132
RESTAURANTS: Governer's Hall, The Terrace, breakfast, lunch & dinner Mon–Sun
EST 1993 (producing wine since 1996). Area under vine: 50ha (80% red). Annual production: 400 tons, 225 000 litres
FLAGSHIP: Classic

LISTED WINES

Five Heirs
RED: Cab Franc (W); Merlot (W); Pinotage (W)

Lanzerac
D WHITE: Chardonnay (W);

S Blanc (U)
RED: Cab (W); Classic (W, Merlot & Cab & Cab Franc & Malbec); Merlot (W); Pinotage (W)

HIGHLY RATED

Lanzerac Cabernet Sauvignon
1999 – W3½ 02
1998 – W3½ 01
1995 – W4 98
Lanzerac Classic
2000 – W4 01, VD 01
Lanzerac Merlot
1995 – W4 98
Lanzerac Pinotage
2000 – W3½ 02
1998 – PT 00

LA PETITE FERME

FRANSCHHOEK MAP
P CELLAR: Franschhoek mountain pass
PO Box 55, Franschhoek 7690
TEL: 021 876-3016
FAX: 021 876-3624
lapetite@iafrica.com
OWNERS: Dendy Young family
C'MASTER: Mark Dendy Young
V'CULTURIST: John Dendy Young
SALES: Mon–Sun 08:30–16:00
TASTINGS: no fee
ACCOM: guest suites
RESTAURANT: Mon–Sun 12:00–16:00
EST 1794 (producing wine since 1996). Area under vine: 6ha (75% white). Annual production: 60 tons, 30 000 litres
FLAGSHIP: S Blanc
SINGLE VYD WINE: Shiraz

LISTED WINES

D WHITE: Blanc Fumé (W, S Blanc); Chardonnay (W); Chenin (W); S Blanc (U); Sémillon (W); Unwooded Chardonnay
OD WHITE: Nectar Du Val (U, Muscat d A & Chenin)
RED: Cab Merlot (W); Merlot (W); Pinotage (W); Shiraz (W)
DESSERT: La Petite Sieste (W, Chenin, Natural Sweet)

L'AUBERGE DU PAYSAN

P PRODUCER: off R44 into Winery Rd, right into Raithby Rd
PO Box 315, Somerset West 7129
TEL: 021 842-2008
FAX: 021 842-2308
www.aubergedupaysan.co.za
OWNERS: Frederick Thermann, Mike Kovensky
MD: F Thermann
W'MAKER: Tjuks Roos (Rust 'n Vrede)
SALES: by appointment only, with meals at restaurant, at retail stores
RESTAURANT: lunch Tues–Sat, dinner Mon–Sat

EST 1995 (producing wine since 1998). Area under vine: 2,5ha (100% red). Annual production: 12,5 tons, 100 000 litres

LISTED WINES

RED: Pinotage (W)

LAUGHING DUCK ▷ LEMBERG
LAUREAT ▷ DISTELL
LAURENTIUS ▷ NEETHLINGSHOF
LAURIES HARVEST ▷ SYLVANVALE
LA VALLEÉ ▷ JC LE ROUX

L'AVENIR

STELLENBOSCH MAP

ESTATE: R44, Stellenbosch
PO Box 1135, Stellenbosch 7599
TEL: 021 889-5001
FAX: 021 889-5258
lavenir@adept.co.za
www.lavenir.co.za
OWNER: Marc Wiehe
C'MASTER: Francois Naudé
V'CULTURIST: M Wiehe
SALES: Mon–Fri 10:00–17:00, Sat 10:00–16:00
TASTINGS: fee of R10
TOURS: by appointment
ACCOM: guest house
PICNICS: welcome to bring
EST 1992. Area under vine: 55ha (50% red). Annual production: 350 tons, 200 000 *l*
FLAGSHIPS: Cab, Chardonnay, Chenin, Pinotage, S Blanc
SINGLE VYD WINES: Chardonnay, Chenin

LISTED WINES

D WHITE: Chardonnay (W); S Blanc (U)
OD WHITE: Vin de Erstelle (U, Colombar & Riesling)
OD PINK: Rosé Maison (U, Cinsaut & various other varieties)
RED: Cab (W); L'Ami Simon (W, Merlot & Cab); Pinotage (W)
DESSERT: Pudding Wine (U, Colombar & Riesling); Vin de Meurveur (W, Colombar & Noble Late Harvest)
PORT: Cape Vintage (W, Cab)

HIGHLY RATED

L'Avenir Chardonnay
1999 – W4 00
L'Avenir Chenin Blanc
1997 – W4 CC 98
L'Avenir Sauvignon Blanc
2001 – W3½ 01
L'Avenir Cabernet Sauvignon
1997 – W4 00
1995 – W3½ 98
L'Avenir CWG Reserve Cabernet Sauvignon
1999 – W4 01
L'Avenir Merlot
2000 – W3½ 02
L'Avenir Pinotage
2000 – W4 01, VD 01, PT 01
1999 – W4 01, SA Trophy 01, PT 00, VD 00
1997 – W4 99, PT 98, VD 98

1996 – W4 98, VD 97, PT 97
L'Avenir CWG Reserve
Pinotage
2000 – W4 01
1996 – W4 98
L'Avenir L'Ami Simon
1997 – W4 00
1996 – VD 97
L'Avenir Vin de Meurveur
1999 – W3½ 00
1997 – PB 98, W4 98
L'Avenir Cape Vintage
1999 – W3½ 02

LE BONHEUR

STELLENBOSCH MAP

ESTATE: R44 between Stellenbosch & Paarl
PO Box 56, Klapmuts 7625
TEL: 021 875-5478
FAX: 021 875-5624
info@lebonheur.co.za
www.lebonheur.co.za
OWNER: Lusan Holdings (Pty) Ltd – see separate listing
DIRECTORS: R Jones (MD); A Diedericks, HA Hëyl, BCM Tan, Prof CJ van Wyk
GM: Sakkie Kotzé
C'MASTER: Sakkie Kotzé
V'CULTURIST: Louis Tshambu
SALES: Mon–Fri 09:00–17:00, Sat 10:00–16:00
TASTINGS: fee of R5
N MONUMENT: manor house (1820)
EST 1820 (producing wine since 1975). Area under vine: 70ha (50% red). Annual production: 7 tons, 67 500 *l*
FLAGSHIP: Cabernet

LISTED WINES

D WHITE: Chardonnay (W); S Blanc (U)
RED: Cab (W); Prima (W, Merlot & Cab)

HIGHLY RATED

Le Bonheur Cabernet Sauvignon
2000 – W4½ 02
1997 – W3½ 00, VD 99
1996 – W4½ 99
1995 – W4 99, VD 98
1994 – W4 99
1993 – W3½ 99
1992 – W3½ 99
Le Bonheur Prima
2000 – W4 02
1998 – W3½ 02
1997 – W3½ 01
1996 – W4 00, VD 98
1995 – W4 98

LE DOMAINE ▷ JC LE ROUX

LEE & JONES

WORCESTER MAP

P CELLAR: Breeland farm, Slanghoek, Rawsonville
PO Box 236, Rawsonville 6845
TEL: 023 349-3137
FAX: 023 349-3137
OWNERS: Lee & Jones families
C'MASTER: Mortimer Lee
V'CULTURIST: Willem Swart
VISITS: not open to the public

🍷 **LISTED WINES**

D WHITE: Chardonnay No 1 (U); Chenin No 1 (U); Sémillon No 1 (U)

LEEF OP HOOP ▷ E LE RICHE

LE GRAND CHASSEUR

ROBERTSON MAP

ESTATE: Le Chasseur, off R60, Robertson
PO Box 439, Robertson 6705
TEL: 023 626-1048 (tastings, sales, cellar); 023 626-1047 (manor house)
FAX: 023 626-1048
lgc@intekom.co.za
OWNER: Albertus Petrus de Wet
C'MASTER: Albertus de Wet
V'CULTURIST: Francois Viljoen
VISITS: by appointment only
N MONUMENT: "Boesman Grot"
EST 1881 (producing wine since 1920). Area under vine: 230ha (70% white). Annual production: 3 200 tons, 2,1m litres
FLAGSHIP: Shiraz
SINGLE VYD WINES: all

🍷 **LISTED WINES**

SPARKLING: Fish Eagle (U, Pinotage)
D WHITE: Chardonnay (W); Fish Eagle White (W, Chardonnay & Colombar); S Blanc (U)
SS PINK: Fish Eagle Rosé (U, Pinotage)
RED: Cab (W); Fish Eagle Red (W, Ruby Cab & Pinotage)

LEIDERSBURG

NÉGOCIANT: PO Box 12505, Stellenbosch 7613
TEL: 021 887-9937
FAX: 021 887-0566
dupwine@intekom.co.za
OWNER: Jan du Preez
W'MAKER: Jan du Preez
SINGLE VYD WINE: Cabernet

🍷 **LISTED WINES**

RED: Cab (W)

LEIPOLDT ▷ SPAR

LEMBERG

TULBAGH MAP

ESTATE: R46 Rd, Straatskerk Rd turn-off (dirt road)
PO Box 317, Tulbagh 6820
TEL: 023 230-0659
FAX: 023 230-0661
schindler@lando.co.za
OWNER: Klaus & Uschi Schindler
C'MASTER: Klaus Schindler
SALES: Mon–Sun 09:00–17:00
TASTINGS: fee of R5
TOURS: yes
ACCOM: guest house
RESTAURANT: light lunches by arrangement
EST 1928 (producing wine since 1995). Area under

vine: 4ha (50% red). Annual production: 5 tons, 9 000 litres

LISTED WINES
D WHITE: Lipovina (U, Hârslevelü); The Laughing Duck (U, S Blanc)
RED: La Féminité (W, Pinot); Phases of the Moon (W, Pinotage)

L'ÉMIGRÉ

STELLENBOSCH MAP
P CELLAR: Stellenboschkloof Rd, Vlottenburg, Stellenbosch
PO Box 14, Vlottenburg 7604
TEL: 021 881-3703 (cellar); 021 881-3702 (manor house)
FAX: 021 881-3773
OWNER: FV Gentis Trust
DIRECTORS: Frans Gentis (MD), Victor Gentis, Emile Gentis
C'MASTER: Emile Gentis
V'CULTURIST: Victor Gentis
SALES: Sat 09:30–16:00
TASTINGS: no fee
ACCOM: De Morgenzon, self-catering
PICNICS: welcome to bring
EST 1920s (producing wine since 1992). Area under vine: 56ha (60% red). Annual production: 400 tons, 250 000 litres
FLAGSHIPS: Nutwood Grove Cab, Nutwood Grove S Blanc
SINGLE VYD WINES: L'Émigré Shiraz, L'Émigré Pinotage

LISTED WINES
L'Émigré
D WHITE: Azure Chardonnay (W)
S WHITE: Glorieux (U, Chenin, Natural Sweet)
RED: Cimmiterre (W, Cinsaut & Pinotage & Malbec & Cab Franc); Pinotage (W); Shiraz (W)
FORTIFIED: Muscat d A (U)
PORT: Port (W, Tinta B & Cab, Vintage)
Nutwood Grove
D WHITE: S Blanc (U)
RED: Cab (W)

L'EMPEREUR ▷ SIGNAL HILL
LE PAVILLON ▷ BOSCHENDAL
LE PLEURS ▷ AVONDALE
LE REVE ▷ SAXENBURG
LE RICHE ▷ ETIENNE LE RICHE
LE SOMMET ▷ MONT DU TOIT
LEUWENBLANC ▷ SEIDELBERG
LIBERTÉ ▷ MEERENDAL
LIEBERSTEIN ▷ DISTELL

LIEVLAND

STELLENBOSCH MAP
ESTATE: R44 Rd, Stellenbosch
PO Box 66, Klapmuts 7625
TEL: 021 875-5226
FAX: 021 875-5213
lievland@icon.co.za
OWNER: Paul Benadé
GM: Jean Pienaar
C'MASTER: Jean Pienaar
SALES: Mon–Thurs

09:00–17:00, Fri 09:00–16:00, Sat 09:00–13:00
TASTINGS: fee of R10pp
PICNICS: welcome to bring
EST 1823 (producing wine since 1982). Area under vine: 65ha (60% red). Annual production: 350 tons
FLAGSHIPS: DVB, Shiraz, Syrah

LISTED WINES

D WHITE: Chardonnay (W); Cheandrie (U, Riesling & Chenin); W Riesling (U)
RED: Cab (W); Cab Franc (W); DVB (W, Cab & Merlot & Cab Franc); Lievlander (W, Cab & Cinsaut & Roobernet & Ruby Cab & Petit Verdot & Shiraz); Shiraz (W); Syrah (W)
DESSERT: Natural Sweet (U, Riesling & S Blanc & Chenin)

HIGHLY RATED

Lievland Cabernet Sauvignon
1994 – VD 98
Lievland Reserve Cabernet Sauvignon
1998 – W3½ 00
Lievland Shiraz
2000 – W4½ 02
1998 – W4 01, VD 99
Lievland Syrah
2000 – W4 02
Lievland DVB
1998 – W3½ 01
1997 – W3½ 00
1995 – W3½ 98
Lievland Reserve Noble Late Harvest
1997 – W4 99, VD 98
Woolworths Riesling (Lievland)
1996 – VD 97

LIFE FROM STONE ▷ SPRINGFIELD
LIMESTONE HILL ▷ DE WETSHOF

LINTON PARK

WELLINGTON MAP

P CELLAR: Slangrivier farm, Wellington
PO Box 1234, Wellington 7654
TEL: 021 873-1625
FAX: 021 873-0851
lpwines@mweb.co.za
www.lintonparkwines.co.za
OWNER: Linton Park Holdings
DIRECTORS: Andrew Siddle (MD), MC Perkins (British), GSG Brown (British), CP Davies
GM: Andrew Siddle
C'MASTER: Hennie Huskisson
V'CULTURIST: TC Botha
VISITS: by appointment only
N MONUMENT: homestead (1809)
EST 1699 (producing wine since 1998). Area under vine: 89ha (60% red). Annual production: 650 tons, 350 000 litres (112 500 litres under own label)
FLAGSHIP: Linton Park Shiraz
SINGLE VYD WINES: Linton Park Cab, Linton Park Chardonnay, Linton Park Merlot, Linton Park Shiraz

LISTED WINES

Capell's Court

D WHITE: Chardonnay (U); S Blanc (U)

RED: Cab (W); Merlot (W); Shiraz (U)

Linton Park

D WHITE: Chardonnay (W)

RED: Cab (W); Merlot (W); Shiraz (W)

HIGHLY RATED

Linton Park Chardonnay 1999 – W4 00

Linton Park Merlot 2000 – W4 02

LIPOVINA ▷ LEMBERG
LISA ▷ AMBELOUI
LIZA ▷ SIMONSIG
LONGITUDE ▷ FLAGSTONE
LONGMARKET ▷ WOOLWORTHS

LONG MOUNTAIN

NÉGOCIANT: 19 Herald St, Stellenbosch – not open to the public

PO Box 1324, Stellenbosch 7599

TEL: 021 880-1688

FAX: 021 880-1691

davies@prsa.co.za

www.longmountain.co.za

OWNER: Pernod Ricard South Africa

DIRECTORS: David de Mardt (MD), Vanda Davies

GM: Jaco Boonzaaier

W'MAKER: Jacques Kruger

SALES: at retail stores only

FLAGSHIPS: Long Mountain wines

LISTED WINES

Gecko Ridge

D WHITE: Chardonnay (U); Chenin Blanc (U); Dry White (U); S Blanc (U)

PINK: Rosé (U)

RED: African Red (W, Ruby Cab & Pinotage); Res Cab (W)

Long Mountain

D WHITE: Chardonnay (U); Chenin (U); S Blanc (U); Sémillon Chardonnay (U)

RED: Cab (W); Merlot Shiraz (W); Pinotage (W); Ruby Cab (W); Shiraz Cab (W)

HIGHLY RATED

Long Mountain Sauvignon Blanc 2000 – W4 00

LONGRIDGE

HELDERBERG MAP

P CELLAR: Eikendal Rd, off R44 between Somerset West & Stellenbosch

PO Box 99, Lyndoch 7630

TEL: 021 855-2004

FAX: 021 881-3690

winecorp@iafrica.com

www.winecorp.co.za

OWNER: Winecorp – see separate listing

MD: Gerrie Wagener

C'MASTER: Christo Versveld
V'CULTURISTS: Johan Smit, Orlando Filander
SALES: Nov–Apr Mon–Fri 09:00–17:00, Sat 09:00–13:00
TASTINGS: fee of R6
EST 1970 (producing wine since 1996). Area under vine: 20ha (60% red). Annual production: 250 tons, 200 000 litres
FLAGSHIP: Chardonnay
SINGLE VYD WINES: Longridge wines

LISTED WINES

Bay View
D WHITE: Chenin (W); S Blanc (U)
RED: Cab (W); Merlot (W); Pinotage (W); Shiraz (W)

Capelands
D WHITE: Chardonnay (W); Chenin (U); Classic Cape White (U)
RED: Classic Cape Red (U); Merlot (U) Pinotage (U)

Longridge
D WHITE: Chardonnay (W)
RED: Cab (W); Merlot (W); Pinotage (W)

HIGHLY RATED

Longridge Brut
1994 – W4 97
Longridge Chardonnay
1999 – JC 00
1998 – W4 99
Longridge Bay View Chenin Blanc
1999 – W4 01
Longridge Cabernet Sauvignon
1999 – W3½ 02, JC 01
1998 – W4 00
1996 – W3½ 99
Longridge Merlot
1998 – W4 00
Longridge Pinotage
1998 – W4 00

LORD NEETHLING ▷ NEETH'HOF

L'ORMARINS

FRANSCHHOEK MAP
P CELLAR: R45, Groot Drakenstein
PO Box 435, Franschhoek Valley 7690
TEL: 021 874-1026
FAX: 021 874-1361
sales@lormarins.co.za
www.lormarins.com
OWNER: Rupert Family Trust
MD: Hein Koegelenberg
C'MASTER: Wrensch Roux
V'CULTURIST: Danie Botha
SALES: Mon–Fri 08:30–16:30, Sat 10:00–15:00
TASTINGS: fee of R5
TOURS: by appointment
N MONUMENTS: cellar (1740s – gable installed 1799), manor house (1811)
PICNICS: available to purchase Nov–Apr
EST 1969 (producing wine since 1985). Area under vine: 170ha (60% white).

Annual production: 1 000 tons, 600 000 litres
FLAGSHIP: Merlot

LISTED WINES

D WHITE: Blanc Fumé (W, S Blanc); Chardonnay (W); Pinot Grigio (U); S Blanc (U)
RED: Cab (W); Merlot (W); Optima (W, Cab & Merlot & Cab Franc)

LOST HORIZONS

NÉGOCIANT: Trading at Simonsvlei, Old Paarl Rd, S Paarl
PO Box 568, S Paarl 7624
TEL: 021 863-3848
FAX: 021 863-3850
horizons@global.co.za
OWNER: NJ Cooper
GM: Jacques P Jordaan
V'CULTURIST: Schalk du Toit (Vinpro)
SALES: at Simonsvlei
EST 1995
FLAGSHIPS: Cab Merlot, Chardonnay

LISTED WINES

Lost Horizons

D WHITE: Chardonnay (U); Classic White (U, Chenin & S Blanc); S Blanc (U)
RED: Cab (U); Cab Merlot (U); Classic Red (U, Ruby Cab & Cab & Merlot); Merlot (U)

Quantum

PINK: Quantum Petillant Rosé (U, Chenin & Muscat & Ruby Cab)

LOUIESENHOF

STELLENBOSCH MAP

P CELLAR: Koelenhof Rd (R304), Stellenbosch
PO Box 2013, Stellenbosch 7601
TEL: 021 865-2632
FAX: 021 865-2613
lhofwine@iafrica.com
OWNER: Stefan Smit
C'MASTER: Stefan Smit
V'CULTURIST: Casper Burger
SALES: Mon–Fri 09:00–17:00, Sat 10:00–18:00, Sun in summer 10:00–18:00
TASTINGS: fee of R5 a wine
RESTAURANT: lunches Mon–Sat, 12:00–15:00
PICNICS: available for purchase by arrangement
EST 1993. Area under vine: 130ha (70% red). Annual production: 600 tons, 400 000 litres
FLAGSHIPS: Marbonne Brandy, Perroquet Cape Tawny Port
SINGLE VYD WINES: all

LISTED WINES

D WHITE: Chardonnay (U); Ltd Release Chardonnay (W); Ltd Release Chardonnay Sur Lie (W); S Blanc (U)
RED: Cab Cab Franc (W); Merlot (W); Perroquet

Rouge (U, Pinotage);
Pinotage (U); Shiraz (W);
Tinta Barocca (U)
FORTIFIED: Red
Muscadel (U)
PORT: Perroquet Cape
Tawny (W, Tinta B)
BRANDY: Marbonne
Potstill (W, Colombar & Ugni
Blanc)

LOUISVALE

STELLENBOSCH MAP

P CELLAR: Devon Valley Rd,
Devon Valley, Stellenbosch
PO Box 542, Stellenbosch 7599
TEL: 021 865-2422
FAX: 021 865-2633
louisval@iafrica.com
www.louisvale.com
OWNERS: Michael A Johnston,
Hendrik Kotze
DIRECTORS: Hendrik
Kotze (MD), Michael A
Johnston, Martin Delaney,
Jonathan Pedley
C'MASTER: Simon Smith
SALES: Mon–Fri 08:00–17:00,
Sat 10:00–13:00
TASTINGS: Mon–Fri
10:00–16:30, Sat 10:00–13:00 –
at a fee of R10
TOURS: by appointment
N MONUMENT: Sir Herbert
Baker manor house
EST 1989. Area under
vine: 26ha (60% red). Annual
production: 260 tons,
182 000 litres
FLAGSHIPS: Chardonnay,
Dominique
SINGLE VYD WINE: Chardonnay

LISTED WINES

LV

D WHITE: Unwooded
Chardonnay
RED: Cab (W); Pinotage (W)

Louisvale

D WHITE: Chardonnay (W);
Chavant Chardonnay (W)
RED: Cab (W);
Dominique (W, Cab &
Merlot); Merlot (W);
Ovation (W, Pinotage &
Merlot)

HIGHLY RATED

Louisvale Chardonnay
2000 – W3½ 02
Louisvale Merlot
1996 – W3½ 99
1994 – W4 97
Louisvale Cabernet
Sauvignon Merlot
1996 – VD 98
Louisvale Dominique
Cabernet Sauvignon Merlot
1999 – W3½ 02

LOURENSFORD

SOMERSET WEST MAP

P CELLAR: under construction,
ready 2003 – Lourensford Rd,
Somerset West
PO Box 16, Somerset West
7129
TEL: 021 886-5641 (Lanzerac,

also owned by Christo Wiese)
FAX: 021 847-1852
linda@lourensford.co.za
www.lourensford.com
OWNER: Christo Wiese
GM: Roux van Zyl
C'MASTER: Wynand Hamman
V'CULTURIST: Kallie Wolfaardt
SALES: details to be announced – cellar and sales office under construction, to be ready from 2003 harvest onwards
VISITS: arrangements not yet finalised; tasting room and cellar will be open to the public in future
EST 2003. Area under fine: 250ha (92% red). Annual production capacity: 2 500 tons, 1,6m litres (when in full production)

LISTED WINES

To be announced

LOURENS RIVER ▷ MORGENSTER
LOUWSHOEK ▷ DASCHBOSCH

LUDDITE

P PRODUCER: wines made at Beaumont
PO Box 656 Bot River 7185
TEL: 028 284-9450
FAX: 028 284-9617
luddite@telkomsa.net
OWNERS: Niels Verburg, Hillie Meyer
W'MAKER: Niels Verburg
SALES: by appointment only

LISTED WINES

RED: Shiraz (W)

LUIPAARDSBERG ▷ D'VILLE HILLS
LUNA DE MIEL ▷ AVONTUUR

LUSAN

PRODUCER-WHOLESALER: head office at Neethlingshof, Polkadraai Rd, Vlottenburg, Stellenbosch
PO Box 104, Stellenbosch 7599
TEL: 021 883-8988
FAX: 021 883-8941 / 8975
nee@mweb.co.za
www.neethlingshof.co.za
OWNERS: Lusan Holdings – 50% Distell, 50% Hans Schreiber
ACTING MD: Richard Jones

LISTED WINES

See separate listings for Alto, Le Bonheur, Neethlingshof, Stellenzicht, Uitkyk

LUSHOF

HELDERBERG MAP

ESTATE: Eikendal Rd, off R44 between Stellenbosch & Somerset West
PO Box 899, Stellenbosch 7599
TEL: 021 855-3134
FAX: 021 855-3623
henni@icon.co.za

www.lushof.co.za
OWNER: Hennie Steyn Trust
DIRECTORS: HJJ Steyn, J du Plessis, K Wessels, S Klymsmith
C'MASTER: Rod Easthope (consultant)
VISITS: by appointment only
EST 1960s (producing wine since 2000). Area under vine: 12ha (66% red). Annual production: 100 tons, 65 000 litres
FLAGSHIP: S Blanc

LISTED WINES

D WHITE: Chardonnay (W); S Blanc (U)
RED: Cab (W); Merlot (W)

LUTZVILLE

OLIFANTS RIVER MAP

COMPANY: R363, Lutzville
PO Box 50, Lutzville 8165
TEL: 027 217-1516
FAX: 027 217-1435
lutzville@kingsley.co.za
www.lutzvillevinyard.com
OWNERS: 109 farmers
DIRECTORS: JM Mostert (chairman), AP van Niekerk (MD), SJ le Roux, HG van Zyl, A Truter, DJ Retief
C'MASTER: Jacques du Toit (Jnr)
V'CULTURIST: Jannie Kotze
SALES: Mon–Fri 08:00–17:00, Sat 08:30–12:00
TASTINGS: no fee
TOURS: by appointment
N MONUMENT: Vlèrmuisklip
PICNICS: available for purchase by arrangement; welcome to bring
EST 1962. Area under vine: 2 100ha (93% white). Annual production: 38 300 tons, 30m litres
FLAGSHIP: Cabernet

LISTED WINES

Lutzville

SS SPARKLING: Sparkling (U, Colombar & Sémillon)
D WHITE: Chardonnay (U); Chardonnay Wooded; Colombar (U); S Blanc (U); Sémillon (U)
OD WHITE: Chenin (U)
RED: Cab (W); Merlot (W); Pinotage (W); Ruby Cab (U)

Fleermuisklip

SS WHITE: Bouquet Blanc (U, Morio Muscat); Late Harvest (U, Chenin); Somersoet (U, Colombar)
PINK: Blanc de Noir (U, Ruby Cab)
RED: Robyn (U, Ruby Cab)
DESSERT: Muscadel (U)

Friends

D WHITE: Misty Morning (U, Chenin & Colombar)
SS WHITE: Sunny Day (U, Colombar)
RED: Ruby Sunset (U, Ruby Cab & Cab)

HIGHLY RATED

Fleermuisklip Robyn
2000 – W3½ 01

LV ▷ LOUISVALE

LYNGROVE

STELLENBOSCH MAP

NÉGOCIANT: 15 Termo Ave, Technopark, Stellenbosch – winery not open to the public
PO Box 7275, Stellenbosch 7600
TEL: 021 880-1221 (tastings, sales); 021 842-2116 (manor house)
FAX: 021 880-0851
chris@baarsma.co.za
www.baarsma.com
OWNER: Baarsma
DIRECTORS: Chris Rabie (MD), Sip Baarsma, Peter Wagenaar, Wim van der Eyk, Tjeerd van der Hoek, Danie Zeeman
C'MASTER: Danie Zeeman
V'CULTURIST: Pikkie Grobbelaar
SALES: at retail stores only
ACCOM: guest house, www.lyngrove.co.za
EST 1985. Area under vine: 76ha (70% red). Annual production: 400 tons, 230 000 litres
FLAGSHIP: Reserve wines

LISTED WINES

D WHITE: Collection Chardonnay (U); Collection S Blanc (U)
RED: Collection Cab (U); Collection Pinotage (U); Collection Shiraz (U); Res Pinotage (W); Res Shiraz Pinotage (W)

LYRIC ▷ NEDERBURG

MADONNA'S KISSES ▷ APPELSDRIFT
MAGIA ▷ MÔRESON

MAIDEN WINES

NÉGOCIANT: 2km outside Vredendal
PO Box 185, Gordon's Bay 7151
TEL: 082 554-9395
FAX: 021 852-8693
mwines@mweb.co.za
www.maidenwines.com
OWNERS: Danie Hattingh, Peter Venter
C'MASTER: Napoleon Stouman
VISITS: by appointment only
FUTURES (EN PRIMEUR): yes
EST 2000. Annual production: 13 500 litres
FLAGSHIP: Private Reserve

LISTED WINES

D WHITE: Chardonnay (W, Chardonnay & Chenin)
RED: Private Res (W, Cab & Shiraz & Pinotage)

MAIN STREET

P CELLAR: 52 Main Rd, Paarl (Schoongezicht)
PO Box 2709, Paarl 7620
TEL: 021 872-3006
FAX: 021 872-3006

mainstreet@mweb.co.za
OWNER: Marais de Villiers
VISITS: by appointment only
FLAGSHIP: Shiraz
SINGLE VYD WINES: Merlot, Shiraz

LISTED WINES

RED: Cab (W); Merlot (W); Shiraz (W); Shiraz Cab (W)

MAISON ROUGE ▷ WOOLWORTHS

MAKRO

NÉGOCIANT-RETAILER: stores throughout the country
Private Bag X4, Sunninghill 2157
TEL: 011 797-0000 (HO, general); 011 797-0664 (wine division)
FAX: 011 803-5833 (HO, general); 011 803-8695 (wine division)
cbarton@makro.co.za
www.makro.co.za
OWNER: Masstores SA
CEO: Gareth E Owens
WINE BUYER: Carolyn Barton
SALES: at retail branches Mon–Fri 09:00–17:30, Sat 08:30–16:00
TASTINGS: at retail branches Sat 10:00–14:00 – no fee

LISTED WINES

@

D WHITE: S Blanc @ Blaauwklippen (U)
RED: Pinot @ Blaauwklippen (W); Shiraz @ Blaauwklippen (W); Zinfandel @ Blaauwklippen (W)

Babbling Brook

RED: Cab (W); Merlot (W); Pinotage (W)

Misty

D WHITE: Mountain (U, Chardonnay & S Blanc)

Mont d'Or

D SPARKLING: Brut (U); Vin Sec (U)
S SPARKLING: Vin Doux (U)

Thomas Kipling

RED: Cab (W); Pinotage (W); Shiraz (W)

Turtle Creek

D WHITE: Dry White (U, Chenin & Colombard & Muscadel); Light (U, Chenin & Colombard & Muscadel)
SS WHITE: Semi Sweet (U, Riesling & Muscadel)
SS PINK: Rosé (U, Muscat de F)
RED: Dry Red (W, Pinotage & Cinsaut)

Vintner's Choice

WHITE: Grand Cru; Light
OD WHITE: Johannisberger
SS WHITE: Stein
PINK: Rosé
RED: Dry Red

Other:

RED: Bouwland Res Cab (W); La Bri Cab Merlot (W); Landskroon Merlot Cab Franc (W); Mischa Res

Cab (W); Overgaauw Touriga Cab (W); Trillennium (W, Cab)
PORT: Boplaas Private Bin

HIGHLY RATED

Babbling Brook Cabernet Sauvignon
1997 – W3½ 00
1995 – W3½ 00
Babbling Brook Merlot
1999 – W4 01
Yellowwood Ridge Chardonnay Semillon
1999 – W3½ 00
Yellowwood Ridge Semillon Chenin Blanc
1997 – W3½ 98

MALAN ▷ SIMONSIG
MALVERNE ▷ CLOS MALVERNE
MAMREWEG ▷ DARLING CELLARS
MANUEL ▷ SAXENBURG
MARBLE ▷ SIMONSVLEI
MARGUERITE ▷ BEAUMONT
MARIE CHRISTINE ▷ AVONDVREDE
MARLBROOK ▷ KLEIN CONSTANTIA
MASKAM ▷ WEST CORP
MAS NICOLAS ▷ WILHELMSHOF
MASON'S HILL ▷ JA CLIFT
MASSWINE ▷ CITRUSDAL
MASTERS ▷ BOPLAAS, D KRANS, A HILL

McGREGOR

ROBERTSON MAP

CO-OP: Vrolijkheid, McGregor
Private Bag X619, McGregor 6708
TEL: 023 625-1741
FAX: 023 625-1829
mcg@intekom.co.za
www.mcgregor-wines.co.za
DIRECTORS: Pieter Le Roux Carstens (MD), PF Malan, JH Cloete, JH Malherbe, JG Fouché, AB Naudé
C'MASTER: Pieter le Roux Carstens
V'CULTURIST: Anton Laas
SALES: Mon–Thurs 08:00–12:30 & 13:30–17:00, Fri 08:00–17:00, Sat 09:00–14:00
TASTINGS: no fee
TOURS: by appointment
PICNICS: welcome to bring
EST 1948. Annual production: 10 000 tons
FLAGSHIP: Village Red
SINGLE VYD WINES: Cab, Res S Blanc

LISTED WINES

D WHITE: Chardonnay (W); Chenin (U); Colombar (U); Colombar Chardonnay (U); S Blanc (U)
RED: Cab (W); Pinotage (W); Res Cab (W); Village Red (W, Ruby Cab)
DESSERT: Red Muscadel (U); White Muskadel (U)

MEANDER ▷ GROOT EILAND
MEDALLION ▷ NAPIER

MEERENDAL

DURBANVILLE MAP

ESTATE: Vissershok Rd (M48), Durbanville

PO Box 2, Durbanville 7551
TEL: 021 975-1655 (tastings, sales, cellar); 021 976-1915 (manor house)
FAX: 021 975-1657
meerendal@mweb.co.za
www.meerendal.co.za
OWNER: JCF Starke Trust
MD: Christa Starke
C'MASTER: Soon Potgieter
V'CULTURIST: Jan Marais
SALES: Mon–Fri 09:00–17:00, Sat 09:00–13:00
TASTINGS: fee of R10pp
EST 1716 (producing wine since 1952). Area under vine: 150ha (70% red). Annual production: 1 000 tons, 480 000 litres

LISTED WINES

D WHITE: Blanc de Blanc (U, Chardonnay & S Blanc); Gewürz (U); Liberté Blanc (W, Chardonnay & Gewürz); S Blanc (U)
RED: Cab (W); Cabochon (W, Cab & Merlot); Merlot (W); Pinotage (W); Shiraz (W)
DESSERT: Natural Sweet (U, Gewürz)

HIGHLY RATED

Meerendal Pinotage
1996 – W3½ 98
Meerendal Cabochon
1998 – W3½ 00
Meerendal Natural Sweet
1999 – W3½ 02
1998 – W4 99

MEERHOF

SWARTLAND MAP

P CELLAR: R46 between Malmesbury & Riebeek Kasteel/Hermon
PO Box 1229, Malmesbury 7550
TEL: 082 786-1843
FAX: 021 421-6191
cobusk@meerhof.co.za
OWNER: Cobus Kotzé, Mountain View partners
DIRECTORS: Cobus Kotzé (MD), Dave Thompson, Brad Harris, Willem Malherbe
C'MASTER: Frikkie Botes
V'CULTURIST: Paul Wallace
SALES: Mon–Fri 09:00–17:00, Sat 09:00–13:00
TASTINGS: no fee
TOURS: by appointment
FUTURES (EN PRIMEUR): yes
N MONUMENT: stone statue with bronze plate commemorating Pieter van Meerhof's first visit in 1661
RESTAURANT: Meerhof
PICNICS: available for purchase; welcome to bring
EST 18th Century (producing wine since 2001). Area under vine: 75ha (80% red). Annual production: 300 tons, 180 000 litres
FLAGSHIP: De la Caille wines

LISTED WINES

D WHITE: De la Caille Chardonnay (W)

RED: De la Caille Pinotage (W); De la Caille Cab (W); De la Caille Shiraz (W); Rolling Hills (U, Cinsaut & Shiraz); Wild Kwagga (U, Cinsaut & Shiraz)

MEERLUST

HELDERBERG MAP

P CELLAR: Baden Powell Drive (R310), Exit 33 from the N2
PO Box 15, Faure 7131
TEL: 021 843-3587
FAX: 021 843-3274
info@meerlust.co.za
www.meerlust.co.za
OWNER: Hannes Myburgh
GM: Trudy Dempsey
C'MASTER: Giorgio Dalla Cia
V'CULTURIST: Roelie Joubert
SALES: Mon–Thurs 09:00–17:00, Fri 09:00–16:30
TASTINGS: by appointment only, at a fee of R60
TOURS: by appointment, at a fee of R60
N MONUMENT: estate (1693)
EST 1693. Area under vine: 150ha (80% red). Annual production: 350 000 litres
FLAGSHIP: Rubicon

LISTED WINES

Dalla Cia

D WHITE: Chardonnay (U); S Blanc (U)

Meerlust

D WHITE: Chardonnay (W)
RED: Merlot (W); Res Pinot (W); Rubicon (W, Cab & Merlot & Cab Franc)
GRAPPA: Cab Merlot (U); Pinot Chardonnay (U); Res Cab Merlot (W)

HIGHLY RATED

Meerlust Merlot
1995 – W4 98
Meerlust Reserve Pinot Noir
1998 – PB 01
Meerlust Rubicon
1995 – W4 00
1994 – W4 98

MEIFORT ▷ BUITENVERWACHTING

MEINERT

STELLENBOSCH MAP

P CELLAR: Devon Valley, Stellenbosch
PO Box 7221, Stellenbosch 7599
TEL: 021 865-2363
FAX: 021 865-2363
info@meinertwines.com
www.meinertwines.co.za
OWNER: Martin Meinert
C'MASTER: Martin Meinert
VISITS: not open to the public
TASTINGS: at 96 Winery Road Restaurant, off Stellenbosch-Somerset West Rd, Mon–Sun 09:00–12:30 & 15:00–17:00
EST 1987 (producing wine since 1991). Area under vine: 11ha (100% red). Annual production: 24 500 litres

LISTED WINES
RED: Cab (W); Merlot (W); Synchronicity (W, Cab & Merlot & Pinotage & Cab Franc)
HIGHLY RATED
Meinert Merlot
1997 – W3½ 00
Meinert Cabernet Sauvignon Merlot
1997 – W4 00

MELCK'S ▷ MURATIE

MELLASAT

PAARL MAP

P CELLAR: Dekkersvlei, Klein Drakenstein, Paarl
PO Box 7169, Paarl 7623
TEL: 021 862-4525
FAX: 021 862-4525
mellasat@mweb.co.za
www.mellasat.com
OWNER: Mellasat Wine & Fruit cc
DIRECTORS: Stephen Richardson (MD), Alison Richardson, Poena Malherbe
C'MASTER: Stephen Richardson
V'CULTURIST: Poena Malherbe
VISITS: by appointment only
EST 1996 (producing wine since 1999). Area under vine: 8ha (80% red). Annual production: 30 tons, 21 000 litres
FLAGSHIP: Shiraz
LISTED WINES
D WHITE: Res Chenin (U); Res S Blanc (U)
RED: Cab (W); Pinotage (W); Res Cab (W); Res Pinotage (W); Shiraz (W)
HIGHLY RATED
Mellasat Pinotage
2000 – W3½ 02
Mellasat Dekkersvlei Reserve Pinotage
1999 – W3½ 00

MELLOW RED ▷ CAPE BAY
MEMORANTI ▷ AFRICAN WINES
MÉNAGE À TROIS ▷ LA COURONNE
MERINDOL ▷ SIMONSIG

MERWESPONT

ROBERTSON MAP

CO-OP: Bonnievale/Swellendam Rd
PO Box 68, Bonnievale 6730
TEL: 023 616-2800
FAX: 023 616-2734
merwespont@lando.co.za
GM: Dirk Cornelissen
C'MASTER: Esmarie Smuts
V'CULTURIST: Brian Stipp
SALES: Mon–Fri 09:00–12:30 & 13:30–17:00; Sat–Sun by appointment
TASTINGS: no fee
TOURS: during harvest, by appointment
ACCOM: Merwensien B&B, Tel 023 616-2806
PICNICS: welcome to bring
EST 1956. Area under vine: 700ha (33% red). Annual

production: 8 300 tons, 5m litres
FLAGSHIP: Cabernet

LISTED WINES

Agulhas
OD WHITE: White (U, S Blanc & Colombar & Chenin)
SS WHITE: Late Harvest (U); Stein (U)
RED: Red (U, Ruby Cab & Pinotage)

Merwespont
D WHITE: Chardonnay (W); S Blanc (U)
S WHITE: Special Late Harvest (U, Morio Muscat)
RED: Cab (W); Shiraz (W)

MERWIDA

WORCESTER MAP

P CELLAR: Rawsonville
PO Box 4, Rawsonville 6845
TEL: 023 349-1144
FAX: 023 349-1953
wines@merwida.com
www.merwida.com
OWNERS: Schalk & Pierre van der Merwe
GM: Magnus Kriel
C'MASTER: Magnus Kriel
SALES: Mon–Fri 08:00–12:30 & 13:30–17:00, Sat 09:00–12:00
TASTINGS: no fee
ACCOM: country lodge, Tel 023 349-1435
EST 1840. Area under vines: 600ha (70% red). Annual production: 8 000 tons, 5,6m litres

LISTED WINES

Riverstone
D WHITE: Chardonnay Viognier (W); Chenin Viognier (U); S Blanc (U); Sémillon S Blanc (U); Viognier (W)
RED: Cab (W); Shiraz (W); Shiraz Cab (W)

MERWO ▷ MERWIDA
MEURVEUR ▷ L'AVENIR

MIDDELVLEI

STELLENBOSCH MAP

ESTATE: Flamingo St, Stellenbosch
PO Box 66, Stellenbosch 7599
TEL: 021 883-2565 (tastings, sales); 021 883-4516 (cellar)
FAX: 021 883-9546
info@middelvlei.co.za
www.middelvlei.co.za
OWNERS: Momberg family
DIRECTORS: JH Momberg (MD), MLN Momberg, BJS Momberg
C'MASTER: Tinnie Momberg
V'CULTURIST: Ben Momberg
SALES: Mon–Sat 10:00–16:30
TASTINGS: fee of R3
TOURS: Feb–Mar, by appointment
PICNICS: welcome to bring
EST 1918 (producing wine since 1942, under own label since 1973). Area under vine: 130ha (65% red). Annual

production: 1 000 tons, 400 000 litres
FLAGSHIP: Pinotage

LISTED WINES

Hagelsberg (export only)
RED: Merlot Pinotage (W); Robyn (U, Tinta B & Pinotage)

Middelvlei
D WHITE: Chardonnay (W)
RED: Cab (W); Pinotage (W); Pinotage Merlot (W); Shiraz (W)

HIGHLY RATED

Middelvlei Pinotage
1999 – W3½ 01
Middelvlei Pinotage Merlot
1996 – W4 99, VD 98

MIGRATION

NÉGOCIANT: PO Box 12505, Stellenbosch 7613
TEL: 021 887-9937
FAX: 021 887-0566
dupwine@intekom.co.za
www.migrationwines.co.za
OWNER: Jan du Preez
W'MAKER: Jan du Preez

LISTED WINES

D WHITE: S Blanc (U)
RED: Cab Merlot (W); Pinotage (W); Shiraz (W)

MIJN BURG

PAARL MAP

P CELLAR: off Klipheuwel Rd (R304), off N1 Klipdam, PO Box 191, Klapmuts 7625
TEL: 021 884-4892
FAX: 021 884-4882
mijnwyne@new.co.za
www.mijnburg.co.za
OWNER: PR Myburgh
DIRECTORS: Paul Roux, Wilma Myburgh
C'MASTER: Chris Williams
VISITS: by appointment only
EST 1999. Area under vine: 135ha (66% red). Annual production: 199 tons, 135 000 litres

LISTED WINES

D WHITE: Chenin (U)
RED: Pinotage (W)

MILESTONE ▷ DOMINION
MILLENNIUM ▷ LA MOTTE
MILL RACE ▷ VERGELEGEN
MILLSTONE ▷ STETTYN
MINE D'OR ▷ DE WETSHOF
MIRANDA ▷ AMBELOUI

MISCHA

WELLINGTON MAP

ESTATE: Voorgroenberg Rd, outside Wellington
PO Box 183, Wellington 7654
TEL: 021 864-1019
FAX: 021 864-2312
mischa@webmail.co.za
OWNER: John H Barns
GM: Gareth Barns
C'MASTER: Andrew Barns
VISITS: by appointment only
EST 1972 (producing wine

since 1999). Area under vine: 21ha (60% red). Annual production: 50 tons, 30 000 litres

LISTED WINES

Eventide

RED: Cab (W); Shiraz (W)

Mischa

RED: Cab (W); Shiraz (W); Shiraz Americana (W)

HIGHLY RATED

Eventide Cabernet Sauvignon
2000 – W4 02
Mischa Shiraz
2000 – W3½ 02, JC 02
1999 – W3½ 01

MISSIONVALE ▷ B FINLAYSON
MISTY MORNING ▷ LUTZVILLE
MISTY MOUNTAIN ▷ MAKRO
MISTY POINT ▷ BARRYDALE
MONDIAL ▷ AFRICAN WINES
MONIS ▷ DISTELL
MONRO ▷ VILLIERA

MONS RUBER

LITTLE KAROO MAP

ESTATE: N12 from Oudtshoorn/Beaufort West via Meiringspoort
PO Box 1585, Oudtshoorn 6620
TEL: 044 251-6550 (tastings); 044 251-6812 (manor house)
FAX: 044 251-6550
monsr@lantic.net
OWNERS: CJ & CE Meyer
DIRECTORS: Coenraad Jesephus Meyer (MD), Erhard Meyer
C'MASTER: Radé Meyer
SALES: Mon–Fri 09:00–17:00, Sat 09:00–13:00
TASTINGS: no fee
N MONUMENT: tasting room in restored hotel of late 19th Century fame
ACCOM: self-catering
EST 1850. Area under vine: 38ha (80% white). Annual production: 500 tons, 375 000 litres
FLAGSHIP: Elegantia

LISTED WINES

RED: Conari (U, Cab)
PORT: Elusivo (W, Muscadel, White Port); Port (W, Cab)
FORTIFIED: Elegantia (U, Cab, Jerepigo); Hanepoot Jerepigo (U); Red Muscadel Jerepigo (U); Regalis (U, Muscadel); Sultana Jerepigo (U)

MONTAGU

ROBERTSON MAP

CO-OP: Bath St, Montagu
PO Box 29, Montagu 6720
TEL: 023 614-1125
FAX: 023 614-1793
mkwkelder@lando.co.za
GM: SB Malan
C'MASTER: SB Malan
V'CULTURIST: C Wright
SALES: Mon–Fri 08:00–12:30 & 13:30–17:50, Sat 09:00–12:00

TASTINGS: no fee
TOURS: by appointment only
EST 1942. Area under vine: 727ha (81% white). Annual production: 12 000 tons, 9m litres
FLAGSHIP: Rooi Muskadel

LISTED WINES

SS SPARKLING: Vin Doux (U, White Muscadel)
D WHITE: Chardonnay (U); Chenin (U)
OD WHITE: Colombar (U); Mont Blanc (U)
SS WHITE: Late Harvest (U, Chenin)
RED: Merlot Ruby Cab (U)
DESSERT: Rooi Muskadel (U); Wit Muskadel (U)

MONTAGU ▷ RIETRIVIER
MONT BLANC ▷ MONTAGU

MONT DESTIN

PAARL MAP

P CELLAR: Valley Rd (R44), Klapmuts, Stellenbosch
PO Box 1237, Stellenbosch 7599
TEL: 021 875-5040 (sales); 021 875-5870 (manor house)
FAX: 021 875-5040
wine@montdestin.co.za
www.montdestin.co.za
OWNERS: Ernst & Samantha Bürgin
V'CULTURIST: André van den Berg
VISITS: to the cellar by appointment only
ACCOM: guest apartments
RESTAURANT: Rosemarie's, Mon & Wed–Fri 08:30–17:00, Fri & Sat 19:30–21:30, Tel 021 875-5443
PICNICS: available for purchase
EST 1996. Area under vine: 18ha (70% red). Annual production: 80 tons, 56 000 litres
FLAGSHIP: 1482
SINGLE VYD WINES: Chenin, Pinotage

LISTED WINES

D WHITE: Stone Pine White (U, S Blanc)
OD WHITE: Chenin (U)
RED: 1482 (W, Cab & Merlot & Shiraz); Pinotage (W)

HIGHLY RATED

Mont Destin Passioné 1998 – W3½ 00

MONT D'OR ▷ MAKRO

MONT DU TOIT

WELLINGTON MAP

P CELLAR: Blouvlei Rd, Wellington
PO Box 704, Wellington 7654
TEL: 021 873-7745
FAX: 021 864-2737
kelder@montdutoit.co.za
www.montdutoit.co.za
OWNER: Stephan du Toit
W'MAKERS: Bernd Philippi,

Bernhard Breuer, Pieter-Niel Rossouw
V'CULTURIST: Alwyn Myburgh
VISITS: by appointment only
EST 1996. Area under vine: 28ha (100% red). Annual production: 120 tons, 72 000 litres

LISTED WINES

RED: Hawequas (W, Cab & Merlot & Shiraz); Le Sommet (W, various cultivars); Mont du Toit (W, Cab & Merlot & Shiraz)

HIGHLY RATED

Mont du Toit
2000 – JC 02

MONTE VISTA ▷ MOOIUITSIG
MONTINO ▷ RIEBEEK
MONT NOIR ▷ KANGO

MONTPELLIER

TULBAGH MAP

P CELLAR: R46, Tulbagh
PO Box 24, Tulbagh 6820
TEL: 023 230-0631
FAX: 023 230-0631
jstheron@intekom.co.za
OWNERS: Jan & Gerd Theron
C'MASTERS: Jan & Gerd Theron
SALES: Mon–Fri 09:00–17:00, Sat 09:00–12:00 – from Dec 2002
TASTINGS: no fee
TOURS: from Dec 2002, for a fee
EST 1857 (producing wine since 1880s – commercially since 1950s). Area under vine: 40ha (75% white). Annual production: 450 tons, 270 000 litres
FLAGSHIP: Pinot Noir
SINGLE VYD WINES: all

LISTED WINES

MCC SPARKLING: Tethrippon (U, Riesling)
D WHITE: S Blanc (U)
OD WHITE: Chenin (U); Huiswyn (U, Colombar & Pinot Gris & Clairette Blanche); Private Bin 43 (U, Riesling)
PINK: Pinot Blanc de Noir (W)
RED: Artemis (W, Merlot); Cab (W); Pinot (W)

MONT ROCHELLE

FRANSCHHOEK MAP

P CELLAR: Dassenberg Rd, Franschhoek
PO Box 334, Franschhoek 7690
TEL: 021 876-3000; 082 901-5076 (cellarmaster)
FAX: 021 876-2362
info@montrochelle.co.za
www.montrochelle.co.za
OWNER: Miko Rwayitare
C'MASTER: Justin Hoy
V'CULTURIST: Alwyn Geldenhuys
SALES: Mon–Sat 11:00–17:00, Sun 11:00–13:00
TASTINGS: fee of R10

TOURS: 11:00, 12:30, 15:00 on weekdays at a fee of R10; by appointment for large groups
PICNICS: available for purchase by arrangement
EST 1992 (producing wine since 1996). Area under vine: 9ha (50% red). Annual production: 200 tons, 140 000 litres
FLAGSHIP: Cabernet

LISTED WINES

D WHITE: Natural Chardonnay (U); Oak Matured Chardonnay; S Blanc (U)
RED: Cab (W); Merlot (W)

HIGHLY RATED

Mont Rochelle Oak Matured Chardonnay
1999 – W4 00, JC 00
Mont Rochelle Cabernet Sauvignon
1996 – VD 97
Mont Rochelle De Villiers Pinotage
1999 – W4 01

MOOIPLAAS

STELLENBOSCH MAP

ESTATE: Bottelary Rd (M23)
PO Box 104, Koelenhof 7605
TEL: 021 903-6273
FAX: 021 903-3474
info@mooiplaas.co.za
www.mooiplaas.co.za
OWNER: Mooiplaas Trust
OWNERS: Roos Family
C'MASTER: Louis Roos
V'CULTURIST: Tielman Roos
SALES: Mon–Fri 09:00–16:00, Sat 10:00–14:00
TASTINGS: no fee
TOURS: by appointment
N MONUMENT: manor house & cellar (1833)
EST 1963. Area under vine: 120ha (75% red). Annual production: 700 tons, 200 000 litres

LISTED WINES

D WHITE: S Blanc (U)
RED: Cab (W); Pinotage (W); Shiraz (W)

HIGHLY RATED

Mooiplaas Cabernet Sauvignon
1999 – W4 02
1997 – W3½ 00
Mooiplaas Pinotage
1999 – W3½ 01
1998 – W3½ 00, PT 99
Mooiplaas Shiraz
2000 – W4 02

MOOIUITSIG

ROBERTSON MAP

CO-OP: on Robertson/ Bonnievale Rd
PO Box 15, Bonnievale 6730
TEL: 023 616-2143
FAX: 023 616-2675
info@mooiuitsig.co.za
OWNER: Mooiuitsig Boere (Pty) Ltd
DIRECTORS: NL Jonker (MD), AJ Jonker, FA Claasen, JJ Claasen,

W'MAKER: Carel Hugo
V'CULTURISTS: C Mathee
SALES: not open to the public
TOURS: by appointment
EST 1947. Area under vine: 200ha (95% white). Annual production: 4 000 tons, 2,8m *l*
FLAGSHIPS: African Wine Adventure, Rusthof

LISTED WINES

African Wine Adventure
RED: (W, Cab & Merlot)

Clemence Creek
D WHITE: Chardonnay

Mooiuitzicht
S SPARKLING: Vin Doux (U, S Blanc)
D WHITE: S Blanc (U)
RED: Cab
BRANDY: Buchu; Da Gama; Spes Bona Hanepoot
OTHER SPIRITS: Gin, Vodka, Witblits

Oude Rust
DESSERT: Rooi Muskadel; Soet Hanepoot; Wit Muskadel

Rusthof
D WHITE: Chardonnay S Blanc Colombar (U)
RED: Dry Red (W, Merlot & Cab & Pinotage)

Other:
Bonwin Ruby, Marsala, Mooiuitsig Soet Hanepoot, Monta Vista Sherries, Nagmaal Wine, Oulap se Rooi, Overberg Club Sherry, Overberg Hanepoot, Ports, Rooi Jerepigo, Rooi Mustana, Trawler's Old Brown Sherry, Wit Jerepigo

MORAND ▷ DOUGLAS GREEN

MÔRESON

FRANSCHHOEK MAP
P CELLAR: Happy Valley Rd, La Motte, Franschhoek
PO Box 114, Franschhoek 7690
TEL: 021 876-3055
FAX: 021 876-2348
sales@moreson.co.za
www.moreson.co.za
www.pinehurst.co.za
OWNER: Richard Friedman
GM: Anton Beukes
C'MASTER: Christo Versfeld
V'CULTURIST: Anton Beukes
SALES: Mon–Sun 11:00–17:00
TASTINGS: fee of R10pp
RESTAURANT: Bread & Wine, Wed–Sun 12:00–16:00, Tel 021 876-3692

LISTED WINES

Môreson
MCC SPARKLING: Soleil du Matin (U, Chardonnay & Chenin)
D WHITE: Premium Chardonnay (W); S Blanc (U)
OD WHITE: Chenin (U)
RED: Cab (W); Magia (W, Cab & Merlot & Cab Franc); Merlot (W); Pinotage (W)

Morningside
D WHITE: White (U, S Blanc

& Chardonnay)
RED: Red (W, Cab & Merlot)

Pinehurst

D WHITE: Chardonnay (W); S Blanc (U)
OD WHITE: Chenin (U)
RED: Cab (W); Pinotage (W)

HIGHLY RATED

Môreson Premium Chardonnay
2000 – W3½ 02
1997 – VD 99

Môreson Sauvignon Blanc
2000 – JC 00

Môreson Cabernet Sauvignon
1998 – W4 02
1996 – W3½ 99
1995 – VD 97

Môreson Merlot
1999 – W3½ 02
1997 – W3½ 00
1995 – VD 97

Môreson Pinotage
2000 – TG 02, W5 02, PT 01
1999 – W4 01

Môreson Magia
1999 – W3½ 02, PB 01
1998 – W3½ 02

Morningside Cabernet Sauvignon Merlot
1999 – W4 02

Pinehurst Cabernet Sauvignon
1998 – W3½ 01
1997 – W4 99

Pinehurst Pinotage
2000 – W4 02

MORGENHOF

STELLENBOSCH MAP

ESTATE: Klapmuts Rd (R44), Stellenbosch
PO Box 365, Stellenbosch 7599
TEL: 021 889-5510
FAX: 021 889-5266
info@morgenhof.com
www.morgenhof.com
OWNER: Anne Cointreau-Huchon
C'MASTER: Rianie Strydom
V'CULTURIST: Pieter Haasbroek
SALES: May–Oct Mon–Fri 09:00–16:30 Sat–Sun 10:00–15:00; Nov–Apr Mon–Fri 09:00–17:30 Sat–Sun 10:00–17:00
TASTINGS: fee of R10
TOURS: yes; R17,50 fee for group tastings and viewing of underground cellar
RESTAURANT: Gazebo, May–Oct 12:00–14:30, Nov–Apr 12:00–15:00; breakfasts at Dame Jeanne Coffee Shop, 09:00–11:30
EST 1968 (producing wine since 1984). Area under vine: 67ha (50% red). Annual production: 250 tons, 170 000 litres
FLAGSHIP: Première Selection

LISTED WINES

MCC SPARKLING: Brut (W, Chardonnay & Pinot)
D WHITE: Chenin (W); S Blanc (U); S Blanc

Chenin (U)
RED: Cab (W); Merlot (W); Pinotage (W); Première Selection (W, Merlot & Cab Franc & Cab); Res Merlot (W)
PORT: Late Bottled Vintage (W, Tinta B)

HIGHLY RATED

Morgenhof Brut Reserve
1997 – VD 00
1995 – W3½ 99
Morgenhof Res Centenaire Brut
1997 – W4 00, JC 01
Morgenhof Chardonnay
1999 – W4 02
Morgenhof Chenin Blanc
2001 – W4 02
1999 – W4 01
Morgenhof Sauvignon Blanc
2001 – 01
Morgenhof Cabernet Sauvignon
2000 – W4 02
1999 – W3½ 02
1998 – JC 01
1996 – W3½ 98
Morgenhof Reserve Cabernet Sauvignon
1998 – W4 01, JP 01, VD 00
Morgenhof Merlot
1999 – JC 02
1998 – W4½ 00
Morgenhof Reserve Merlot
2000 – W3½ 02
Morgenhof Pinotage
2000 – W3½ 02
1999 – W4 01, JC 01
1997 – W3½ 99
Morgenhof Cabernet Franc Cabernet Sauvignon
2000 – W4 01
Morgenhof Cuvée Centenaire
1998 – PB 01
Morgenhof Merlot Malbec
1999 – W4 01
Morgenhof Première Selection
1998 – W4 02
1997 – W4 01
1996 – W4 99
1995 – JP 00
Morgenhof Noble Late Harvest
1998 – W4 01
Morgenhof Cape LBV
1997 – W3½ 02
1994 – W4 99
1993 – W4 98
Morgenhof Cape Vintage
1998 – VD 01
Morgenhof Centenaire Cape Vintage Reserve
1998 – VD 00, W4 00

MORGENSTER

SOMERSET WEST MAP

ESTATE: Vergelegen Rd, off Lourensford Rd, Somerset West
PO Box 1616, Somerset West 7129
TEL: 021 852-1738
FAX: 021 852-1141
wine@morgenster.co.za
www.morgenster.co.za
OWNER: Giulio Bertrand
C'MASTER: Marius Lategan
CONSULTANT: Pierre

Lurton (Cheval Blanc)
V'CULTURIST: Basie Fisher
SALES: Mon–Fri 10:00–17:00
TOURS: by appointment
N MONUMENT: manor house (1786)
EST 1993 (producing wine since 1998). Area under vine: 41ha (100% red). Annual production: 150 tons, 100 000 litres

LISTED WINES

Lourens River Valley

RED: Lourens River Valley (W, Merlot & Cab & Cab Franc)

MORGENTHAU ▷ BERNHEIM
MORNINGSIDE ▷ MÔRESON
MOSAIC ▷ BOWE JOUBERT
MOSCATO ▷ DISTELL
MOSSOPS ▷ AXE HILL
MOTIF ▷ STEENBERG
MOULIN ROUGE ▷ BERNHEIM
MOUNTAIN ▷ CORDOBA
MOUNTAINSIDE ▷ RUITERSVLEI
MOUNT CLAIRE ▷ CORDOBA
MOUNT MARBLE ▷ SIMONSVLEI
MOUNT MASKAM ▷ WEST CORP

MOUNT ROZIER

HELDERBERG MAP
ESTATE: Sir Lowry's Pass Village
PO Box 784, Somerset West 7129
TEL: 021 858-1130
FAX: 021 858-1131
winesbyjf@hotmail.com
www.mountrozier.co.za
OWNERS: D Lyddell, MJ Rubin, PA Loebenberg
MD: D Lyddell
C'MASTER: Jacques Fourie
V'CULTURIST: Johan Wiese
VISITS: by appointment only
EST 1997. Area under vine: 45ha (60% white). Annual production: 25 tons, 16 250 litres
FLAGSHIP: Cabernet

LISTED WINES

D WHITE: S Blanc (U)
RED: Annie Rozier (W, Cab & Merlot & Pinotage); Cab (W); Merlot (W)

MULDERBOSCH

STELLENBOSCH MAP
P CELLAR: Koelenhof Rd (R304)
PO Box 548, Stellenbosch 7599
TEL: 021 885-2488
FAX: 021 885-2351
info@mulderbosch.co.za
www.mulderbosch.co.za
OWNER: Hydro Holdings
MGR: Ben Truter
C'MASTER: Mike Dobrovic
V'CULTURIST: Johann Schloms
SALES: Mon–Fri 08:00–17:00
TASTINGS: by appointment only, at a fee of R25
EST 1989 (producing wine since 1991). Area under vine: 27ha (70% white). Annual production: 300 tons,

200 000 litres
FLAGSHIPS: Chardonnay Barrel Fermented, S Blanc

LISTED WINES

D WHITE: Chardonnay (W); Chardonnay Barrel Fermented; S Blanc (U); S Blanc Barrel Fermented; Steen op Hout (W, Chenin)
RED: Faithful Hound (W, Cab & Merlot & Malbec & Cab Franc)

HIGHLY RATED

Mulderbosch Chardonnay
2001 – W3½ 02
1998 – PB 99
1997 – PB 98
Mulderbosch Barrel Fermented Chardonnay
2000 – TT 02, W5 02
1999 – W4 00
1998 – W4 99, PB 99
1997 – W4½ 99
Mulderbosch Steen-op-Hout
2001 – W3½ 02
1998 – W4 99
1997 – W4 98
Mulderbosch Sauvignon Blanc
1997 – W4 98
Mulderbosch Barrel Fermented Sauvignon Blanc
2000 – W3½ 01
Mulderbosch Faithful Hound
2000 – W3½ 02
1999 – W3½ 01
1997 – W3½ 00

MURATIE

STELLENBOSCH MAP
ESTATE: Knorhoek Rd, off R44, Koelenhof
PO Box 133, Koelenhof 7605
TEL: 021 865-2330 (tastings, sales, cellar); 021 865-2906 (manor house)
FAX: 021 865-2790
muratie@kingsley.co.za
www.muratie.co.za
OWNERS: Melck family
DIRECTORS: APC Melck, RHM Melck, AJM Melck, AC Melck
GM: AC Melck
C'MASTER: Mark Carmichael-Green
V'CULTURIST: Paul Wallace
SALES: Mon–Fri 09:00–17:00, Sat–Sun 10:00–16:00
TASTINGS: fee of R2 per serving
EST 1920s. Area under vine: 35ha (98% red). Annual production: 200 tons, 130 000 litres
FLAGSHIP: Ansela van de Caab

LISTED WINES

D WHITE: Isabella (W, Chardonnay)
RED: Ansela van de Caab (W, Merlot & Cab & Shiraz); Cab (W); Melck's Dry Red (W, Cab & Merlot); Merlot (W); Pinot (W); Shiraz (W)
DESSERT: Amber (U, Muscat d A)
PORT: Ruby (W, Tinta B &

Tinta Röritz & Souzão); Vintage (W, Tinta B & Tinta Röritz & Souzão)

🍷 **HIGHLY RATED**

Muratie Shiraz
2000 – JP 02, W4 01
Muratie Vintage Port
1997 – W3½ 00

MUSCATHEER ▷ JONKHEER
MUSIC ROOM ▷ FLAGSTONE
MUSTIQUE ▷ SIMONSIG
MYRIAD ▷ BWC

NADINE ▷ RUPERT & ROTHSCHILD
NAKED TRUTH ▷ PICARDI REBEL
NALEDI ▷ SAVANHA
NAMAQUA ▷ WEST CORP

NAPIER

WELLINGTON MAP

P CELLAR: Off Bainskloof Rd
PO Box 638, Wellington 7654
TEL: 021 873-7829 (sales, cellar); 021 864-1231 (manor house)
FAX: 021 864-2728
napierwines@iafrica.com
OWNER: GRT Farming & Financial
DIRECTORS: Dr Christoph Hermann Kuhn (MD), L Desfarges, C Bertie, C Loubser
C'MASTER: Coenraad Leon Bester
VISITS: by appointment only
EST 1989 (producing wine since 1994). Area under vine: 21ha (60% white).
Annual production: 120 tons, 40 000 litres
FLAGSHIPS: Chardonnay, Red Medallion

🍷 **LISTED WINES**

D WHITE: Chardonnay (W); Chenin Blanc (U)
RED: Red Medallion (W, Cab & Merlot & Cab Franc)

NARINA ▷ SPAR
NARRABOS ▷ SPAR
NATURAL CORP ▷ CAPE CLASSICS
NECTAR DE PROV ▷ LOUWSHOEK
NECTAR DU VAL ▷ LA PETITE FERME

NEDERBURG

PAARL MAP

P CELLAR: Sonstraal Rd, Daljosafat, Paarl
Private Bag X3006, Paarl 7620
TEL: 021 877-5123 (tastings); 021 877-5123 (sales); 021 877-5145 (cellar); 021 877-5124 (manor house)
FAX: 021 862-4887 (public relations); 021 862-0878 (sales)
nedwines@distell.co.za
www.nederburg.co.za
OWNER: Distell
MD: Jan Scanell
C'MASTER: Razvan Macici
V'CULTURISTS: Hannes van Rensburg
SALES: Mon–Fri 08:30–17:00, Sat 10:00–14:00 – Nov–Mar Sat

10:00–16:00, Sun 11:00–16:00
TASTINGS: fee of R10
TOURS: by appointment, for a fee of R17 (weekdays), R25 (Sat), R35 (Sun)
N MONUMENT: manor house (1800)
PICNICS: available for purchase Nov–Mar
EST 1800. Under vine: 750ha (65% red). Annual production: 12 000 tons, 7,2m litres
FLAGSHIPS: Nederburg Auction wines
SINGLE VYD WINES: Reserve wines

LISTED WINES

Reserve

RED: Res Malbec (W); Res Petit Verdot (W)

Standard Range

MCC SPARKLING: Blanquette (W, Chardonnay)
D SPARKLING: Kap Sekt (W, Riesling & Crouchen & Chardonnay); Premier Cuvée Brut (U, Chenin & Crouchen & S Blanc)
D WHITE: Chardonnay (W); Paarl Riesling (U, Crouchen); Prelude (W, S Blanc & Chardonnay); Premier Grand Cru (U, Chenin & Crouchen & S Blanc); S Blanc (U)
OD WHITE: Elegance (U, Muscat & Riesling); Lyric (U, S Blanc & Crouchen & Chardonnay); R Riesling (U)
SS WHITE: Stein (U, Chenin & Riesling & Gewürz)
S WHITE: Special Late Harvest (U, Chenin & Riesling & Gewürz)
ROSÉ: Rosé (U, Cinsaut)
RED: Cab (W); Baronne (W, Dab & Shiraz); Duet (U, Pinot & Cab); Edelrood (W, Cab & Merlot); Pinotage (W)
DESSERT: Noble Late Harvest (U, Chenin & Muscat & Sémillon)
BRANDY: NV (W, S Blanc)

Auction Wines

RED: Private Bin R103 Cab Shiraz (W); Private Bin R108 Merlot (W); Private Bin R121 Shiraz (W); Private Bin R115 Shiraz Cab (W)
DESSERT: Edelkeur (U, Chenin, NLH); Eminence (U, Muscat, NLH)

HIGHLY RATED

Nederburg Blanquette
1993 – W3½ 98

Nederburg Reserve Chenin Blanc
2000 – W3½ 01

Nederburg Chenin Blanc
2002 – JC 02

Nederburg Bin 250 Auction Reserve Pinot Blanc
1998 – W3½ 01, PB 00

Nederburg Sauvignon Blanc
2001 – W4 01

Nederburg Private Bin 181 Merlot

2000 – JC 02
Nederburg Noble Late Harvest
1999 – W3½ 01
1997 – W4 99
Nederburg Edelkeur
Chenin Blanc
1998 – W3½ 02
1995 – W4 98
Nederburg Eminence Muscadel
2000 – JC 02
1999 – W3½ 01
1996 – W4 02
Nederburg Private Bin S333
Special Late Harvest
1997 – VD 98

NEETHLINGSHOF

STELLENBOSCH MAP

ESTATE: M12 between Stellenbosch & Kuils River
PO Box 104, Stellenbosch 7599
TEL: 021 883-8988
FAX: 021 883-8941
nee@mweb.co.za
www.neethlingshof.co.za
OWNER: Lusan Holdings
DIRECTORS: R Jones (MD); A Diedericks, HA Hëyl, BCM Tan, Prof CJ van Wyk
C'MASTER: Philip Costandius
V'CULTURISTS: Dawid Saayman, Eben Archer
SALES: Mon–Fri 09:00–17:00, Sat–Sun 10:00–16:00
TASTINGS: fee of R10
TOURS: by appointment
FUTURES (EN PRIMEUR): yes
N MONUMENT: manor house (1814)
RESTAURANT: Lord Neethling, Tel 021 883-8966, Fax 021 883-9803
PICNICS: welcome to bring
EST 1800. Area under vine: 210ha (70% red). Annual production: 1 200 tons, 900 000 litres
FLAGHIPS: Lord Neethling
SINGLE VYD WINES: Lord Neethling Cab, Lord Neethling Cab Franc, Lord Neethling Pinotage

LISTED WINES

D WHITE: Chardonnay (W); S Blanc (U)
OD WHITE: Gewürz (U)
RED: Cab (W); Cab Merlot (W); Lord Neethling Cab (W); Lord Neethling Cab Franc (W); Lord Neethling Laurentius (W, Cab & Merlot & Cab Franc); Lord Neethling Pinotage (W); Merlot (W); Pinotage (W); Shiraz (W)
DESSERT: W Riesling Noble Late Harvest (U)

HIGHLY RATED

Lord Neethling Cabernet Franc
1999 – W4 02
Lord Neethling Pinotage
1998 – W4½ 00, PT 01
1997 – W4 00, PT 00, VD 99
Lord Neethling Laurentius
1997 – W3½ 01
Neethlingshof Gewürztraminer
2001 – W3½ 02

Neethlingshof Cabernet Franc
1996 – W3½ 00
Neethlingshof Merlot
1996 – W3½ 01
Neethlingshof Weisser Riesling Noble Late Harvest
1998 – W4 01, JP 00
1997 – JP 99
1997 – VD 97
Neethlingshof Special Late Harvest
1998 – W3½ 99
1997 – W4 99

NEIL ELLIS

STELLENBOSCH MAP

P CELLAR: Jonkershoek Rd
PO Box 917, Stellenbosch 7599
TEL: 021 887-0649
FAX: 021 887-0647
tasting@neilellis.com
www.neilellis.com
OWNERS: Neil Ellis Wines
C'MASTER: Neil Ellis
SALES: Mon–Fri 09:30–16:30, Sat 10:00–14:00
TASTINGS: fee for groups

LISTED WINES

D WHITE: Elgin Chardonnay (W); Groenekloof S Blanc (U); Stellenbosch Chardonnay (W)
RED: Cab (W); Cab Merlot (W, Cab & Merlot & Cab Franc); Coastal Pinotage (W); Stellenbosch Shiraz (W); Vyd Selection Cab (W); Vyd Selection Shiraz (W)

HIGHLY RATED

Neil Ellis Chardonnay
1997 – W3½ 98
Neil Ellis Elgin Chardonnay
2001 – W4 02
1999 – VD 00
1998 – JP 00
1997 – W3½ 98
Neil Ellis Stellenbosch Chardonnay
2000 – W3½ 01
Neil Ellis Groenekloof Sauvignon Blanc
1999 – W4 99
1998 – W4 98
1997 – W4 98
1996 – W3½ 98
1993 – W3½ 98
Neil Ellis Cabernet Sauvignon
1998 – JC 00
1997 – W3½ 99, VD 99
1996 – W4 99
Neil Ellis Vineyard Selection Cabernet Sauvignon
1999 – W4 02
1998 – VD 01, W4½ 01
1997 – JP 00
Neil Ellis Swansong Pinotage
1998 – W3½ 00, PT 99
Neil Ellis Stellenbosch Shiraz
1999 – W4½ 01, SA Trophy 01
1998 – SA Trophy 00, W3½ 99
Neil Ellis Vineyard Selection Shiraz
1999 – W3½ 02, VD 01
1998 – W3½ 01, JC 00
Neil Ellis Cabernet

Sauvignon Merlot
1999 – W3½ 01
Woolworths Cabernet
Sauvignon Merlot (Neil Ellis)
1999 – W4 01

NELSON'S CREEK

PAARL MAP

ESTATE: R44, Windmeul
PO Box 2009, Windmeul 7630
TEL: 021 869-8453 (tastings, sales, cellar); 021 869-8463 (manor house)
FAX: 021 869-8424
info@nelsonscreek.co.za
www.nelsonscreek.co.za
OWNER: AJ Nelson
C'MASTER: Richard Kershaw
V'CULTURIST: Walter Schlomms
SALES: Mon–Fri 09:00–17:00, Sat 09:00–14:00
TASTINGS: fee of R5
TOURS: at a fee of R15
PICNICS: available for purchase; welcome to bring
EST 1987 (producing wine since 1994). Area under vine: 50ha (60% red). Annual production: 300 tons

LISTED WINES

D WHITE: Chardonnay (W); Dry White (U, Sémillon & Chardonnay & S Blanc & Chenin); S Blanc (U)
SS WHITE: Marguerite (U, Chenin & Muscat)
RED: Albenet (W, Merlot & Ruby Cab & Cab); Cab (W); Merlot (W); Pinotage (W); Shiraz (W)

NEW BEGINNINGS

WINES MADE AT NELSON'S CREEK
Wittenberg Farm, R44
Windmeul
PO Box 2009, Windmeul 7630
TEL: 021 869-8612
FAX: 021 869-8612
nbwines@iafrica.com
OWNERS: 16 families
CHAIRMAN: Pieter Jacobs
C'MASTER: Matewis Thabo
V'CULTURIST: Sollie Skippers
SALES: by appointment only
EST 1988. Under vine: 10,5ha (80% red). Annual production: 35 000 litres

LISTED WINES

PINK: Rosé (U, Palomino & Cinsaut)
RED: Classic Dry Red (U, Cinsaut & Cab); Pinotage (U)

NEWTON JOHNSON ▷ CAPE BAY
NICOLE'S HAT ▷ ASHANTI

NIEL JOUBERT

PAARL MAP

P CELLAR: Klein Simonsvlei, Klapmuts – on the R45 between Klapmuts & Simondium
PO Box 17, Klapmuts 7625
TEL: 021 875-5936 (tastings, sales, cellar); 021 875-

5419 (manor house)
FAX: 021 875-5936
wine@nieljoubert.co.za
www.nieljoubert.co.za
OWNERS: DM Joubert Snr & Jnr
GM: DM Joubert Jnr
C'MASTER: Ernst Leicht
V'CULTURIST: DM Joubert Jnr
VISITS: by appointment only
EST 1898. Area under vine: 250ha (50% red). Annual production: 2 000 tons, 1,2m litres

LISTED WINES

D WHITE: Chardonnay (W); S Blanc (U)
OD WHITE: Chenin (U)
RED: Cab (W); Merlot (W); Pinotage (W); Shiraz (W)

HIGHLY RATED

Niel Joubert Cabernet Sauvignon
1996 – W3½ 99
Niel Joubert Merlot
1999 – W3½ 01
Niel Joubert Shiraz
1999 – W3½ 01

NIETVOORBIJ

STELLENBOSCH MAP

P CELLAR, RESEARCH INST: Klapmuts Rd (R44), Stellenbosch
Private Bag X5026, Stellenbosch 7599
TEL: 021 809-3084 (tastings, sales); 021 809-3091 (cellar)
FAX: 021 809-3202
shirley@infruit.agric.za
www.nietvoorbij.co.za
OWNER: Agricultural Research Council Infruitec – Nietvoorbij
MD: Dr Johan van Zyl
GM: Dr Ockert Augustyn
C'MASTER: Adéle Louw
V'CULTURIST: Willem Laubscher
SALES: Mon–Fri 09:00–15:00
TASTINGS: fee of R1 per wine if no purchase made
EST 1960s. Area under vine: 30ha (50% red). Annual production: 120 tons, 45 000 *l*
FLAGSHIPS: Kwartet, Pinotage

LISTED WINES

D WHITE: Chardonnay (W); Droë Wit (U, Chenin & Gewürz & Colombar); S Blanc (U)
RED: Cab (W); Cab Franc (W); Dry Red (U, Cab & Ruby Cab & Cab Franc & Pinotage); Kwartet (W, Merlot & Cab & Cab Franc & Malbec); Merlot (W); Pinotage (W)

HIGHLY RATED

Nietvoorbij Pinotage
1999 – W4 01

NIGHT NECTAR ▷ TJG
NIKELA ▷ GRANGEHURST

NITIDA

DURBANVILLE MAP

P CELLAR: Tygerbergvalley Rd, Durbanville

PO Box 1423, Durbanville 7551
TEL: 021 976-1467
FAX: 021 976-1467
nitida@mweb.co.za
www.nitida.co.za
OWNERS: Veller family
MD: Bernhard Veller
C'MASTER: Bernhard Veller
SALES: Mon–Fri 09:00–17:00, Sat 09:30–13:00
TASTINGS: no fee
TOURS: by appointment
EST 1992 (producing wine since 1995). Area under vine: 13ha (50% red). Annual production: 132 tons, 90 000 litres
FLAGSHIPS: Calligraphy, S Blanc

LISTED WINES

D WHITE: S Blanc (U); Sémillon (W)
RED: Cab (W); Calligraphy (W, Merlot & Cab & Cab Franc); Pinotage (W); Shiraz (W)

HIGHLY RATED

Nitida Chardonnay
1998 – W4 99, VD 98
Nitida Sauvignon Blanc
2001 – W4 01
2000 – W3½ 00
1999 – W3½ 99
Nitida Sémillon
2001 – W4 02
Nitida Cabernet Sauvignon
1997 – W4 99
Nitida Calligraphy
2000 – W4 02
1999 – W4½ 01
Nitida Pinotage
2001 – W4 02
2000 – W3½ 01
Nitida Shiraz
1999 – W3½ 01
1998 – W4 00
1997 – W4½ 99

NOBLE HILL ▷ COWLIN
NOON GUN ▷ FLAGSTONE

NORDALE

ROBERTSON MAP

CO-OP: 22 Waterkant St, Bonnievale
PO Box 105, Bonnievale 6730
TEL: 023 616-2050
FAX: 023 616-2192
nordalewines@yebo.co.za
www.nordale.co.za
GM: Emile Schoch
C'MASTER: Simon Basson
V'CULTURISTS: S Basson, Brian Stipp
SALES: Mon–Thurs 08:00–12:30 & 13:30–17:00, Fri 07:30–12:30 & 13:30–16:00
TASTINGS: no fee
TOURS: yes; by appointment for groups
FUTURES (EN PRIMEUR): yes
PICNICS: welcome to bring
EST 1950. Area under vine: 510ha (77% white). Annual production: 7 000 tons, 5m *l*
FLAGSHIP: Captains' Drift

SINGLE VYD WINE: Captains' Drift Shiraz

LISTED WINES

D WHITE: Colombar Chardonnay (U); S Blanc (U)
RED: Cab Merlot (W); Cab Shiraz (W); Captains' Drift Shiraz (W); Double Cab (U, Cab & Ruby Cab); Shiraz (U)
FORTIFIED: Red Muskadel Jerepigo (U); Wit Muskadel Jerepigo (U, Muscat de F)

NUANCE ▷ SEIDELBERG
NUTWOOD GROVE ▷ L'ÉMIGRÉ

NUY

WORCESTER MAP

CO-OP: off R60 between Worcester & Robertson
PO Box 5225, Worcester 6849
TEL: 023 347-0272
FAX: 023 347-4994
wines@nuywinery.co.za
GM: Wilhelm Linde
C'MASTER: Wilhelm Linde
SALES: Mon–Fri 08:30–16:30, Sat 08:30–12:30
TASTINGS: fee for groups larger than ten
EST 1963. Annual production: 9 000 tons

LISTED WINES

SPARKLING: Sparkling (U, S Blanc)
D WHITE: Chardonnay (W); Chant de Nuit (U, Chenin & Colombard & Ferdinand de Lesseps); Colombard Dry (U); Riesling (U); S Blanc (U)
OD WHITE: Fernão Pires (U)
SS WHITE: Colombard Semi-Sweet (U)
RED: Cab (W); Rouge de Nuy (W, Cab & Merlot)
FORTIFIED: Red Muscadel (U); White Muscadel (U)

HIGHLY RATED

Nuy Red Muscadel
1999 – VD 01
1998 – VD 00
1997 – VD 99
1996 – W4 98
Nuy White Muscadel
1999 – VD 01
1998 – VD 00
1996 – W5 98

NYATHI ▷ SPAR

OAKRIDGE ▷ SHOPRITE CHECKERS
OCCASIONAL ▷ HARTENBERG
OCULUS ▷ QUOIN ROCK
ODDBINS ▷ SHOPRITE CHECKERS
OLD PADDOCK ▷ ZANDVLIET
OLD ROAD ▷ GRAHAM BECK

OLD VINES

P PRODUCER: 50 Liesbeek Rd, Rosebank, Cape Town 7700 – wines made elsewhere
TEL: 021 685-6428
FAX: 021 685-6446
fran@oldvines.co.za

www.oldvines.co.za
OWNERS: Irina von Holdt, Françoise Botha
GM: Françoise Botha
W'MAKER: Irina von Holdt
VISITS: by appointment only
EST 1995. Annual production: 220 tons, 132 000 litres
FLAGSHIP: Old Vines wines

LISTED WINES

Blue White
D WHITE: Chenin (U)

Cheerful White
D WHITE: Cheerful White (U, Chenin)

Old Vines
D WHITE: Barrel Res Chenin; Chenin (U)

Spring Valley
D WHITE: Chenin S Blanc (U)
D PINK: Rosé (U, Pinotage)
RED: Shiraz Merlot (W)

HIGHLY RATED

Blue White Chenin Blanc
1998 – W3½ 02
1997 – W4 01

ONDERKLOOF

HELDERBERG MAP

P CELLAR: 17 Sir Lowry's Pass Rd
PO Box 90, Sir Lowry's Pass 7133
TEL: 021 858-1538
FAX: 021 858-1536
wine@onderkloofwines.co.za
www.onderkloofwines.co.za
OWNERS: Danie Truter, Beat Musfeld
GM: Danie Truter
C'MASTER: Danie Truter
VISITS: by appointment only
EST 1950 (producing wine since 1999). Area under vine: 31ha (60% white). Annual production: 120 tons, 80 000 litres
FLAGSHIPS: Cab, Chardonnay
SINGLE VYD WINES: Cab, Chardonnay, Chenin, Pinotage, S Blanc

LISTED WINES

D WHITE: Chardonnay (W); Chenin (U); S Blanc (U)
OD WHITE: Floreal Blanc de Blanc (U, Chenin & Crouchen & Muscat d A)
RED: Cab (W); Pinotage (W)

ONVERWACHT ▷ STORMBERG
ONYX ▷ DARLING CELLARS
OOM PAGEL ▷ CHARLES BACK
OOM TAS ▷ DISTELL

OPSTAL

WORCESTER MAP

ESTATE: off R101, Slanghoek, Rawsonville
PO Box 27, Rawsonville 6845
TEL: 023 344-3001
FAX: 023 344-3002
opstal@lando.co.za
www.opstal.co.za
OWNER: Stanley Louw
C'MASTER: Stanley Louw
V'CULTURIST: Kobus Theron

SALES: Mon–Fri 08:30–17:00, Sat 10:00–13:00
TASTINGS: no fee
TOURS: by appointment
EST 1930s. Area under vine: 103ha (70% white). Annual production: 1 500 tons
FLAGSHIP: Cabernet

🍷 LISTED WINES

D WHITE: Carl Everson S Blanc Sémillon (U); Chardonnay (W); Chenin (U)
RED: Cab (W); Carl Everson Classic Red (W, Pinotage & Ruby Cab & Cinsaut)

OPTIMA ▷ L'ORMARINS

ORANJERIVIER

NORTHERN CAPE

CO-OP: Industria Rd, Upington
PO Box 544, Upington 8800
TEL: 054 337-8800
FAX: 054 332-4408
marketing@owk.co.za
www.owk.co.za
DIRECTORS: JAC Burger (MD), AJJ van Rensburg, FP de Wet, JC Eggers, JA Husselman, CJ Nortje, CFJ van Rensburg, LE van Zyl
GM: Herman Cruywagen
C'MASTER: Jannie Engelbrecht
V'CULTURIST: Dirk Malan
SALES: Mon–Fri 08:00–17:00, Sat 09:00–12:00
TASTINGS: no fee
TOURS: Jan–Mar, by appointment for groups of 15 or more
EST 1968. Area under wine vines: 3 232ha (97% white). Annual production: 171 000 tons, 130m litres
SINGLE VYD WINES: Chenin, Muscadels, Pinotage, Ruby Cab

🍷 LISTED WINES

D WHITE: Chenin (U); Colombard (U); Grand Cru (U, Chenin & Colombard)
SS WHITE: Late Harvest (U, Chenin)
S WHITE: Nouveau Blanc (U, Chenin or Colombard, Natural Sweet); Special Late Harvest (U, Chenin)
SS PINK: Rosé (U, Ruby Cab & Pinotage)
RED: Pinotage (W); Ruby Cab (W)
FORTIFIED: Cream Sherry (U, Sultana); Red Muscadel (U); White Muscadel (U)
PORT: Red Port (W, Ruby Cab); White Port (W, Chenin)

ORGANIC ▷ AFRICAN TERROIR
ORIGINAL ▷ RAATS

OUDEKLOOF

P PRODUCER: Tulbagh – wine made at Montpellier
PO Box 191, Tulbagh 6820
TEL: 023 230-1925
OWNERS: Paul O'Riordan,

Francois Rozon
GM: Arnold Hugo
W'MAKER: Gert Theron (Montpellier)
V'CULTURIST: Arnold Hugo
VISITS: not open to the public
EST mid-1960s (producing wine since 2000). Area under vine: 22ha (75% white). Annual production: 65 tons, 37 000 litres
FLAGSHIP: Oudekloof Res Chenin
SINGLE VYD WINE: Foregon Cinsaut Pinotage

LISTED WINES

OD WHITE: Oudekloof Chenin (U); Oudekloof Res Chenin (W)
RED: Foregon (W, Cinsaut & Pinotage)

OUDE RUST ▷ MOOIUITSIG

OUDE WELLINGTON

WELLINGTON MAP

ESTATE: Bainskloof Rd (R303), 2km outside Wellington towards Bainskloof pass
PO Box 622, Wellington 7654
TEL: 021 873-1008 (tastings, sales); 021 873-2262 (cellar, manor house)
FAX: 021 873-4639
info@kapwein.com
www.estate.kapwein.com
OWNER: Dr Rolf Schumacher
DIRECTORS: Dr Rolf Schumacher (MD), Vanessa Simkiss
C'MASTER: Vanessa Simkiss
VISITS: to the cellar by appointment only
FUTURES (EN PRIMEUR): yes
N MONUMENT: farm-house (1790)
ACCOM: guest house
RESTAURANT: River Café, dinner Tues–Sat (Sun in summer), lunch Sun, Tel 021 873-2009
PICNICS: available for purchase in summer
EST 1995. Area under vine (16ha (50% red). Annual production: 80 tons, 20 000 litres
FLAGSHIP: Cabernet
SINGLE VYD WINES: Chardonnay, Currant Abbey, Gewürz, Ruby Cab

LISTED WINES

D WHITE: Chardonnay (U); Chardonnay (W); Chenin
OD WHITE: Gewürz (U)
BLANC DE NOIR: Rose Petal (U, Ruby Cab)
RED: Cab (W); Currant Abbey (Ruby Cab & Chenin); Rubignon (Ruby Cab & Cab); Ruby Cab (W); Shiraz
SPIRITS: Brandy; Grappa

OUDE WELTEVREDEN ▷ WELTEVREDE
OUMA SE WYN ▷ WELTEVREDE
OUPA SE WYN ▷ WELTEVREDE

OUT OF AFRICA ▷ AFRICAN TERROIR
OVATION ▷ LOUISVALE
OVATION ▷ WELTEVREDE

OVERGAAUW

STELLENBOSCH MAP

ESTATE: Stellenboschkloof Rd, Vlottenburg
PO Box 3, Vlottenburg 7604
TEL: 021 881-3815
FAX: 021 881-3436
info@overgaauw.co.za
OWNER: Braam van Velden
GM: Chris Joubert
C'MASTER: Chris Joubert
V'CULTURIST: Johan Pienaar
SALES: Mon–Fri 08:30–12:30 & 14:00–17:00, Sat 10:0–12:30
TASTINGS: fee of R5pp
TOURS: by appointment, at R5pp
ACCOM: guest cottage
EST 1905 (producing wine since 1910). Area under vine: 75ha (70% red). Annual production: 400 tons, 135 000 litres
FLAGSHIPS: Merlot, Tria Corda

LISTED WINES

D WHITE: Chardonnay (W); S Blanc (U)
OD WHITE: Sylvaner (U)
RED: Cab (W); Merlot (W); Pinotage Cab Franc (W); Shiraz Cab (W); Touriga N Cab (W); Tria Corda (W, Cab & Merlot & Cab Franc)
PORT: Cape Vintage (W, Tinta B & Tinta F & Souzão)

HIGHLY RATED

Overgaauw Cabernet Sauvignon
1998 – W3½ 00
Overgaauw Merlot
1999 – W4 01
1996 – W4 99
1995 – W4 98
Overgaauw Pinotage Cabernet Franc
1998 – W3½ 00
1997 – W3½ 99
Overgaauw Tria Corda
1998 – W4 01
Overgaauw Cape Vintage
1997 – W3½ 01
1995 – W4 99
1994 – W3½ 99
Overgaauw 2000 Cape Vintage Reserve
1998 – W4½ 00

OVERMEER ▷ DISTELL
OVERTURE ▷ BONFOI

PAARL PERLÉ ▷ DISTELL

PADDAGANG

NÉGOCIANT: 23 Church St, Tulbagh
PO Box 303, Tulbagh 6820
TEL: 023 230-0394
FAX: 023 230-0433
paddagang@mweb.co.za
OWNER: Paddagang Wines (Pty) Ltd

DIRECTORS: PF Theron, JG du Plessis, Nicky Krone, PW Theron, PF Theron (Jnr)
GM: Marius Burger
C'MASTER: Stephan Smit
SALES: Mon–Sun 09:00–16:00
TASTINGS: fee of R5
RESTAURANT: 08:00–17:00, Tel 023 230-0242
EST 1987. Annual production: 27 000 litres

LISTED WINES

D WHITE: Paddadundee (W, Chardonnay); Paddasang (U, S Blanc)
OD WHITE: Paddaspring (U, Chenin & Colombar)
S WHITE: Platanna (U, Chenin & Hanepoot, Natural Sweet)
RED: Paddajolyt (W, Pinotage); Paddarotti (W, Merlot)
PORT: Brulpadda (U, Pinotage & Ruby Cab)

PAINTED LADY ▷ BERGSIG
PAPILLON ▷ VAN LOVEREN
PAPKUILSFONTEIN ▷ TUKULU

PARADYSKLOOF

STELLENBOSCH MAP

P CELLAR: Paradyskloof Rd, Stellenbosch
PO Box 155, Stellenbosch 7559
TEL: 021 880-0284
FAX: 021 880-1503
info@paradys.co.za
www.paradys.co.za
OWNER: Jan Boland Coetzee, Chardonnay Wine Estates
C'MASTER: Jan Boland Coetzee
VISITS: by appointment only
EST 1982. Area under vine: 16,3ha (90% red). Annual production: 130 tons, 90 000 litres
FLAGSHIP: Vriesenhof Kallista
SINGLE VYD WINES: Talana Hill Chardonnay, Talana Hill Royale

LISTED WINES

Paradyskloof

D WHITE: S Blanc Chardonnay (W)
RED: Cab Merlot (W); Pinotage (W)

Talana Hill

D WHITE: Chardonnay (W)
RED: Royale (W, Merlot & Cab & Cab Franc)

Vriesenhof

D WHITE: Chardonnay (W)
RED: Cab (W); Kallista (W, Cab & Merlot & Cab Franc); Pinot (W); Pinotage (W)

HIGHLY RATED

Paradyskloof Cabernet Sauvignon Merlot
1998 – W3½ 00
Talana Hill Royale
1998 – JC 00
Vriesenhof Cabernet Sauvignon
1998 – W4 01
1993 – W4 99
1991 – W3½ 99
Vriesenhof Special Selection Cabernet Sauvignon

1995 – W4 99
Vriesenhof Kallista
1998 – W3½ 01, JC 00

PARAGON ▷ SIMONSVLEI

PAUL CLUVER

ELGIN MAP

ESTATE: off the N2 (Kromco turn-off) between Cape Town & Hermanus
PO Box 48, Grabouw 7160
TEL: 021 844-0605
FAX: 021 844-0150
info@cluver.co.za
www.cluver.com
OWNER: Dr Paul Clüver
GM: Paul Clüver (Jnr)
C'MASTER: Andries Burger
V'CULTURIST: Wayne Voigt
SALES: Mon–Fri 08:00–17:00, Sat 09:00–13:00
TASTINGS: fee of R10 for groups of more than eight
ACCOM: guest house
PICNICS: available for purchase; welcome to bring to music concerts
EST 1986 (producing wine since 1997). Area under vine: 62ha (50% red). Annual production: 150 tons, 100 000 litres

LISTED WINES

D WHITE: Chardonnay (W); S Blanc (U); Wooded S Blanc
OD WHITE: Gewürz (U); W Riesling (U)
SS WHITE: W Riesling Special Late Harvest (U)
RED: Cab (W); Pinot (W)
DESSERT: W Riesling Noble Late Harvest (U)

HIGHLY RATED

Paul Cluver Weisser Riesling
1998 – W3½ 00
1997 – W3½ 99
Paul Cluver Cabernet Sauvignon
1999 – W4 02
Paul Cluver Pinot Noir
2000 – JC 02
1998 – W3½ 99
Paul Cluver Weisser Riesling Noble Late Harvest
2001 – JC 02
1999 – W4 00
Paul Cluver Weisser Riesling Special Late Harvest
1999 – W4 00

PAUL DE VILLIERS ▷ LANDSKROON
PAULINA'S ▷ RICKETY BRIDGE
PAUL SAUER ▷ KANONKOP
PAVILLON ▷ BOSCHENDAL
PEAK ▷ ASHANTI
PEARLY BAY ▷ KWV
PECAN STREAM ▷ WATERFORD
PEGLEG ▷ CHARLES BACK

PERDEBERG

PAARL MAP

CO-OP: off the R44, Boland Agricultural School Rd, Vryguns, Windmeul
PO Box 214, Paarl 7620

TEL: 021 869-8112 (tastings); 021 869-8244 (sales, cellar)
FAX: 021 869-8245
pwynk@iafrica.com
GM: Kobus de Kock
C'MASTER: Kobus de Kock
V'CULTURIST: Stefan Joubert
SALES: Mon–Fri 08:00–12:30 & 13:30–17:00
TASTINGS: no fee
TOURS: by appointment (seasonal)
EST 1941. Area under vine: 2 500ha (50% red). Annual production: 18 000 tons, 14m litres
FLAGSHIP: Chenin Dry

LISTED WINES

D WHITE: Chenin Dry (U); S Blanc (U)
RED: Cab Merlot (W); Cinsaut (U); Pinotage (U); Res Pinotage (W); Shiraz (U)

HIGHLY RATED

Perdeberg Reserve Pinotage
2000 – W3½ 02
1997 – VD 97

PERKY PINK ▷ WOOLWORTHS
PERNOD RICARD ▷ LONG MOUNTAIN
PEROLD ▷ KWV
PERROQUET ▷ LOUIESENHOF
PETER BARLOW ▷ RUSTENBERG
PETIT ▷ KEN FORRESTER
PETITE ▷ RIETRIVIER
PETTILANT FRONTE ▷ DE WET
PHASES OF THE MOON ▷ LEMBERG
PHILIP JONKER ▷ WELTEVREDE

PHILIP JORDAAN

P PRODUCER: wines made at Du Toitskloof – R101 Rd, Hercules farm, Rawsonville
PO Box 55, Rawsonville 6845
TEL: 023 349-1601
FAX: 023 349-1581
philipjordaan@intekom.co.za
www.dutoitskloof.com
OWNER: Philip Jordaan
W'MAKER: Philip P du Toit Jordaan
V'CULTURIST: Schalk du Toit
SALES: by appointment only
EST 1998. Area under vine: 1ha (red). Annual production: 7 tons, 5 000 litres

LISTED WINES

RED: Cab Franc (W)

PHILLIPS ▷ BERNHEIM
PHYLLIS HANDS ▷ HELDERKRUIN

PICARDI REBEL

NÉGOCIANT: retail stores countrywide
PO Box 18130, Wynberg 7824
TEL: 021 700-5500 (head office)
FAX: 021 700-5515
info@picardirebel.co.za
OWNERS: Seawind Investments 10 (Pty) Ltd
DIRECTORS: Brian McIntyre & Darron Swersky (joint MDs), Helio Tomaz, Ivan Anderson,

Shona Anderson, Jonathan January, Marius Jute
WINE CONSULTANT: Mark Norrish
SALES: at retail branches
EST 1920

OWN LABELS

Hippo Creek
D WHITE: Chardonnay (W); Chenin (U); S Blanc (U)
RED: Cab (W); Merlot (W); Pinotage (U); Shiraz (W)

Naked Truth
D SPARKLING: Brut
S SPARKLING: Vin Doux
D WHITE: White
S WHITE: Soetes
RED: Dry Red

PICK 'N PAY

NÉGOCIANT: national supermarket chain
Boston Circle, Airport Industria 7490
TEL: 021 936-8400
FAX: 021 934-2639
egray@pnp.co.za
www.picknpay.co.za
OWNERS: Pick 'n Pay Retailers (Pty) Ltd – public company with Ackerman family holding 51%
DIRECTORS: RD Ackerman (chairman), SR Summers (CEO), NP Badminton (MD), JG Ackerman, DJ Boshoff, PJ Connellan, DG Cope, HR Herzfeld, A Jakoet, IJ Joubert, I Motaung, M Pulik, NR Quirk, C Reed, LA Saacks, FP van der Walt, JC van Rooyen
WINE BUYER: David Smith
EST 1967
SALES: at retail branches countrywide

OWN LABELS

Selection of "No Name" box wines

PIERRE JOURDAN ▷ CABRIÈRE
PIK 'N WYNTJIE ▷ BONNIEVALE
PINAZ ▷ BERNHEIM
PINEAU ▷ LABORIE
PINEAU DES CHARANTES ▷ BWC
PINEHURST ▷ MÔRESON
PINNO ▷ GRAHAM BECK
PINTA ▷ DOMEIN DOORNKRAAL
PK MORKEL ▷ BELLEVUE

PLAISIR DE MERLE

FRANSCHHOEK MAP
P CELLAR: R45, Simondium – between Paarl & Franschhoek
PO Box 121, Simondium 7670
TEL: 021 874-1071/2 (tastings & sales); 021 874-1488 (cellar)
FAX: 021 874-1689 (general); 021 874-1488 (cellar)
nbester@distell.co.za
www.plaisirdemerle.co.za
OWNER: Distell
MD: Jan Scanell
C'MASTER: Niel Bester
V'CULTURIST: Freddie le Roux

SALES: Mon–Fri 08:30–17:00, Sat 10:00–14:00 (May–Sept) 10:00–16:00 (Oct–Apr)
TASTINGS: fee of R20
TOURS: by appointment, for a fee of R20 (includes tastings)
EST 1964 (producing wine since 1993). Area under vine: 400ha – 300ha for Distell wines, 100ha for Plaisir de Merle wines (60% red). Annual production: 800 tons, 550 000 litres (balance sold to Distell)
FLAGSHIPS: Cab, Grand Plaisir
SINGLE VYD WINE: Cellarmaster's Release Cab

LISTED WINES

D WHITE: Chardonnay (W); S Blanc (U)
RED: Cab (W, Cab & Shiraz & Petit Verdot & Merlot); Cellarmaster's Release Cab (W); Merlot (W, Merlot & Shiraz)

HIGHLY RATED

Plaisir de Merle Chardonnay 2000 – W3½ 01
Plaisir de Merle Sauvignon Blanc 2000 – W3½ 01
Plaisir de Merle Cabernet Sauvignon 1998 – W3½ 01
Plaisir de Merle Cellarmaster's Release Cabernet Sauvignon 1999 – W4 01
Plaisir de Merle Shiraz 1998 – W4 99

PLATANNA ▷ PADDAGANG
PLOUGHMAN ▷ AVONDALE
POETRY COLLECTION ▷ FLAGSTONE
POLLA'S RED ▷ DU PREEZ EST
PONGRÁCZ ▷ JC LE ROUX
PORCUPINE ▷ BOEKENHOUTSKLOOF

PORTERVILLE

SWARTLAND MAP

CO-OP: 1 Voortrekker St, Porterville
PO Box 52, Porterville 6810
TEL: 022 931-2170
FAX: 022 931-2171
portwyn@xsinet.co.za
CHAIRMAN: Johan Conradie
GM: NW de Jongh
C'MASTER: André Oberholzer
V'CULTURIST: Johan Viljoen (Vinpro)
SALES: Mon–Fri 08:00–13:00 & 14:00–17:00, Sat 08:00–11:00
TASTINGS: no fee
TOURS: Jan–Feb, by appointment for groups larger than five
PICNICS: welcome to bring
EST 1941. Area under vine: 1 559ha (13% red). Annual production: 14 000 tons, 9,4m litres
FLAGSHIP: Res Pinotage

LISTED WINES

D WHITE: Chardonnay (U); Premier Grand Cru (U,

Colombar); S Blanc (U)
OD WHITE: Blanc de Blanc (U, Chenin & Colombar); Chenin (U)
SS PINK: Rosé (U, Colombar & Shiraz)
RED: Enigma (W, Shiraz & Cab & Merlot & Pinotage); Res Pinotage (W)

HIGHLY RATED

Porterville Chenin Blanc 2001 – PB 01
Porterville Pinotage 1999 – W3½ 99

POST HOUSE

HELDERBERG MAP

P CELLAR: Raithby Rd, Raithby, Somerset West
PO Box 5635, Helderberg 7135
TEL: 021 842-2409
FAX: 021 842-2409
OWNER: Nicholas Gebers
C'MASTER: Nicholas Gebers
VISITS: by appointment only
EST 1995. Area under vine: 35ha (60% red). Annual production: 25 tons, 15 000 litres
FLAGSHIP: Cabernet

LISTED WINES

D WHITE: Chenin (W)
RED: Cab (W); Merlot (W)

HIGHLY RATED

Post House Cabernet Sauvignon 1998 – W3½ 01

POUR MES AMIS ▷ CAPE BAY
POTJIE ▷ MOOIUITSIG
PRELUDE ▷ NEDERBURG
PREMIÈRE SEL ▷ MORGENHOF
PRESTIGE ▷ THEUNISKRAAL
PRIMA ▷ LE BONHEUR
PRIMO ▷ CHARLES BACK
PRIVÈ DU BOIS ▷ WELTEVREDE
PUDDING ▷ L'AVENIR

QF ▷ RUSTENBERG

QUANDO

ROBERTSON MAP

P CELLAR: R317 between Bonnievale & Robertson
PO Box 82, Bonnievale 6730
TEL: 023 616-2752 (cellar); 082 926-0805 (tastings, sales)
FAX: 023 616-2752
quandowines@mweb.co.za
OWNER: Fanus & Martin Bruwer
C'MASTER: Fanus Bruwer
V'CULTURIST: Martin Bruwer
VISITS: by appointment only
EST early 1900s (producing wine since 2001). Area under vine: 65ha (80% white). Annual production: 15 tons, 9 000 litres
SINGLE VYD WINE: S Blanc

LISTED WINES

D WHITE: S Blanc

QUANTUM ▷ LOST HORIZONS

QUOIN ROCK

STELLENBOSCH MAP

P CELLAR: Knorhoek Rd, off R44, Klapmuts side of Stellenbosch (previously named Glenhurst)
PO Box 1193, Stellenbosch 7599
TEL: 021 888-4740
FAX: 021 888-4744
OWNER: Talacar Holdings
C'MASTER: Rod Easthope
V'CULTURIST: Jaco van der Westhuizen
VISITS: by appointment only
EST 1999 (producing wine since 2001). Area under vine: 65ha (70% red). Annual production: 300 tons, 10 000 *l*

LISTED WINES

D WHITE: Chardonnay (W); Oculus; S Blanc (U)
RED: Merlot (W)

RAATS

P PRODUCER: wines made at Delaire
PO Box 2068, Stellenbosch 7601
TEL: 021 885-1756
FAX: 021 885-1270
braats@mweb.co.za
www.raats.co.za
OWNERS: Bruwer & Jasper Raats
W'MAKER: Bruwer Raats
V'CULTURIST: Jasper Raats
SALES: by arrangement
EST 2000. Area under vine: 6ha (70% white). Annual production: 20 tons, 15 000 litres
FLAGSHIP: Cabernet, Chenin

LISTED WINES

D WHITE: Chenin (W); Original Chenin (U)
RED: Cab Franc (W)

RAILROAD ▷ GRAHAM BECK
R&R ▷ RUPERT & ROTHSCHILD
RAOUL'S ▷ BEAUMONT
RAPHAEL ▷ DOUGLAS GREEN
RATAFIA ▷ CABRIÈRE
RAWSON'S ▷ CAPE VINEYARDS
REBATTET ▷ SIGNAL HILL
REDHILL ▷ SIMONSIG
RED MEDALLION ▷ NAPIER
RED ZIN ▷ KLEINE ZALZE
REGALIS ▷ MONS RUBER

REMHOOGTE

STELLENBOSCH MAP

ESTATE: R44, Koelenhof
PO Box 2032, Dennesig 7602
TEL: 021 889-5005
FAX: 021 889-6907
remhoogte@adept.co.za
OWNER: Murray Boustred
DIRECTORS: Murray Boustred (MD), Michel Rolland
C'MASTER: Murray Boustred
V'CULTURIST: Hendrik De Beer
VISITS: by appointment only
EST 1812 (producing wine since 1995). Area under

vine: 30ha (90% red). Annual production: 80 tons, 50 000 *l*
FLAGSHIP: Merlot

LISTED WINES
RED: Cab (W); Merlot (W); Pinotage (W)

HIGHLY RATED
Remhoogte Cabernet Sauvignon 1999 – W3½ 01
Remhoogte Merlot 1998 – W3½ 00
Remhoogte Pinotage 1997 – W4 99

REMINISCE ▷ BERNHEIM
RESOLUTE ▷ FLAGSTONE
RETREAT ▷ ROBERTSON CO-OP

REYNEKE

STELLENBOSCH MAP

P CELLAR: Uitzicht Farm, Polkadraai Rd (M12), Stellenbosch
PO Box 61, Vlottenburg 7604
TEL: 021 881-3517
FAX: 021 881-3451
wine@reynekewines.co.za
www.reynekewines.co.za
OWNERS: Johan Reyneke, James Farquharson
C'MASTER: James Farquharson
VISITS: to the cellar by appointment only
FUTURES (EN PRIMEUR): yes
ACCOM: B&B, Tel 021 881-3261
EST 1998. Area under vine: 20ha (80% red). Annual production: 30 tons, 22 000 litres
FLAGSHIP: Reyneke Estate
ORGANIC WINE: Pinotage

LISTED WINES
D WHITE: S Blanc (W)
RED: Cab Merlot Shiraz (W); Pinotage (W); Reyneke Estate (W, Shiraz & Cab & Merlot)

RHEBOKSKLOOF

PAARL MAP

ESTATE: Agter-Paarl
PO Box 7141, N Paarl 7623
TEL: 021 869-8386
FAX: 021 869-8504
info@rhebokskloof.co.za
www.rhebokskloof.co.za
OWNER: Rhebokskloof Farming & Trading (Pty) Ltd
DIRECTORS: Tracey Thornycroft (MD), HW Jenkins, LS Tregoning
C'MASTER: Daniel Langenhoven
V'CULTURIST: Werner de Villiers
SALES: Mon–Sun 09:00–17:00
TASTINGS: fee of R6pp
TOURS: by appointment, at R10pp
FUTURES (EN PRIMEUR): yes
RESTAURANT: lunch from 12:00, teas 09:30–14:30, dinner Thurs–Mon, Sun lunch buffet, TEL: 021 869-8606
EST 1976 (producing wine since 1989). Area under

vine: 77ha (55% white). Annual production: 450 tons, 50 000 *l*
FLAGSHIPS: Grande Res Chardonnay, Shiraz
SINGLE VYD WINE: Shiraz

LISTED WINES

Rhebok

D WHITE: Chardonnay (W); Dry White (U, Chardonnay & Riesling)
OD WHITE: Bouquet Blanc (U, Chenin); Dry Red (W, Gamay Noir & Pinotage & Merlot & Cab)

Rhebokskloof

S SPARKLING: Tamay (U, Chardonnay & Riesling & Hanepoot)
D WHITE: Chardonnay Grande Res (W); Chardonnay Sur Lie (W); W Riesling (U)
RED: Cab (W); Cab Merlot (W); Gamay Noir (U); Merlot (W); Pinotage (W); Shiraz (W)

HIGHLY RATED

Rheboksskloof Grande Reserve Chardonnay
1999 – W3½ 00
Rheboksskloof Sur Lie Chardonnay
1999 – W3½ 00
Rheboksskloof Weisser Riesling
1997 – W3½ 01
Rheboksskloof Merlot
1998 – W4 01
Rheboksskloof Pinotage
2000 – W4 02
Rheboksskloof Shiraz
2000 – W4 02
Rheboksskloof Cabernet Sauvignon Merlot
2000 – W3½ 02

RHINOFIELDS ▷ DURBANVILLE HILLS
RHONA ▷ GRAHAM BECK

RICKETY BRIDGE

FRANSCHHOEK MAP

P CELLAR: off Main Rd, Franschhoek
PO Box 455 Franschhoek 7690
TEL: 021 876-2129 (sales, cellar); 021 876-3669 (tastings)
FAX: 021 876-3486
onaghan@mweb.co.za
OWNER: Duncan Spence
DIRECTORS: D Spence (MD), Milton Smith, Bruce Birrell
GM: Butch McEwan
C'MASTER: Wilhelm van Rooyen
SALES: Mon–Fri 10:00–17:00
TASTINGS: fee of R5
TOURS: at a fee of R10
PICNICS: welcome to bring
EST 1992 (producing wine since 1995). Area under vine: 16ha (65% red). Annual production: 120 tons, 68 000 *l*
FLAGSHIP: Paulina's Res

LISTED WINES

Duncan's Creek

D WHITE: Classic White (U, Chenin & S Blanc & Sémillon)

RED: Classic Red (W, Cab & Merlot & Malbec & Cab Franc)

Rickety Bridge

D WHITE: Chardonnay (W); S Blanc (U); Sémillon (W)

RED: Cab (W); Merlot (W); Paulina's Res (W, Cab & Merlot); Shiraz (W)

HIGHLY RATED

Rickety Bridge Chardonnay
1999 – W3½ 00

Rickety Bridge Sauvignon Blanc
2001 – W3½ 01
2000 – W4 00

Rickety Bridge Sémillon
1999 – W3½ 01
1998 – W3½ 99

Rickety Bridge Cabernet Sauvignon
1998 – PB 01
1997 – W3½ 99

Rickety Bridge Merlot
1999 – W3½ 01
1998 – W4½ 02

Rickety Bridge Paulina's Res
1996 – VD 98

Rickety Bridge Shiraz
1999 – W4 01
1998 – MG 99
1997 – W4½ 99

RIDGE ▷ GRAHAM BECK

RIEBEEK

SWARTLAND MAP

COMPANY: Pieter Cruythoff St, Riebeek Kasteel
PO Box 13, Riebeek Kasteel 7307
TEL: 022 448-1213
FAX: 022 448-1281
riebeek@mweb.co.za
www.riebeekcellars.co.za
GM: Zakkie Bester
C'MASTER: Eric Saayman
V'CULTURIST: Hanno van Schalkwyk
SALES: Mon–Fri 08:00–17:00, Sat 09:00–14:00
TASTINGS: no fee
TOURS: by appointment
EST 1850 (producing wine since 1941). Area under vine: 1 400ha (65% white). Annual production: 14 000 tons, 10,5m litres
FLAGSHIPS: "A Few Good Men" wines

LISTED WINES

A Few Good Men

RED: Cab (W); Merlot (W); Pinotage (W); Shiraz (W)

Montino

PETILLANT: Montino (U, Muscat de F)

Riebeek

D SPARKLING: Brut (U, S Blanc)

RED: Cab (W); Dry Red (U, Cinsaut & Pinotage); Merlot (W); Pinotage (W); Shiraz (W)

FORT: Red Jerepigo (U, Cab)

PORT: Cape Ruby (W, Tinta B)

Other:

Cape Table (U, Chardonnay);

Cellar Gold (U, Chenin & Hanepoot); Cellar Light (U); Cellar Red (U, Pinotage & Merlot); Cellar Rosé (U, Pinotage); Cellar White (U, Chenin & Colombar)

HIGHLY RATED

A Few Good Men Shiraz 2000 – W3½ 02

Riebeek Cape Ruby NV – W3½ PS 01, DT 00

RIDGEBACK

PAARL MAP

P CELLAR: Langverwacht Farm, 10km from Paarl on Noorde Agter Paarl Rd

PO Box 2076, Windmeul 7630

TEL: 021 869-8068

FAX: 021 869-8068

lesleyc@mweb.co.za

OWNER: Vansha Farming (Pty) Ltd

MD: Jeremy Parker

MGR: Toit Wessels

C'MASTER: Cathy Marshall

V'CULTURIST: Paul Wallace

SALES: Mon–Sat 08:00–17:00

TASTINGS: no fee

TOURS: by appointment

FUTURES (EN PRIMEUR): yes

EST 1997 (producing wine since 2001). Area under vine: 28ha (70% red). Production: 160 tons, 110 000 litres

LISTED WINES

Ridgeback

RED: Cab (W); Merlot (W); Shiraz (W)

Vansha

D WHITE: S Blanc (U)

RIETRIVIER

LITTLE KAROO MAP

CO-OP: on R62, 20km from Montagu on Barrydale Rd

PO Box 144, Montagu 6720

TEL: 023 614-1705

FAX: 023 614-1705

rietrivier@lando.co.za

CHAIRMAN: Retief Uys

GM: Chris-Willem de Bod

C'MASTER: Chris-Willem de Bod

V'CULTURIST: Willem Botha

SALES: Mon–Thurs 08:00–17:00, Fri 08:00–15:00

TASTINGS: no fee

TOURS: by appointment

EST 1962 (producing wine since 1968). Area under vine: 300ha (95% white). Annual production: 6 000 tons, 4,2m litres

FLAGSHIP: Montagu Rooi Muskadel

LISTED WINES

Montagu

FORTIFIED: Hanepoot (U); Rooi Muskadel (U); Wit Muskadel (U)

Petite

OD WHITE: Blanc (U, Colombar & Chardonnay & S Blanc)

RED: Rouge (U, Pinotage &

Shiraz)

Rietrivier

SPARKLING: Sparkling (U, S Blanc)

RED: Shiraz (W)

RIETVALLEI

ROBERTSON MAP

ESTATE: Klaasvoogds – 8km outside Robertson on the way to Ashton

PO Box 386, Robertson 6705

TEL: 023 626-3596 (cellar); 023 626-4147 (manor house)

FAX: 023 626-4514

info@rietvallei.co.za

www.rietvallei.co.za

OWNER: Johnny Burger

C'MASTER: Johannes Marthinus Burger

VISITS: by appointment only

EST 1864. Area under vine: 142ha (82% white). Annual production: 2 500 tons, 2m litres

FLAGSHIP: Muscadel

LISTED WINES

D WHITE: Chardonnay (W); S Blanc (U)

RED: Cab (W)

FORTIFIED: Muscadel (U)

HIGHLY RATED

Rietvallei Sauvignon Blanc 2001 – W3½ 02

Rietvallei Red Muscadel 1998 – W3½ 00

RIGGTON ▷ BONNIEVALE

RIJCKSHOF ▷ KANGO

RIJK'S

TULBAGH MAP

P CELLAR: outside Tulbagh

PO Box 400, Tulbagh 6820

TEL: 023 230-1622

FAX: 023 230-1650

wine@rijks.co.za

www.rijks.co.za

OWNER: Neville Dorrington

C'MASTER: Pierré Wahl

V'CULTURIST: Johan Wiese

SALES: Mon–Fri 08:00–17:00, Sat 10:00–16:00

TASTINGS: fee of R15

TOURS: daily

ACCOM: country hotel, Tel 023 230-1006, Fax 023 230-1125, PO Box 340 Tulbagh 6820, bookings@rijks.co.za

RESTAURANT: breakfast 07:30–10:00, lunch 12:00–14:00, dinner 19:00–21:00, Tel 023 230-1006

EST 1997 (producing wine since 2000). Area under vine: 26ha (70% red). Annual production: 150 tons, 100 000 litres

FLAGSHIPS: Pinotage, S Blanc

SINGLE VYD WINES: Cab, Chardonnay, Chenin, Merlot, Pinotage, S Blanc, Sémillon, Shiraz

LISTED WINES

D WHITE: Chardonnay (W); Chenin (W); Iceberg

White (U, S Blanc);
S Blanc (U); Sémillon (W)
RED: Cab (W); Merlot (W);
Pinotage (W); Shiraz (W)

HIGHLY RATED

Rijk's Chardonnay
2000 – PB 01, W3½ 01
Rijk's Chenin Blanc
2001 – W3½ 02
2000 – W3½ 02
Rijk's Sauvignon Blanc
2001 – MG 01
2000 – W4 01, VD 00
Rijk's Sémillon
2000 – W3½ 02
Rijk's Cabernet Sauvignon
2000 – W3½ 02
Rijk's Pinotage
2000 – W4½ PC 02, VD 01, MG Trophy 01, PT 01
Rijk's Shiraz
2000 – W3½ 01

RIVER RED ▷ VAN LOVEREN
RIVERAIN ▷ VAN ZYLSHOF
RIVER GRANDEUR ▷ VILJOENSDRIFT
RIVER'S EDGE ▷ WELTEVREDE
RIVERSTONE ▷ MERWIDA
ROBERT ALEXANDER ▷ TRAVINO
ROBERT ALEXANDER ▷ WILHELMSHOF
ROBERT FULLER ▷ AFRICAN WINES

ROBERTSON

ROBERTSON MAP

CO-OP: 1 Constitution St, Robertson
PO Box 37, Robertson 6705
TEL: 023 626-8017 (tastings, sales); 023 626-3059 (cellar)
FAX: 023 626-2926
info@robertsonwine.co.za
www.robertsonwinery.co.za
GM: Bowen Botha
C'MASTER: Lollie Lourens
V'CULTURIST: Anton Laas
SALES: Mon–Thurs 08:00–17:00, Fri 08:00–18:00, Sat 09:00–15:00
TASTINGS: no fee
TOURS: by appointment
EST 1941. Area under vine: 1 500ha (70% white). Annual production: 25 000 tons, 20m litres
FLAGSHIPS: Wide River Res wines
SINGLE VYD WINES: Almond Grove Noble Late Harvest, Wide River Res S Blanc

LISTED WINES

Fat Bastard
D WHITE: Chardonnay (W)

Robertson
D WHITE: Chardonnay (W); Chenin (U); Colombard (U); S Blanc (U)
OD WHITE: Beaukett (U, Gewürz & Colombard)
S WHITE: Natural Sweet White (U, Chenin & Colombard & Muscadel); Special Late Harvest (U, Gewürz)
S RED: Natural Sweet Red (U, Colombard & Ruby Cab)
S PINK: Natural Sweet Rosé

(U, Colombard & Muscadel & Ruby Cab)
RED: Cab (W); Cab Shiraz (W); Merlot (W); Pinotage (U); Ruby Cab (U); Shiraz (W)

Silver Sands
D WHITE: Chardonnay (W)

Wide River Reserve
D WHITE: S Blanc (U)
OD WHITE: Riesling (U)
RED: Cab (W)
DESSERT: Almond Grove Noble Late Harvest

HIGHLY RATED
Wide River Chardonnay
1998 – W3½ 99
Wide River Reserve Sauvignon Blanc
2000 – W4½ 00
Wide River Reserve Almond Grove Weisser Riesling Noble Late Harvest
2001 – W3½ 02
1999 – W4½ 00

ROBERT'S ROCK ▷	KWV
ROBUSTO ▷	RUDERA
ROBYN ▷	LUTZVILLE
ROCKWOOD ▷	FLAGSTONE
RODEWAGEN ▷	WABOOMSRIVIER
RODINI ▷	GOUDINI
ROGUES ▷	LA COURONNE
ROLAND'S ▷	SEIDELBERG
ROLLING HILLS ▷	MEERHOF

ROMANSRIVIER

WORCESTER MAP

CO-OP: on R303, outside Wolseley
PO Box 108, Wolseley 6830
TEL: 023 231-1070/80
FAX: 023 231-1102
romans@cybertrade.co.za

LISTED WINES
No wines produced in 2002

HIGHLY RATED
Romansrivier Ceres Chardonnay Noble Late Harvest
1997 – W3½ 99

ROODEBERG ▷	KWV
ROODEHOF ▷	DE DOORNS
ROODENDAL ▷	DISTELL

ROODEZANDT

ROBERTSON MAP

CO-OP: Voortrekker St, Robertson
PO Box 164, Robertson 6705
TEL: 023 626-1160
FAX: 023 626-5074
roodez@intekom.co.za
www.roodezandt.co.za
GM: Abé Rossouw
C'MASTER: Christie Steytler
V'CULTURIST: Anton Laas
SALES: Mon–Fri 08:00–17:30, Sat 09:00–12:30
TASTINGS: no fee
TOURS: by appointment
EST 1953. Area under vine: 1 400ha (80% white). Annual production: 23 000 tons, 17m litres
FLAGSHIP: Balthazar wines

LISTED WINES

Balthazar Classic
D WHITE: Chardonnay (U)
RED: Cab (W)

Roodezandt
D WHITE: Colombar (U); S Blanc (U)
SS WHITE: Special Late Harvest (U, White Muscadel & Chenin)
RED: Cab (W); Keizer's Creek (U, Ruby Cab & Merlot & Cab)
DESSERT: Red Muscadel (U); White Muscadel (U)
PORT: Port (W, Ruby Cab)

ROOIBERG

ROBERTSON MAP

CO-OP: on R60 between Worcester & Robertson
PO Box 358, Robertson 6705
TEL: 023 626-1663
FAX: 023 626-3295
rooiberg@wine.co.za
www.rooiberg.co.za
GM: Johan du Preez
C'MASTER: Tommy Loftus
V'CULTURIST: Brian Stipp
SALES: Mon–Fri 08:00–17:30, Sat 09:00–15:00
TASTINGS: no fee
TOURS: by appointment
RESTAURANT: tea garden/farm stall
PICNICS: welcome to bring
EST 1964. Area under vine: 741ha. Annual production: 9 000 tons, 6,3m litres
FLAGSHIP: Henri Roselt wines

LISTED WINES

D SPARKLING: Brut (U, S Blanc)
S SPARKLING: Alcohol Free Doux (U, Chenin & Hanepoot); Flamingo (U, Muscadel); Vin Doux (U, Gewürz & Chenin)
D WHITE: Chardonnay (U); Premier Grand Cru (U, Chenin & Colombar & S Blanc); Riesling (U); S Blanc (U)
OD WHITE: Bukettraube (U); Chenin (U); Colombard (U); R Riesling (U)
S WHITE: Blanc Natural Sweet (U, White Muscadel & Colombard)
S PINK: Rosé Natural Sweet (U, Red Muscadel & Colombard)
S RED: Red Natural Sweet (U, Ruby Cab)
RED: Cab (W); Henri Roselt Cab (W); Henri Roselt Shiraz (W); Merlot (W); Pinotage (W); Roodewyn (W, Cab & Cab Franc & Pinotage); Selected Red (U, Cinsaut & Shiraz & Ruby Cab & Pinot); Shiraz (W)
DESSERT: Hanepoot (W); Red Jerepigo (W, Pinotage); Red Muscadel (W)
PORT: Vintage Port (W, Pinotage & Souzão & Tinta B)

🍷 **HIGHLY RATED**
Rooiberg Pinotage
1995 – PT 97
Rooiberg Red Muscadel
1996 – W3½ 98
Rooiberg Cape Ruby
1995 – W3½ 98
Rooiberg LBV
1993 – VD 97
1992 – DT 00

ROSENLESE ▷ BERGKELDER
ROSE PETAL ▷ OUDE WELLINGTON
ROSETTE ▷ SWARTLAND
ROSSINI ▷ EIKENDAL
ROYAL ▷ CITRUSDAL

ROZENDAL

STELLENBOSCH MAP

P CELLAR: Omega St, Jonkershoek Valley, Stellenbosch
PO Box 160, Stellenbosch 7599
TEL: 021 887-8737 (sales); 021 887-6854 (cellar, manor house)
FAX: 021 883-8738
rozendal@mweb.co.za
www.rozendal.co.za
OWNER: Kurt Ammann
C'MASTER: K Ammann
VISITS: cellar not open to the public
ACCOM: Auberge
RESTAURANT: Auberge Rozendal, lunch by appointment, dinner Mon–Sun
EST 1983. Area under vine: 6ha (100% red). Annual production: 20 000 litres

🍷 **LISTED WINES**
RED: Rozendal (W, Merlot & Cab)

🍷 **HIGHLY RATED**
Rozendal
1998 – W4 00
1995 – W4 98

R62 ▷ JOUBERT-TRADAUW
RUBELLITE ▷ WABOOMSRIVIER
RUBÉR ▷ STONEWALL
RUBICON ▷ MEERLUST
RUBIGNON ▷ OUDE WELLINGTON
RUBY BLUSH ▷ AVONTUUR
RUBY SUNSET ▷ LUTZVILLE

RUDERA

P PRODUCER: wines made at Kanu
PO Box 2868, Somerset West 7129
TEL: 083 461-8111
info@rudera.co.za
www.rudera.co.za
OWNER: Teddy Hall
W'MAKER: Teddy Hall
SALES: at retail stores or by arrangement
EST 1999. Area under vine: 13ha (75% red). Annual production: 45 tons, 27 000 litres
FLAGSHIPS: Rudera Cab, Rudera Chenin

LISTED WINES
D WHITE: Chenin (W); Robusto Chenin (W)
RED: Cab (W); Syrah (W)
DESSERT: Chenin Noble Late Harvest (W)

HIGHLY RATED
Robusto Chenin Blanc
2001 – W4 02, DT 01
2000 – W4½ 02
Rudera Chenin Blanc
2001 – W4 02
Rudera Robusto Chenin Blanc
2001 – W4 02
Rudera Cabernet Sauvignon
2000 – W3½ 02

RUITERSVLEI

PAARL MAP

P CELLAR: Suid-Agter-Paarl Rd, off the R44 from N1
PO Box 532, S Paarl 7624
TEL: 021 863-1517
FAX: 021 863-1443
ruitersv@iafrica.co.za
www.ruitersvlei.co.za
OWNER: John Faure
DIRECTORS: Belinda Faure-Griessel, Nicki de Villiers, Sheryl Faure
C'MASTER: Hein Hesebeck
V'CULTURIST: Kobus Mostert
SALES: Mon–Sun 09:00–17:00
TASTINGS: no fee
TOURS: by appointment
ACCOM: guest rooms
RESTAURANT: Tel 021 863-3959
EST ±1690 (producing wine since 1979). Area under vine: 265ha (60% red). Production: 1 000 tons, 600 000 *l*
FLAGSHIPS: Cab, Merlot, S Blanc

LISTED WINES
MCC SPARKLING: Four Sisters (U, Pinot)
D WHITE: Chardonnay (U); Chenin (U); Chenin Chardonnay (U); S Blanc (U)
PINK: Rosé (U, Chenin & Merlot)
RED: Cab Merlot (W); Cinsaut Cab (W); Merlot (W); Mountainside Red (U, Tinta B & Carignan & Ruby Cab & Merlot); Pinotage (W); Shiraz (W)

HIGHLY RATED
Four Sisters Pinot Noir MCC
1997 – W4 01
Ruitersvlei Cabernet
1996 – VD 97
Ruitersvlei Cab Merlot
1999 – W4 00

RUMOUR MILL ▷ FLAGSTONE

RUPERT & ROTHSCHILD

PAARL MAP

P CELLAR: at Fredericksburg, Klapmuts Rd, Simondium
PO Box 412, Franschhoek Valley 7670
TEL: 021 874-1648 (sales, cellar); 021 874-1640 (tastings)
FAX: 021 874-1802

info@fredericksburg.co.za
www.fredericksburg.co.za
OWNERS: Rupert Family (SA), Benjamin De Rothschild (France)
DIRECTORS: Hein Koegelenberg (MD), BEMA De Rothschild, JP Rupert, BMP Otto, L Saidelli, JJ Borman
C'MASTER: Schalk-Willem Joubert
V'CULTURIST: André Pentz
VISITS: by appointment only
EST 1997. Area under vine: 40ha (95% red). Annual production: 600 tons, 300 000 *l*
FLAGSHIP: Baron Edmond

LISTED WINES

D WHITE: Baroness Nadine (W, Chardonnay)
RED: Baron Edmond (W, Cab & Merlot); Classique (W, Cab & Merlot & Cab Franc)

HIGHLY RATED

Rupert & Rothschild Baroness Nadine Chardonnay
2000 – W4½ 02
1999 – W4 02
Rupert & Rothschild Baron Edmond
1999 – W4 02
1998 – W3½ 01
Rupert & Rothschild Classique
1998 – W3½ 00

RUSTENBERG

STELLENBOSCH MAP
P CELLAR: Lelie Rd, Idas Valley, Stellenbosch (Nooitgedacht cellar, Helderberg, not open to the public)
PO Box 33, Stellenbosch 7599
TEL: 021 809-1200
FAX: 021 809-1219
wine@rustenberg.co.za
www.rustenberg.co.za
OWNER: Simon Barlow
C'MASTER: Adi Badenhorst
V'CULTURIST: Nico Walters
SALES: Mon–Fri 09:00–16:30, Sat 10:00–13:30
TASTINGS: no fee
N MONUMENT: Idas Valley (decl 1974)
EST 1682 (producing wine since 1692). Under vine: 180ha (80% red). Annual production: 650 tons, 455 000 *l*
FLAGSHIP: Peter Barlow
SINGLE VYD WINES: Five Soldiers, Peter Barlow

LISTED WINES

Brampton

D WHITE: Chardonnay (U); S Blanc (U)
RED: Cab Merlot (W); Old Vines Red (W, Cab & Merlot & Carignan & Shiraz)
DESSERT: QF2 Natural Sweet (W, S Blanc, NLH)

Rustenberg

D WHITE: Five Soldiers (W, Chardonnay); Stellenbosch Chardonnay (W)
RED: John X Merriman (W, Merlot & Cab); Peter Barlow (W, Cab)

HIGHLY RATED
Brampton Chardonnay
1997 – W3½ 98
Brampton Cabernet Sauvignon Merlot
1999 – W3½ 01
1997 – W4 98
Brampton QF Natural Sweet
2000 – TT 02, W5 02
1997 – W4 99
Rustenberg Stellenbosch Brut
NV – W4 00
Rustenberg Five Soldiers
1999 – W3½ 01
Rustenberg Stellenbosch Chardonnay
2000 – W3½ 01
Rustenberg John X Merriman
1999 – SA Certificate 02, W4½ 01
Rustenberg Peter Barlow
1999 – W4½ 01
1998 – W3½ 01
1997 – W4½ 00
1996 – JP 99
Rustenberg Stellenbosch
1998 – W3½ 01
1996 – W3½ 99

RUST EN VREDE

STELLENBOSCH MAP

ESTATE: Annandale Rd (off R44), Stellenbosch
PO Box 473, Stellenbosch 7599
TEL: 021 881-3881
FAX: 021 881-3000
info@rustenvrede.com
www.rustenvrede.com
OWNER: Jannie Engelbrecht
GM: Jean Engelbrecht
C'MASTER: Louis Strydom
V'CULTURIST: Wessel Combrink
SALES: Mon–Fri 09:00–17:00, Sat 09:00–15:00
TASTINGS: fee of R10pp
N MONUMENTS: Cape Dutch-style buildings (18th Century)
EST 1979. Area under vine: 36ha (100% red). Production: 300 tons, 250 000 *l*
FLAGSHIP: Estate wine

LISTED WINES

Guardian Peak
RED: Cab Syrah (W); Merlot (W); Shiraz (W); Syrah Mourvèdre Grenache (W); Trinity (W, Merlot & Cab & Tinta B)

Rust en Vrede
RED: Cab (W); Estate (W, Cab & Shiraz & Merlot); Merlot (W); Shiraz (W); Tinta B (W)

HIGHLY RATED
Guardian Peak Cabernet Sauvignon Syrah
2001 – W3½ 02
2000 – W3½ 01
Guardian Peak Merlot Cabernet Sauvignon Tinta Barocca (Trinity)
2000 – W3½ 01
Rust en Vrede Cabernet Sauvignon
1998 – W4½ 01

1997 – W3½ 00
1995 – W3½ 99
Rust en Vrede Estate
1998 – JP 02, W4 01
1997 – VD 00, W3½ 00, DT 99
1996 – PB 99, W4½ 99, VD 99
1995 – VD 00
1993 – VD 97
Rust en Vrede Merlot
2000 – W3½ 01
1999 – W3½ 01
Rust en Vrede Shiraz
1999 – W4 02
1998 – W4 01
1994 – DT 98
Rust en Vrede Tinta Barocca
2000 – W3½ 01

RUSTHOF ▷ MOOIUITSIG
RYK ▷ TULBAGH
RYLANDS GROVE ▷ INT W SERVICES

SAINT ▷ DOUGLAS GREEN
SAINT OMER ▷ BERNHEIM
SANDBERG PURPLE ▷ APPELSDRIFT
SANDOWN BAY ▷ CAPE BAY
SANS SOUFRE ▷ SIGNAL HILL
SATYNROOI ▷ ASHTON
SAUVENAY ▷ BELLINGHAM

SAVANHA

Wines made at Spier
PO Box 99, Lynedoch, 7603
TEL: 021 881-1143 (cellar); 021 809-1143 (sales)
FAX: 021 881-3699
winecorp@iafrica.com
www.winecorp.co.za
OWNER: Winecorp
MD: Gerrie Wagener
W'MAKERS: Frans Smit (Savanha), Alain Moueix (Sejana, Naledi)
V'CULTURIST: Orlando Filander
SALES: at Spier
EST 1998. Annual production: 50 tons, 30 000 litres
FLAGSHIPS: Naledi, Sejana
SINGLE VYD WINES: Naledi, Sejana

LISTED WINES

Naledi
RED: Naledi (W, Cab)

Savanha
D WHITE: S Blanc (U)
RED: Cab (W); Merlot (W); Shiraz (W)

Sejana
RED: Sejana (W, Merlot)

HIGHLY RATED

Savanha Barrel Fermented Chenin Blanc
1999 – W4 98
Savanha Sauvignon Blanc
2001 – VD 01
Savanha Cabernet Sauvignon
1998 – W3½ 99
Savanha Reserve Pinotage
1997 – W4 99
Savanha Shiraz
2000 – W3½ 02
Sejana Merlot
1999 – PB 00
1998 – PB 00

SAVISA ▷ AFRICAN TERROIR

SAXENBURG

STELLENBOSCH MAP

P CELLAR: Polkadraai Rd, Kuils River
PO Box 171, Kuils River 7579
TEL: 021 903-6113
FAX: 021 903-3129
saxfarm@iafrica.com
www.saxenburg.co.za
OWNER: Saxenburg Wine (Pty) Ltd
DIRECTORS: Adrian Bührer (MD), Brigit Bührer
GM: Nico van der Merwe
C'MASTER: Nico van der Merwe
V'CULTURIST: Len Coetzee
SALES: Mon–Fri 09:00–17:00, Sat 10:00–16:00, Sun 10:00–16:00
TASTINGS: fee of R15
TOURS: by appointment
RESTAURANT: Guinea Fowl, lunch Wed–Mon, dinner Wed–Sat, Tel 021 906-5232
EST 1600 (producing wine since 1990). Area under vine: 70ha (80% red). Annual production: 400 tons, 280 000 litres
FLAGSHIPS: Private Collection wines
SINGLE VYD WINE: Shiraz Select (SSS)

LISTED WINES

MCC SPARKLING: Private Collection Brut (U, Chardonnay & Pinot)
D WHITE: Private Coll Chardonnay (W); Private Coll S Blanc (U)
RED: Cab (W); Grand Vin Blanc (U, Chenin & Chardonnay from Ch Capion, France, and Saxenburg); Grand Vin Rouge (W, red from Ch Capion, France, & five cultivars from Saxenburg); Gwendolyn (W, Shiraz & Cab); Manuel (W, Cab & Merlot); Merlot (W); Pinotage (W); Private Coll Cab (W); Private Coll Merlot (W); Private Coll Shiraz (W); Shiraz Select (W)
DESSERT: Private Coll Le Revé (U, S Blanc & Chenin)

HIGHLY RATED

Saxenburg Private Collection Chardonnay
1999 – W3½ 00
1997 – W3½ 98
Saxenburg Private Collection Cabernet Sauvignon
1998 – W3½ 01
1997 – W4 00
1996 – W4 99
1995 – W4 99
1994 – W5 99
Saxenburg Cabernet Sauvignon
1992 – W4½ 99
1991 – W4 99
Saxenburg Gwendolyn
1998 – W3½ 01

1996 – W3½ 98
Saxenburg Private Collection Merlot
1998 – W4 00
1997 – JP 00, W4½ 99
1996 – W4 98
Saxenburg Private Collection Pinotage
1999 – W3½ 01
1997 – W3½ 99
1996 – W4 98
Saxenburg Select Shiraz
1998 – W3½ 01, JP 01
Saxenburg CIWG Shiraz
1997 – JP 00
Saxenburg Private Collection Shiraz
1999 – W3½ 01
1998 – W3½ 00
1997 – DT 98
Saxenburg La Rêve
NV – W3½ 98

SCALI

PAARL MAP

P CELLAR: Schoone Oord farm, off R44 (Voor-Paardeberg exit)
PO Box 7143, N Paarl 7623
TEL: 021 869-8340
FAX: 021 869-8383
info@scali.co.za
www.scali.co.za
OWNERS: Willie & Tania de Waal
C'MASTERS: W & T de Waal
VISITS: by appointment only
EST 1877 (producing wine since 1999). Area under vine: 74ha (over 50% red). Production: 20 tons, 12 400 *l*

LISTED WINES

RED: Pinotage (W); Syrah (W)

SCARBOROUGH ▷ CAPE POINT
SCHANDERL ▷ TJG
SCHOLTZENHOF ▷ KEN FORRESTER
SCHOONE GEVEL ▷ LA MOTTE
SCULPTED ▷ AMANI
SEBASTIAAN ▷ HIGH CONSTANTIA
SEDGWICK'S ▷ DISTELL

SEIDELBERG

PAARL MAP

ESTATE: Suid-Agter Paarl Rd, exit 47 from N1, Paarl
PO Box 505, S Paarl 7624
TEL: 021 863-3495/6
FAX: 021 863-3797
info@seidelberg.co.za
www.seidelberg.co.za
OWNER: Roland Seidel
GM: Nicolaas Rust
C'MASTER: Nicolaas Rust
V'CULTURIST: Conré Groenewald
SALES: Mon–Fri 09:00–17:00, Sat–Sun 10:00–16:00
TASTINGS: fee of R10
TOURS: by appointment for large groups, at a fee of R5
FUTURES (EN PRIMEUR): yes
RESTAURANT: Olive and Vine
PICNICS: available for purchase
EST 1700 (producing wine since 1989). Area under

vine: 93ha (80% red). Annual production: 420 tons, 292 000 litres

FLAGSHIPS: Seidelberg Roland's Res wines

LISTED WINES

De Leuwen Jagt

D WHITE: Chardonnay (U); Leuwenblanc (U, S Blanc & Colombard)

OD WHITE: Nuance (U, Riesling & S Blanc & White Muscat)

SS WHITE: Stein (U, Riesling & S Blanc & White Muscat)

RED: Cab (W); Cab Franc Merlot (U); Merlot (U); Shiraz (U)

FORTIFIED: Muskadel (U, Muscat de F)

Seidelberg

D WHITE: Chardonnay (W); Chenin (U); S Blanc (U); Viognier (W)

OD PINK: Rosé (U, Muscat de F & Pinotage)

RED: Cab (W); Cab Merlot (W); Merlot (W); Pinotage (W); Roland's Res Cab (W); Roland's Res Pinotage (W); Roland's Res Syrah (W); Un Deux Trois (W, Cab & Merlot & Cab Franc)

HIGHLY RATED

Seidelberg Chardonnay
1999 – VD 01

Seidelberg Chenin Blanc
2001 – DT 01, VD 01

Seidelberg Roland's Reserve Merlot
2000 – W3½ 01

Seidelberg Roland's Reserve Shiraz
2000 – W4½ 01

Seidelberg Pinotage
2001 – W4 02

SEJANA ▷ SAVANHA
SEMAPHORE ▷ FLAGSTONE
SEMINAY ▷ TULBAGH
SENTINEL ▷ COPPOOLSE FINLAYSON
SERENDIPITY ▷ BERNHEIM
SERIOUSLY GOOD ▷ V JOHNSON'S
SEUGNET ▷ SOMERBOSCH
SEVEN FALLS ▷ BARRYDALE
SEVERNEY ▷ SIYABONGA
SHAKA ▷ COPPOOLSE FINLAYSON
SHINDIG ▷ FLAGSTONE
SHIP ▷ DISTELL

SHOPRITE CHECKERS

NÉGOCIANT: national supermarket chain

Head Office: cnr William Dabb & Old Paarl roads, Brackenfell

PO Box 215, Brackenfell 7561

TEL: 021 980-4000

FAX: 021 980-4012

seksteen@shoprite.co.za

www.shoprite.co.za

OWNER: public company

MD: Whitey Basson

WINE BUYER: Stephan Eksteen

SALES: at selected stores

OWN LABELS

Oakridge

RED: Cab (W); Merlot (W); Shiraz (W)

Oddbins

S WHITE: No 115 – Natural Sweet (U, Chenin & White Muscadel)

PINK: No 84 – Rosé (U, Merlot); No 109 – Rosé (U, Colombard & Chenin & Ruby Cab)

RED: No 93 (W, Cab); No 68 (W, Shiraz); No 149 (W, Merlot)

SIESTA ▷ LA PETITE FERME

SIGNAL HILL

CAPE TOWN

P CELLAR: Culemborg Park, Foreshore, Cape Town

POSTAL ADDRESS: 24 Scott Street, Cape Town 8001

TEL: 021 461-9590 (sales); 083 675-0280 (cellar)

FAX: 021 465-0342

info@winery.co.za

www.winery.co.za

OWNER: Ridon Family Vineyards (Pty) Ltd

DIRECTORS: Jean Vincent Ridon (MD), Kulekani Buthelezi

C'MASTER: Jean Vincent Ridon

V'CULTURIST: Maritje Marais

VISITS: by appointment only

FUTURES (EN PRIMEUR): yes

EST 1997. Area under vine: 3ha (80% red). Annual production: 45 tons, 30 000 litres

FLAGSHIP: Creme de Tete

SINGLE VYD WINES: Rebattet, Rouge De Constance, Vin de L'Empereur

WINES WITHOUT SULPHUR: Sans Soufré Cab

LISTED WINES

D WHITE: Tete Blanche (W, Chenin)

RED: Antica (W, Cab); Cabeça Tinta (W, Tinta B); Petit Verdot (W); Pinot Barrel; Rebattet (W, Shiraz); Rouge de Constance (W)

DESSERT: Creme de Tete (W, Muscat, Noble Late Harvest); Vin de Glaciere (W, Muscat, Icewine); Vin de L'Empereur (U, Muscat d A, Natural Sweet); Vin de Paille (W)

HIGHLY RATED

Signal Hill Cabernet Sauvignon
1997 – W4 00

Signal Hill Vin de Glaciere
NV – W4 01

Signal Hill Vin de L'Empereur
2000 – W3½ 02
1998 – W4 99

Signal Hill Crème de Tete
2000 – W4 02
1998 – W4 01

SIGNATURE ▷ JEAN DANEEL
SILVER SANDS ▷ ROBERTSON

SIMONSIG

STELLENBOSCH MAP

ESTATE: Kromme Rhee Rd, off R304, Koelenhof, Stellenbosch
PO Box 6, Koelenhof 7605
TEL: 021 888-4900 (sales); 021 888-4915 (tastings); 021 888-4925 (cellar)
FAX: 021 888-4909
wine@simonsig.co.za
www.simonsig.co.za
OWNERS: Malan Brothers
DIRECTORS: Pieter Malan (MD), Francois Malan, Johan Malan, W Sendin
C'MASTER: Johan Malan
V'CULTURIST: Francois Malan
SALES: Mon–Fri 08:30–17:00, Sat 08:30–16:30
TASTINGS: fee of R10
TOURS: Mon–Fri 10:00 & 15:00, at a charge of R15
PICNICS: available for purchase in season; welcome to bring by prior arrangement
EST 1942 (bottling wine since 1968). Area under vine: 205ha (55% red). Annual production: 2 000 tons, 1,4m litres
FLAGSHIP: Tiara
SINGLE VYD WINES: Merindol Syrah, Redhill Pinotage

LISTED WINES

Jean Le Riche

OD SPARKLING: Vin Sec (U, Chenin & Colombar)
S SPARKLING: Vin Doux (U, Chenin & Colombar)

Malan Family Vintners

D WHITE: Adelblanc S Blanc Sémillon (U); Chardonnay (W); S Blanc (U)
RED: Adelberg Cab Merlot (W); Pinotage (U); Rouge Du Cap (U, Cab & Pinotage & Shiraz)

Merindol

RED: Syrah (W)

Redhill

RED: Pinotage (W)

Simonsig

MCC SPARKLING: Cuvée Royale (U, Chardonnay & Pinot); Kaapse Vonkel (U, Pinot & Chardonnay & Pinot Meunier)
D WHITE: Chardonnay (W); Chenin (U); S Blanc (U); Vin Fumé (W, S Blanc & Sémillon)
OD WHITE: Mustique (U, Bukettraube & Muscat Ottonel & Morio Muscat)
S WHITE: Franciskaner (U, Bukettraube & Chenin, Natural Sweet); Gewürz (U, SLH)
RED: Cab (W); Frans Malan Res (W, Pinotage & Cab & Merlot); Pinotage (U); Shiraz (W); Tiara (W, Cab & Merlot & Petit Verdot & Cab Franc)
DESSERT: Vin de Liza (U, Bukettraube, Noble Late Harvest)

PORT: Port (W, Pinotage)

HIGHLY RATED

Merindol Syrah
1999 – PB 01
1998 – W4 01
1997 – W4 00
Redhill Pinotage
1999 – PT 01, W4 01
1998 – W3½ 01
Simonsig Brut Cuvée Royale
1992 – W4 01, VD 01
Simonsig Kaapse Vonkel Brut
1997 – W3½ 01
1996 – W4½ 00
Simonsig Chardonnay
2000 – W4 02
1998 – VD 99, W3½ 99
1996 – W4 97
Simonsig Chenin Blanc
2001 – VD 01
1997 – W3½ 02
Simonsig Sauvignon Blanc
1998 – W3½ 98
1997 – W4 97
Simonsig Adelberg
2000 – W3½ 01
Simonsig Frans Malan Reserve
1998 – JC 02, W3½ 01
1997 – VD 99
Simonsig Pinotage
2000 – W3½ 02
1997 – W3½ 99
1995 – VD 99
Simonsig Shiraz
1997 – DT 98
Simonsig Reserve Shiraz
1997 – W4 99
Simonsig Tiara
2000 – W3½ 02
1999 – W3½ 02
1998 – W4 01
1995 – VD 97
Simonsig Bukettraube
Noble Late Harvest
1996 – W4 98
Simonsig Vin de Liza
2001 – W4 02
Simonsig Morio Muscat
1998 – W3½ 99

SIMONSVLEI

PAARL MAP

COMPANY: R101 (old Paarl Rd), between Klapmuts & Paarl
PO Box 584, S Paarl 7624
TEL: 021 863-3040
FAX: 021 863-1240
info@simonsvlei.co.za
www.simonsvlei.co.za
OWNER: Simonsvlei International
DIRECTORS: GS Scheepers (chairman), FS de Manielle (vice-chairman), JA Louw (MD), JE Lambrechts, DT van der Merwe, EB van Zyl, GF Malherbe, JJ Vermeulen, FJ Joubert, DB Malan, DF Hugo, MT Malan, A de Villiers, IB van der Vyver
C'MASTER: Francois van Zyl
V'CULTURIST: Cobus van Graan
SALES: Mon–Fri 08:00–17:00, Sat 08:00–16:30, Sun 11:00–15:00
TASTINGS: R8 for six wines

TOURS: by appointment, at a fee of R15pp
RESTAURANT: R101, Tues–Sun, Tel 021 863-2486, info@restaurant101.co.za
EST 1945. Area under vine: 1 200ha (70% white). Annual production: 9 000 tons, 5,4m litres
FLAGSHIPS: Hercules Paragon wines

LISTED WINES

Hercules Paragon

D WHITE: S Blanc (U)
RED: Cab (W)

Mount Marble

S SPARKLING: Vin Doux (U, Chenin)
D WHITE: Blanc de Blanc (U, Chenin); Riesling (U)
OD WHITE: Stein (U, Chenin)
SS WHITE: Late Vintage (U, Chenin)

Simonsvlei

D WHITE: Chardonnay (W); Premier Bukettraube (U); Premier Chenin Blanc (U); S Blanc (U)
RED: Cab (W); Cab Merlot (W); Pinotage (W); Shiraz (W); Simonsrood (U, Shiraz & Ruby Cab & Cinsaut & Pinotage)
DESSERT: Humbro Hanepoot (U); Humbro Red Jerepigo (U, Red Muscadel); White Port (W, Chenin); White Muscadel (U)

HIGHLY RATED

Hercules Paragon Sauvignon Blanc
2001 – W4 01
Hercules Paragon Cabernet Sauvignon
1998 – W3½ 99
Hercules Paragon Merlot
1998 – W4½ 01
Hercules Paragon Shiraz
1997 – VD 99, W3½ 99
Simonsvlei Premier Chenin Blanc
2002 – W3½ 02
Simonsvlei Shiraz
1998 – W3½ 99
Simonsvlei Premier White Muscadel
1995 – W4 98
Simonsvlei Port
1994 – W3½ 99
1993 – W3½ 01

SIYABONGA

WELLINGTON MAP

P CELLAR: Ranzadale Rd, Cumminos St, Wellington
PO Box 1209, Wellington 7654
TEL: 021 864-3295
FAX: 021 914-9162
doolhof@mweb.co.za
OWNER: H Investments No 121
DIRECTORS: Graham Knox (MD), Alain Moueix
C'MASTER: Charles Stassen
VISITS: by appointment only
EST 1998. Area under vine: 20ha (70% red). Annual pro-

duction: 100 tons, 60 000 litres

LISTED WINES

D WHITE: Severney (W, Chenin & Chardonnay & Sémillon & S Blanc)
RED: Cab Merlot (W); Pinotage (W)

1682 ▷ STEENBERG

SLALEY

STELLENBOSCH MAP

ESTATE: cnr R44 & Kromme Rhee Roads, Koelenhof
PO Box 119, Koelenhof 7605
TEL: 021 865-2123
FAX: 021 865-2798
info@slaley.co.za
www.slaley.co.za

OWNERS: Hunting family
DIRECTORS: Sue Rowe (MD), Lindsay Hunting
C'MASTER: Thierry Lesne (flying winemaker)
V'CULTURIST: Jaco Mouton
SALES: Mon–Fri 08:30–16:30, Sat & evenings by appointment
TASTINGS: no fee
FUTURES (EN PRIMEUR): yes
N MONUMENTS: boundary markers (1751)
EST 1957 (producing wine since 1997). Area under vine: 65ha (90% red). Annual production: 300 tons, 200 000 litres
FLAGSHIPS: Slaley Shiraz
SINGLE VYD WINES: Slaley Merlot, Slaley Shiraz, Slaley Pinotage, Broken Stone S Blanc

LISTED WINES

Broken Stone
D WHITE: S Blanc (U)
RED: Cab (W); Cab Shiraz (W); Pinotage (W)

Slaley Hunting Family
D WHITE: Chardonnay (W)
RED: Cab (W); Cab Merlot (W); Merlot (W); Pinotage (W); Shiraz (W)

Slaley Reserve
RED: Merlot (W); Shiraz (W)
DESSERT: Noble Late Harvest (W)

HIGHLY RATED

Broken Stone Pinotage
2000 – W4 02
1998 – W4 00
Broken Stone Cabernet Sauvignon Shiraz
2000 – W3½ 01
1998 – W4 00, VD 99
Slaley Chardonnay
2000 – W3½ 01
1999 – VD 01
Slaley Merlot
2000 – W3½ 02
Slaley Shiraz
2000 – W4½ 01, JC 01
1999 – W3½ 01
1998 – JP 00, W5 99, VD 99
1997 – W4 98
Slaley Cabernet

Sauvignon Merlot
1999 – W3½ 01

SLANGHOEK

WORCESTER MAP

CO-OP: Slanghoek Valley, off the old W1 between Worcester & Rawsonville
PO Box 75, Rawsonville 6845
TEL: 023 344-3026/7/8
FAX: 023 344-3157
slanghoek@lando.co.za
www.slanghoek.co.za
OWNERS: 25 farmers
GM: Peter Carstens
C'MASTER: Christo Pienaar
V'CULTURIST: Francois Nel
SALES: Mon–Fri 08:00–12:30 & 13:30–17:30, Sat 10:00–13:00
TASTINGS: no fee
TOURS: by appointment
PICNICS: welcome to bring
EST 1951.
Area under vine: 1 500ha (75% white). Annual production: 24 000 tons, 21m litres
FLAGSHIPS: Cab, Noble Late Harvest, S Blanc
SINGLE VYD WINES: Cabernet, S Blanc

LISTED WINES

D WHITE: Chardonnay (W); Chenin (U); Riesling Sémillon (U); S Blanc (U)
OD WHITE: Chardonnay S Blanc (U); Vinay (U, Chenin & Colombar & S Blanc)
S WHITE: Special Late Harvest (U, Hanepoot)
D PINK: Vinay Rosé (U, Pinotage)
RED: Cab (W); Camerca (W, Cab & Merlot & Cab Franc); Merlot (W); Pinotage (W); Shiraz (W); Vinay Red (U, Pinotage)
DESSERT: Noble Late Harvest (U, Chenin); Noble Late Harvest (W, Chenin)
FORTIFIED: Red Muscadel (U); Rooi Jerepigo (U, Pinotage); Soet Hanepoot (U)
PORT: Cape Ruby (U, Cinsaut)

HIGHLY RATED

Slanghoek Noble Late Harvest
2000 – W4½ 02, MG 01
1999 – W4 00
Slanghoek Hanepoot
1995 – VD 97
Slanghoek Red Jerepigo
1997 – W3½ 99
Slanghoek Red Jerepigo
2000 – VD 01
Slanghoek Cape Ruby
1998 – DT 00

SMV ▷ CHARLES BACK
SOCIABLE ▷ BLAAUWKLIPPEN
SOLEIL ▷ APPELSDRIFT
SOLEIL DU MATIN ▷ MÔRESON
SOLITUDE ▷ CHARLES BACK

SOMERBOSCH

HELDERBERG MAP

P CELLAR: on the R44, halfway between Somerset West & Stellenbosch
PO Box 12181, Die Boord 7613
TEL: 021 855-3615
FAX: 021 855-4457
enquiries@somerbosch.co.za
www.somerbosch.co.za
OWNER: WJD Roux
GM: Walter Marius Roux
W'MAKERS: M Roux, JPR Roux
SALES: Mon–Fri 09:00–17:00, Sat 09:00–13:00
TASTINGS: fee of R10pp
EST 1950 (producing wine since 1995). Area under vine: 90ha (60% red). Annual production: 650 tons, 3,7m *l*
FLAGSHIP: Ltd Release Cab Merlot

LISTED WINES

D WHITE: Chardonnay (W); S Blanc (U); Seugnet Blanc (U, Chenin)
RED: Cab (W); Ltd Rel Cab Merlot (W); Merlot (W); Pinotage (W); Seugnet Rouge (W, Cinsaut & Cab)
DESSERT: Ltd Rel Chenin Natural Sweet (U)
PORT: Port (W, Cab)

HIGHLY RATED

Somerbosch Cabernet Sauvignon
1998 – W4 00

Somerbosch Reserve Pinotage
1999 – W4 01

SOMERSOET ▷ LUTZVILLE
SONNET ▷ VILLIERA
SONNIGDAL ▷ CITRUSDAL

SONOP

PAARL MAP

P CELLAR: Windmeul, Paarl
PO Box 1142, Stellenbosch 7599
TEL: 021 887-2409 (office at African Terroir)
FAX: 021 886-4838 (office at African Terroir)
office@africanterroir.co.za
www.africanterroir.co.za
OWNER: Jacques Germanier Group (SAVISA)
GM: Sebastian Dayer
C'MASTER: Alain Cajeux
VISITS: by appointment
ACCOM: guest house, Tel 021 869-8723
ANNUAL PROD: 1 300 tons

LISTED WINES

See African Terroir

SOOPJESHOOGTE ▷ WELGEMEEND

SOUTHERN RIGHT

WALKER BAY MAP

P CELLAR: Hemel-en-Aarde Valley, R320, Hermanus
PO Box 158, Hermanus 7200
TEL: 028 312-3595
FAX: 028 312-1797

hrv@hermanus.co.za
OWNERS: Anthony Hamilton-Russell (MD), Kevin Grant
C'MASTER: Kevin Grant
VISITS: not open to the public
EST 1995. Area under vine: 5ha (90% red). Annual production: 175 tons, 116 500 litres

LISTED WINES

D WHITE: S Blanc (U)
RED: Pinotage (W)

HIGHLY RATED

Southern Right Sauvignon Blanc
1999 – PB 99
Southern Right Pinotage
1999 – W3½ 00
1998 – W3½ 00
1997 – W4 99, PT 98

SOUTH SOUTHWEST ▷ AFRICAN W

SPAR

NÉGOCIANT: national supermarket chain
Head Office: 22 Chanecry Lane, Pinetown, Durban
PO Box 1589, Pinetown 3600
TEL: 031 719-1900
FAX: 031 719-1991
saexvino@mweb.co.za
OWNER: franchise operator
CEO: Peter Hughes
GRP LIQUOR MANAGER: Ray Edwards
WINE CONSULTANT: Tinus van Niekerk, Tel 012 998-4737, Fax 012 998-4738
SALES: from retail chain's branches countrywide

OWN LABELS

Country Cellars

D WHITE: Blanc de Blanc (U, Chenin & Colombard); Chardonnay (U); Dry White (U, Chenin & Colombard); Lite (U, Chenin); S Blanc (U); Vin Classique Blanc (U, Chenin & Colombard & Chardonnay)
SS WHITE: Late Harvest (U, Chenin); Stein (U, Chenin); Vin Classique Stein (U, Chenin)
PINK: Caresse (U, Chenin & Ruby Cab); Rosé (U, Chenin & Pinotage); Vin Classique Rosé (U, Chenin & Pinotage)
RED: Bohlôkwa Pinotage (W); Cab (W); Classic Red (W, Pinotage & Ruby Cab); Merlot (U); Pinotage (U); Ruby Cab (W); Vin Classique Rouge (W, Pinotage & Ruby Cab)

SPATZENDRECK ▷ DELHEIM
SPES BONA ▷ VAN LOVEREN

SPICE ROUTE

SWARTLAND MAP

P CELLAR: Klein Amoskuil Farm, Kalbaskraal Rd, outside Malmesbury – see separate listing for wines of Charles Back

PO Box 583, S Paarl 7642
TEL: 021 863-2450 (sales); 022 487-7139 (cellar)
FAX: 021 863-2591 (office); 022 487-7169 (cellar)
fairwine@mweb.co.za
OWNER: Charles Back – see also separate listing for Charles Back
C'MASTER: Charl du Plessis
V'CULTURISTS: Dawie Krynauw, Andrew Teubus
VISITS: not open to the public – sales at Fairview, together with other Wines of Charles Back
EST 1998. Area under vine: 120ha (90% red)
FLAGSHIPS: Flagship Merlot, Flagship Pinotage, Flagship Syrah

LISTED WINES

See separate listing under Charles Back

SPIER

STELLENBOSCH MAP

P CELLAR: Annandale Rd, Lynedoch
PO Box 1078, Stellenbosch 7660
TEL: 021 809-1100 (tastings, sales)
FAX: 021 809-1144
info@spier.co.za
www.spier.co.za
OWNER: Winecorp – see separate listing
MD: Gerrie Wagener
C'MASTER: Frans Smit
V'CULTURISTS: Johan Smit, Orlando Filander
SALES: Mon–Sun 09:00–17:00
TASTINGS: fee of R12 (conducted tasting)
TOURS: by appointment, for a fee of R17,50
N MONUMENTS: manor house (1884), cellar (1767)
ACCOM: Village at Spier Hotel, Tel 021 809-1100
RESTAURANT: Jonkershuis 12:30–15:30 & 18:00–22:00; Taphuis 12:30–15:30; Toto's 12:30–15:30, Figaro's 06:00–10:30 & 18:00–22:00, Riverside Pub 11:00–18:00 &18:00–21:00 (drinks only Sun–Wed); Farm Stall 09:00–17:00; Tel 021 809-1100
PICNICS: available for purchase
EST 1970 (producing wine under current ownership since 1995). Area under vine: 106ha (80% red). Annual production: 2 800 tons, 2m litres
FLAGSHIPS: Private Collection
SINGLE VYD WINES: Private Collection wines

LISTED WINES

D WHITE: Chardonnay (W); Chenin (W); Private Coll S Blanc (U); S Blanc (U)
RED: Cab (W); Merlot (W); Pinotage (W); Private Coll

Cab (W); Private Coll Merlot (W)
DESSERT: Private Coll Noble Late Harvest (W, Bukettraube)

HIGHLY RATED

Spier Chardonnay
2000 – W4 01
Spier Private Collection Sauvignon Blanc
2001 – W3½ 01, JC 01
Spier Private Collection Merlot
1998 – W4 00, VD 99
Spier Private Collection Cabernet Sauvignon
1998 – W4 00
Spier Private Collection Noble Late Harvest
1998 – W4 01
1996 – W3½ 99, MG 98, VD 97

SPIER GROUP

Office at Spier resort, R310, Stellenbosch
PO Box 1078, Lynedoch 7603
TEL: 021 809-1100
FAX: 021 881-3302
info@spier.co.za
www.spier.co.za
OWNER: Capricorn
PROPRIETOR: Dick Enthoven
CHIEF W'MAKER: Frans Smit
V'CULTURIST: Johan Smit

LISTED WINES

See separate listings for Spier, Longridge, Savanha, Winecorp

SPITZ ▷ BELLINGHAM

SPRINGFIELD

ROBERTSON MAP

ESTATE: on the R317 (Bonnievale Rd), 4km from Robertson
PO Box 770, Robertson 6705
TEL: 023 626-3661
FAX: 023 626-3664
admin@springfieldestate.com
www.springfieldestate.com
OWNERS: Bruwer family
DIRECTORS: Abrie & Jeanette Bruwer
C'MASTER: Abrie Bruwer
SALES: Mon–Fri 08:00–17:00, Sat 09:00–16:00
TASTINGS: no fee
TOURS: during business hrs
FUTURES (EN PRIMEUR): yes
PICNICS: available for purchase in summer
EST 1902. Area under vine: 55% red
FLAGSHIPS: Life from Stone S Blanc, Méthode Ancienne Chardonnay, Special Cuvée S Blanc, Whole Berry Cab
SINGLE VYD WINES: Life from Stone S Blanc, Méthode Ancienne Chardonnay, Special Cuvée S Blanc

LISTED WINES

D WHITE: Life from Stone S Blanc (U); Méthode Ancienne Chardonnay (W); Special Cuvée S Blanc (U); Wild Yeast

Chardonnay (U)
RED: Méthode Ancienne Cab (W); Whole Berry Cab (W)

HIGHLY RATED

Springfield Méthode Ancienne Chardonnay
1999 – JC 01
Springfield Wild Yeast Chardonnay
1999 – W3½ 01
Springfield Life from Stone Sauvignon Blanc
2001 – JC 01
Springfield Special Cuvée Sauvignon Blanc
2000 – W3½ 00
Springfield Whole Berry Cabernet Sauvignon
2000 – W3½ 02

SPRING VALLEY ▷ OLD VINES
SPRITZENLESE ▷ BERGKELDER
SPRITZIGER ▷ BERGKELDER

SPRUITDRIFT

OLIFANTS RIVER MAP
CO-OP: on R363 between Klawer & Vredendal
PO Box 129, Vredendal 8160
TEL: 027 213-3086
FAX: 027 213-2937
spruitdrift@kingsley.co.za
OWNER: West Corp – see separate listing
GM: Izak Visagie
C'MASTER: Len Knoetze
V'CULTURIST: PD Koegelenberg
SALES: Mon–Fri 08:00–17:00, Sat 08:30–12:30
TOURS: Feb–Apr, 10:00 & 15:00
PICNICS: welcome to bring
EST 1968 (producing wine since 1972)

LISTED WINES

See separate listing under West Corp – reds made at Spruitdrift, whites made at Vredendal

SSS ▷ SAXENBURG
SSW ▷ AFRICAN WINES
ST (SAINT) ▷ DOUGLAS GREEN

STARK-CONDÉ

STELLENBOSCH MAP
P CELLAR: Oude Nektar farm, Jonkershoek, Stellenbosch
PO Box 389, Stellenbosch 7599
TEL: 021 887-3665
FAX: 021 887-4340
wine@thecapeco.com
OWNER: Schröder Family Trust
MD: HP Schröder
W'MAKERS: José Condé, Pieter Smit
V'CULTURIST: Pieter Smit
VISITS: by appointment only
EST 1998. Area under vine: 40ha (70% red). Annual production: 20 tons, 13 000 litres
FLAGSHIPS: Condé Cab, Stark Cab
SINGLE VYD WINES: Condé Cab

LISTED WINES
RED: Condé Cab (W); Pinot (W); Pinotage (W); Stark Cab (W)
HIGHLY RATED
Condé Cabernet Sauvignon 2000 – TT 02, W5 02
Starke Cabernet Sauvignon 1998 – W4 02, JP 02

STARKE

STELLENBOSCH MAP

P CELLAR: Muldersvlei farm, Muldersvlei Rd, Muldervlei, Stellenbosch
PO Box 66, Muldersvlei 7607
TEL: 021 884-4433 (manor house); 083 631-2229 (sales)
FAX: 021 883-4324
wine@muldersvlei.co.za
OWNER: Muldersvlei Estates (Pty) Ltd
MD: Julian Starke
C'MASTER: Ian Starke
V'CULTURISTS: Julian & Ian Starke
VISITS: by appointment only
ACCOM: The Cellars Mulderbosch B&B, info@muldersvlei.co.za
EST 1889 (producing wine since 1999). Area under vine: 35ha (100% red). Annual production: 50 tons, 24 500 litres
FLAGSHIP: Res Pinotage, Shiraz
LISTED WINES
RED: Juliette (W, Pinotage & Shiraz); Pinotage (W); Res Pinotage (W); Shiraz (W)

STEENBERG

CONSTANTIA MAP

P CELLAR: Steenberg Estate, Steenberg Rd, Tokai
PO Box 224, Steenberg 7947
TEL: 021 713-2211
FAX: 021 713-2201
info@steenbrg.co.za
www.steenberg-vineyards.co.za
OWNER: Adrian Gardiner – The Mantis Collection
DIRECTORS: Adrian Gardiner (MD), G Savoi Diz (Uruguay), PF Retief, FP Fleck, RK Jackson, F Scelza Martinez, D Malan
GM: Herman Hanekom
C'MASTER: John Loubser
V'CULTURIST: Herman Hanekom
SALES: Mon–Fri 09:00–16:30, Sat 09:00–13:00
TASTINGS: fee of R5
TOURS: by appointment
N MONUMENTS: hotel 'werf' area incl manor house, jonkershuis, barn (1682)
ACCOM: Country Hotel, Tel 021 713-2222, Fax 021 713-2221, reservations@steenberghotel.com, www.steenberghotel.com
RESTAURANT: breakfast 07:00–10:00, lunch 12:00 onwards, dinner 19:00–22:00,

contact details/address as for hotel
EST 1990 (producing wine since 1996). Area under vine: 63ha (60% white). Annual production: 430 tons, 301 000 litres
FLAGSHIPS: Catharina Red, Res S Blanc

LISTED WINES

Motif
D WHITE: Blanc Sec (U, Sémillon & S Blanc & Chardonnay)
RED: Rouge (W, Cab & Cab Franc & Merlot)

Steenberg
MCC SPARKLING: 1682 (U, Chardonnay & Pinot)
D WHITE: Res S Blanc (U); S Blanc (U); Wooded Sémillon
RED: Cab (W); Catharina Red (W, Cab & Merlot & Shiraz); Merlot (W); Nebbiolo (W); Shiraz (W)

HIGHLY RATED

Motif Rouge
W33½ 99
Steenberg Sauvignon Blanc
1999 – SA Trophy 00, PB 99
1997 – VD 97
Steenberg Reserve Sauvignon Blanc
2001 – SA Trophy 02, VD 01
2000 – W3½ 00
1999 – W4 00
1998 – W4 98
1997 – W4 97
Steenberg Unwooded Sémillon
2001 – W3½ 02
1998 – W4 99
Steenberg Wooded Sémillon
2002 – W3½ 02
2000 – W3½ 02
1999 – W4 00
1998 – W4 99
Steenberg Cabernet Sauvignon
1996 – W4 99
Steenberg Merlot
1999 – JC 00
1997 – W3½ 99
1996 – W4 99
Steenberg Motif Rouge
NV – W3½ 99
Steenberg Shiraz
2000 – W4 01

STEEN OP HOUT ▷ MULDERBOSCH

STELLEKAYA

STELLENBOSCH MAP
P CELLAR: KWV Park, Distillery Rd, Stellenbosch
PO Box 12426, Die Boord 7613
TEL: 021 883-3873
FAX: 021 883-2536
stellekya@iafrica.com
www.stellekaya.com
OWNER: Dave Lello
C'MASTER: Petrus Jacobus le Roux
VISITS: by appointment only
EST 1999. Annual production: 60 tons, 30 000 litres

LISTED WINES
RED: Cab (W); Merlot (W)

STELLENBOSCH VINEYARDS

COMPANY: Stellenbosch Vineyards, based at Welmoed Winery, R310, Lynedoch, Stellenbosch – incorporating Welmoed, Helderberg, Eersterivier
PO Box 465, Stellenbosch 7599
TEL: 021 881-3870
FAX: 021 881-3102
info@stellvine.co.za
www.stellvine.co.za
DIRECTORS: JG Carinus (chairman), Hermann Böhmer (MD), D Smollen, M le Roux, G van der Westhuizen, Willem Louw, T Roos, F Albertyn, D Stanford
W'MAKER: Chris Kelly
V'CULTURIST: Francois de Villiers
SALES: at Welmoed
EST 1997. Area under vine: 2 000ha (70% red). Annual production: 15 000 tons, 9,7m litres
FLAGSHIPS: Genesis, Kumkani
SINGLE VYD WINES: Genesis Chardonnay, Genesis Merlot, Genesis S Blanc, Genesis Shiraz

LISTED WINES

Genesis
D WHITE: Chardonnay (W)
RED: Cab (W); Merlot (W); Shiraz (W)

Infiniti
MCC SPARKLING: Brut (U, Chardonnay & Pinot)

Kumkani
D WHITE: S Blanc (U)
RED: Merlot Cab Franc (W); Pinotage (W); Shiraz (W); Shiraz Cab (W)

Versus
OD WHITE: White (U, Chenin & S Blanc)
RED: Red (W, Merlot & Cab & Shiraz)

HIGHLY RATED
Genesis Chardonnay
2001 – JC 02
2000 – W3½ 01
Genesis Cabernet Sauvignon
2000 – W4½ 02, JC 02
Genesis Merlot
1999 – W3½ 02
Genesis Shiraz
2000 – W4 02
1997 – DT 98
Infiniti Brut
1999 – W3½ 01
Kumkani Sauvignon Blanc
1999 – W4 00
Kumkani Shiraz Cabernet Sauvignon
1998 – DT 99
Kumkani Cabernet Sauvignon Merlot
1999 – JC 00
Kumkani Pinotage
1999 – W3½ 02

Kumkani Shiraz
2000 – W3½ 02

STELLENDRIFT

STELLENBOSCH MAP

P CELLAR: on the M23, off the R304 (former Bottelary co-op)
PO Box 6340, Uniedal, Stellenbosch 7612
TEL: 021 865-2204 (tastings); 021 887-6561 (sales); 082 372-5180 (cellar); 082 417-5467 (manor house)
FAX: 021 887-6561
fcilliers@wam.co.za
OWNER: Fanie Cilliers
C'MASTER: Fanie Cilliers
VISITS: to the cellar by appointment only
RESTAURANT: Bottelary Kos & Wyn, 09:00–17:00, Tel 021 865-2204
PICNICS: available for purchase
EST 1960s (producing wine since 1996). Area under vine: 12ha (100% red). Annual production: 30 tons, 20 000 litres
FLAGSHIPS: Stellendrift Cab, Stellendrift Merlot

LISTED WINES

Cilliers
RED: Cab (W); Jako's Pinotage (W)

Stellendrift
RED: Cab (W); Merlot Cab (W)

STELLENRYCK ▷ BERGKELDER

STELLENZICHT

STELLENBOSCH MAP

P CELLAR: Stellenrust Rd, Stellenbosch
PO Box 104, Stellenbosch 7599
TEL: 021 880-1103
FAX: 021 880-1107
nee@mweb.co.za
www.stellenzicht.co.za
OWNER: Lusan Holdings (Pty) Ltd
DIRECTORS: R Jones (MD); A Diedericks, HA Hëyl, BCM Tan, Prof CJ van Wyk
C'MASTER: Guy Webber
V'CULTURISTS: Dawid Saayman, Eben Archer
SALES: Mon–Fri 09:00–17:00, Sat–Sun 10:00–16:00
TASTINGS: fee of R10
TOURS: by appointment
FUTURES (EN PRIMEUR): yes
EST 1988. Area under vine: 138ha (70% red). Annual production: 800 tons, 500 000 litres
FLAGSHIP: Stellenzicht Syrah
SINGLE VYD WINE: Stellenzicht Syrah

LISTED WINES

Hill & Dale
D WHITE: Chardonnay (W); S Blanc (U)
RED: Cab Shiraz (W); Stellenbosch Red (W, Cab &

Pinotage & Shiraz & Merlot)

Stellenzicht

D WHITE: Golden Triangle Chardonnay (W); Golden Triangle S Blanc (U); Res Sémillon (W)

RED: Golden Triangle Cab (W); Golden Triangle Merlot (W); Golden Triangle Pinotage (W); Golden Triangle Shiraz (W); Stellenzicht (W, Cab & Merlot & Cab Franc & Malbec); Syrah (W)

HIGHLY RATED

Stellenzicht Chardonnay
1997 – SA Trophy 99, PB 98, W4 98
1996 – W4 97
Stellenzicht Hans Schreiber Chardonnay
1998 – JP 00
Stellenzicht Chenin Blanc
1996 – W4 98
Stellenzicht Sauvignon Blanc
1997 – W4 97
Stellenzicht Sémillon
2001 – W3½ 02
Stellenzicht Reserve Sémillon
2000 – W4 02
1998 – W4½ 99
1997 – PB 98, W4 98
Stellenzicht
1998 – W3½ 02
Stellenzicht Bin 404 Cabernet Franc
1995 – W4 98
Stellenzicht Cabernet Sauvignon
1994 – VD 97
Stellenzicht Malbec
1999 – W3½ 01
Stellenzicht Merlot
1995 – VD 97
Stellenzicht Merlot Cabernet Franc
1995 – W4 99, VD 98
Stellenzicht Pinotage
1998 – W4 00
Stellenzicht Syrah
1999 – W4 02, JC 02
1998 – W4½ 01, JP 00, VD 99
1997 – W4 99
1994 – MG 97
Stellenzicht Golden Triangle Pinotage
2000 – JC 02
Stellenzicht Golden Triangle Shiraz
2000 – W4 02
Stellenzicht Shiraz
1997 – W3½ 00
1996 – VD 97
Stellenzicht Weisser Riesling Noble Late Harvest
1998 – W5 01

STETTYN

STELLENBOSCH MAP

CO-OP: R43, 16km from Villiersdorp
PO Box 1520, Worcester 6849
TEL: 023 340-4220 (cellar); 023 340-4972 (manor house)
FAX: 023 340-4220
stettyn@xsinet.co.za
CHAIRMAN: George Botha

GM: Albie Treurnicht
C'MASTER: Albie Treurnicht
V'CULTURIST: Schalk du Toit
VISITS: by appointment only
N MONUMENT: Cape Dutch homestead and wheat mill (1777, restored 1977)
EST 1965. Area under vine: 300ha (85% white). Annual production: 4 000 tons, 3,1m *l*
FLAGSHIPS: Millstone wines

LISTED WINES

D WHITE: Millstone S Blanc (U)
RED: Millstone Pinotage (W); Shiraz Cab (W)
DESSERT: Vin de Paille (W, Muscat d A)

STEYTLER ▷ KAAPZICHT
STONECROSS ▷ DEETLEFS
STONE PINE ▷ MONT DESTIN

STONEWALL

HELDERBERG MAP

P CELLAR: R44 between Somerset West & Stellenbosch
PO Box 5145, Helderberg 7135
TEL: 083 310-2407 (tastings); 021 855-3675 (manor house)
FAX: 021 855-2206
stonewall@mweb.co.za
OWNER: De Waal Koch
W'MAKER: Ronell Wiid (consultant)
SALES: Dec–Jan Mon–Fri 09:00–17:00 & Sat 09:00–13:00; otherwise by appointment
TASTINGS: no fee
TOURS: by appointment
EST 1980 (producing wine since 1997). Area under vine: 80ha (90% red). Annual production: 50 tons, 30 000 *l*
FLAGSHIP: Rubér

LISTED WINES

D WHITE: Chardonnay (W); S Blanc (U)
RED: Cab (W); Pinotage (W); Rubér (W, Cab & Merlot)

HIGHLY RATED

Stonewall Chardonnay
1997 – W3½ 98
Stonewall Pinotage
2000 – W4 01
Stonewall Rubér
1999 – W4 01
1998 – W3½ 00

STONY BROOK

FRANSCHHOEK MAP

P CELLAR: right at monument at end of main road through Franschhoek, look for sign on the left ±2,5km up the road
PO Box 22, Franschhoek 7690
TEL: 021 876-2182
FAX: 021 876-2182
mcnaught@iafrica.com
OWNER: Dr Nigel & Mrs Joy McNaught
GM: Estelle Goschen
C'MASTER: Nigel McNaught
SALES: Mon–Sat 09:00–13:00
TASTINGS: fee of R5pp, refundable on wine purchase

FUTURES (EN PRIMEUR): yes
EST 1995. Area under vine: 14ha (60% red). Annual production: 85 tons, 55 000 litres

LISTED WINES

D WHITE: Chardonnay (W); S Blanc (U); Sémillon (U)
RED: Cab (W); Merlot (W); Reserve (W, Cab & Merlot & Malbec & Cab Franc & Petit Verdot); Res Shiraz (W); Shiraz (W)

HIGHLY RATED

Stony Brook Chardonnay
1997 – W4 98
Stony Brook Res Sémillon
1997 – W4 98
Stony Brook Reserve Cabernet Sauvignon
1998 – W4 01, VD 99
Stony Brook Pinotage
2000 – VD 01
1999 – W3½ 00
1998 – W4 99
Stony Brook Reserve
2000 – W4 01
Stony Brook Shiraz
2001 – W3½ 02
1999 – W3½ 01
Stony Brook Reserve Shiraz
2001 – W3½ 02

STORMBERG

WELLINGTON MAP

NÉGOCIANT: Onverwacht, Addy Rd, Wellington
PO Box 2907, Paarl 7620
TEL: 082 776-3540
FAX: 021 872-3940
kooskit@mweb.co.za
OWNER: JB Jordaan
C'MASTER: Koos Jordaan
VISITS: not open to the public
EST 1998. Annual production: 50 tons, 30 000 litres

LISTED WINES

RED: Barrel Select (W, Merlot & Cab & Cab Franc); Cab Shiraz (W)

STOUMANN'S

OLIFANTS RIVER MAP

P CELLAR: Excelsior farm, Vredendal
PO Box 307, Vredendal 8160
TEL: 027 213-2323
FAX: 027 213-2323
stoutmanns@kingsley.co.za
OWNERS: Stoumann & Sons
DIRECTORS: Napoleon Christiaan Stoumann (MD), GAJ Stoumann, CG Stoumann
C'MASTER: NC Stoumann
SALES: Mon–Sat 08:30–12:30 & 14:00–17:00
TASTINGS: no fee
TOURS: by appointment
PICNICS: welcome to bring
EST 1984 (producing wine since 1998). Area under vine: 1,3ha (100% red). Annual production: 9,8 tons, 5 684 litres
FLAGSHIP: Vin de la Tortue

LISTED WINES
RED: Cab (W); Shiraz (W); Vin de la Tortue (W, Cab & Shiraz & Pinotage)
FORTIFIED: Red Muscadel (W); White Jerepico (W, Hanepoot)

STRATA ▷ FLAGSTONE
SUIDER TERRAS ▷ BLOEMENDAL

SUMARIDGE

WALKER BAY MAP

P CELLAR: Hemel en Aarde Valley, Hermanus
PO Box 1413, Hermanus 7200
TEL: 028 312-1097
FAX: 028 312-2824
sumaridge@itec.co.za
OWNER: Brenda Sumaris
GM: Jasper Raats
C'MASTER: Jasper Raats
SALES: Dec–Jan Mon–Fri 08:00–17:00, Feb–Nov Mon–Fri 08:00–17:00 & Sat 09:00–13:00
TASTINGS: no fee
TOURS: by appointment
FUTURES (EN PRIMEUR): yes
ACCOM: guest house
RESTAURANT: opening soon
PICNICS: available for purchase
EST 1996 (producing wine since 2000). Area under vine: 20ha (60% red). Annual production: 110 tons, 70 000 litres
FLAGSHIP: Syrah
SINGLE VYD WINE: Syrah

LISTED WINES
D WHITE: Chardonnay (W); S Blanc (U)
PINK: Rosé (U, Shiraz & Pinot & Merlot & Pinotage)
RED: Dry Red (U, Pinot); Merlot (W); Pinot (W); Syrah (W)

SUNDAY BEST ▷ V JOHNSON'S
SUNNY DAY ▷ LUTZVILLE
SUNSET BLUSH ▷ ASHANTI

SWARTLAND

SWARTLAND MAP

CO-OP: outside Malmesbury on the R45 (Paarl-Wellington Rd)
PO Box 95, Malmesbury 7299
TEL: 022 482-1134
FAX: 022 482-1750
swartland@sswines.co.za
www.sswines.co.za
MD: Kobus Basson
GM: Marius Erasmus
C'MASTER: Andries Blake
V'CULTURIST: Johannes Mellet
SALES: Mon–Fri 08:00–17:00, Sat 09:00–12:00
TASTINGS: no fee
TOURS: by appointment
EST 1948. Area under vine: 3 200ha (65% white). Production: 20 000 tons, 15m *l*
FLAGSHIPS: Icon wines (to be launched in 2003)

LISTED WINES
Botella
D WHITE: Blanc (U, Chenin

& Colombard)
SS WHITE: Bouquet (U, Bukettraube & Chenin)
PINK: Rosé (U, Pinotage & Chenin)
RED: Rouge (U, Shiraz & Tinta B & Pinotage)

D'Vine
D WHITE: Dry White (U, Chenin & Colombard)
SS WHITE: Semi Sweet (U, Bukettraube & Chenin)
RED: Dry Red (U, Pinotage & Shiraz & Tinta B)

Indalo
D WHITE: Chenin (U); S Blanc
RED: Cab (W); Cab Merlot (W); Pinotage (W); Shiraz (W)

Swartland
D SPARKLING: Brut (U, S Blanc)
SS SPARKLING: Demi Sec (U, S Blanc & Hanepoot)
S SPARKLING: Vin Doux (U, S Blanc & Hanepoot)
D WHITE: Blanc de Blanc (U, Colombard & Chenin); Chardonnay (W); Colombard (U); Grand Cru (U, Chenin & Colombard & Clairette Blanche); Light (U, Fernão Pires); Riesling (U); S Blanc (U); Steen (U, Chenin)
OD WHITE: Bouquet Blanc (U, Bukettraube & Muscat d A) & Colombar)
SS WHITE: Bukettraube (U); Late Vintage (U, Chenin); Natural Sweet (U, Chenin); Stein (U, Chenin)
OD PINK: Blanc de Noir (U, Pinotage)
SS PINK: Rosé (U, Pinotage & Chenin); Rosette (U, Pinotage & Hanepoot & Chenin)
RED: Cab (W); Cinsaut (U); Dry Red (U, Tinta B & Cab & Pinotage); Merlot (W); Pinotage (U); Shiraz (W); Tinta Barocca (U)
FORTIFIED: Hanepoot (U); Red Jerepigo (U, Pinotage); Wit Jerepiko (U, Chenin)
PORT: Port (U, Tinta B); Vintage Port (W, Tinta B & Shiraz & Cab)

HIGHLY RATED

Swartland Cabernet Sauvignon
2000 – W3½ 00
1994 – W4 97

Swartland Indalo Pinotage
2001 – W3½ 02

Swartland Reserve Pinotage
1996 – VD 97, PT 97

Swartland Shiraz
1999 – W4 01
1997 – W4 98
1996 – VD 97

Swartland Rooi Jerepiko
NV – W4 02

Swartland Ruby Port
NV – W4 02

Swartland Vintage

1999 – W4 02, VD 01
1997 – W4 99

SYDNEY BACK ▷ BACKSBERG

SYLVANVALE

STELLENBOSCH MAP

P PRODUCER: Devon Valley Hotel, Stellenbosch – wines made at Meinert cellar
PO Box 68, Stellenbosch 7599
TEL: 021 865-2012
FAX: 021 865-2610
devon@iafrica.com
OWNER: David Nathan-Maister
W'MAKER: Martin Meinert
V'CULTURIST: Lorna Hughes
SALES: at Devon Valley Hotel, Mon–Sun 07:00–21:30
TASTINGS: fee of R5
ACCOM: Devon Valley Hotel
RESTAURANT: The Vineleaf, Mon–Sun 07:00–21:30
EST 1972 (producing wine since 1998). Area under vine: 4ha (66% red). Annual production: 40 tons, 5 000 *l*
FLAGSHIP: Res Pinotage
SINGLE VYD WINES: all

LISTED WINES

D WHITE: Lauries Harvest Chenin (U)
D PINK: Cab Rosé (U)
RED: Res Pinotage (W)
D DESSERT: Vine Dried Chenin (W); Vine Dried Pinotage (W)

HIGHLY RATED

SylvanVale Laurie's Harvest Chenin Blanc
1998 – W3½ 01
SylvanVale Reserve Pinotage
1999 – W3½ 01
SylvanVale Vine Dried Pinotage
2000 – W3½ 02

SYNCHRONICITY ▷ MEINERT

T ▷ KEN FORRESTER
TALANA HILL ▷ PARADYSKLOOF
TAMAY ▷ RHEBOKSKLOOF
TASHEIMER ▷ DISTELL
TASSENBERG ▷ DISTELL

TENFIFTYSIX

FRANSCHHOEK MAP

P CELLAR: Robertsvlei Rd, Franschhoek
PO Box 244, Milnerton 7435
TEL: 021 551-2284
FAX: 021 551-2487
charbrey@new.co.za
OWNER: Michael Falkson
C'MASTER: Gerda Willers
V'CULTURIST: Tol Malherbe
VISITS: by appointment only
EST 1989 (producing wine since 1996). Area under vine 2,7ha (50% red). Annual production: 13 500 litres

LISTED WINES

D WHITE: Chardonnay (W); S Blanc (U)
RED: Cab (W)

TERROIR TRIAL ▷ CHARLES BACK

TESLAARSDAL

WALKER BAY MAP

P CELLAR: Teslaarsdal village, between Hermanus & Caledon
POSTAL ADDRESS: 6 Stemmet St, Hermanus 7200
TEL: 083 258-0952
FAX: 028 312-3591
OWNERS: Jan & Willie Malherbe
C'MASTER: W Malherbe
VISITS: to the cellar by appointment only
ACCOM: The Cottage, Tel 028 312-3591
EST 1998 (producing wine since 2001). Area under vine: 2ha (100% red). Annual production: 10 tons
FLAGSHIP: Cuvée Rouge

LISTED WINES

RED: Cuvée Rouge (W, Cab & Shiraz); Grand Rouge (U, Shiraz & Merlot & Cinsaut)

TETE BLANCHE ▷ SIGNAL HILL
TÊTE DE CUVÉE ▷ B FINLAYSON
TETHRIPPON ▷ MONTPELLIER

THANDI

P PRODUCER: wines made at Paul Cluver
TEL: 021 844-0605
FAX: 021 844-0605
thandi@cluver.co.za
www.cluver.com
OWNER: Dr Paul Cluver
GM: Karin Voigt
W'MAKER: Patrick Kraukamp
V'CULTURIST: Wayne Voigt
SALES: at Paul Cluver
EST 1996. Area under vine: 28ha (52% white). Annual production: 8 tons, 700 litres
FLAGSHIPS: Chardonnay, Pinot

LISTED WINES

D WHITE: Chardonnay (W)
RED: Cab (W); Pinot (W)

THELEMA

STELLENBOSCH MAP

P CELLAR: Helshoogte Pass (R310), Stellenbosch
PO Box 2234, Stellenbosch 7601
TEL: 021 885-1924
FAX: 021 885-1800
wines@thelema.co.za
www.thelema.co.za
OWNERS: McLean & GH Webb Family Trust
DIRECTORS: David McLean (MD), Barbara Webb, Jennifer de Tolly
GM: Gyles Webb
C'MASTER: Gyles Webb
V'CULTURIST: Mathew Castle
SALES: Mon–Fri 09:00–17:00, Sat 09:00–13:00
TASTINGS: no fee
EST 1983 (producing wine

since 1988). Area under vine: 50ha (50% red). Annual production: 350 tons, 315 000 litres
FLAGSHIP: Cabernet

LISTED WINES

D WHITE: Chardonnay (W); R Riesling (U); S Blanc (U)
OD WHITE: Muscat de F (U)
RED: Cab (W); Merlot (W); Res Merlot (W); Shiraz (W)
DESSERT: Muscadel (W)

HIGHLY RATED

Thelema Chardonnay
2000 – W3½ 01
1999 – W4 00
1998 – W3½ 99
1997 – W4 98
Thelema Reserve Chardonnay
1997 – JP 99
Thelema Sauvignon Blanc
1997 – W3½ 98
1994 – W3½ 98
Thelema Cabernet Sauvignon
1999 – W3½ 02
1998 – W4 01
1997 – JP 01, W4½ 00
1996 – W4 99
1995 – JP 99, W4 98
1994 – W4 97
Thelema Merlot
1999 – W3½ 02
1998 – W4 01
1995 – W4 98
1994 – W4 97
Thelema Reserve Merlot
1999 – W4 02, JP 02
Thelema Pinotage
1998 – W4 00
Thelema Shiraz
2000 – W3½ 02

THERONA ▷ WELTEVREDE

THEUNISKRAAL

TULBAGH MAP

ESTATE: 7km north of Tulbagh, Winterhoek Rd
PO Box 34, Tulbagh 6820
TEL: 023 230-0688
FAX: 023 230-1504
tkraal@lando.co.za
www.theuniskraal.co.za
OWNERS: Jordaan Brothers
DIRECTORS: Ernst Gabriel Jordaan, Jacobus Nicolaas Petrus Jordaan
C'MASTER: Andries Jordaan
VISITS: by appointment only
EST 1927 (producing wine since 1930s). Area under vine: 120ha (90% white). Annual production: 1 800 tons
FLAGSHIP: Riesling

LISTED WINES

D WHITE: Riesling (U, Crouchen); Sémillon Chardonnay (W)
S WHITE: Natural Sweet (U, Bukettraube & Muscat Ottonel)
RED: Prestige (W, Cab & Ruby Cab)

THIRTY-NINE ▷ TJG
33° ▷ WAMAKERSVALLEI

THOMAS KIPLING ▷ MAKRO
THORNHILL ▷ VEENWOUDEN
THREE CAPE LADIES ▷ WARWICK
THREE VINES ▷ DELLRUST
TIARA ▷ SIMONSIG
TICKLED PINK ▷ D DOORNKRAAL
TJ ▷ TWEE JONGE GEZELLEN

TOKARA

STELLENBOSCH MAP

P CELLAR: Helshoogte Pass, Stellenbosch
TEL: 021 808-5959
FAX: 021 808-5959
wine@tokara.com
OWNER: GT Ferreira
MD: Gyles Webb
W'MAKER: Miles Mossop
V'CULTURIST: Aidan Morton
VISITS: to the cellar by appointment only
RESTAURANT: lunch & dinner Tues–Sat, Tel 021 808-5959, Fax 021 808-5966, eat@tokara.com
EST 1995 (producing wine since 2000). Area under vine: 39ha (60% red). Annual production: 350 tons, 262 000 litres
FLAGSHIP: Tokara (still to be released)

LISTED WINES

Zondernaam

D WHITE: Chardonnay (W); S Blanc (U)
RED: Cab (W)

TOMORROW

P CELLAR: Tomorrow Wine Company, 42A Main Rd, Paarl
PO Box 9601, Paarl 7620
TEL: 021 863-1471
FAX: 021 863-1473
info@twc.co.za
www.tomorrowwinecompany.co.za
DIRECTORS: Viljoen de Kock (MD), Kosie Möller
C'MASTER: Kosie Möller
SALES: open to the public from end-2002, days and times t.b.a.
TASTINGS: no fee
RESTAURANT: to be opened in 2002/2003
EST 1998 (new cellar; first harvest to be 2003)
FLAGSHIPS: Bushman's Creek wines

LISTED WINES

Alphen Hill

D WHITE: Chardonnay (U); S Blanc (U)
RED: Cab (W); Merlot (W); Pinotage (W); Shiraz (W)

Bushman's Creek

D WHITE: Chardonnay (W); S Blanc (U)
RED: Cab (W); Cab Merlot (W); Merlot (W); Pinotage (W); Shiraz (W)

HIGHLY RATED

Bushman's Creek Merlot 1999 – W3½ 01, JC 02

Bushman's Creek Shiraz 1998 – JC 01

TOPAZ

SOMERSET WEST MAP

P CELLAR: not open to the public
6 Topaz St, Somerset West 7130
TEL: 021 855-4275
FAX: 021 855-5086
topazwines@mweb.co.za
OWNERS: Tanja Beutler, Clive Torr
GM: Tanja Beutler
W'MAKERS: Clive Torr, Tanja Beutler
SALES: by arrangement
FUTURES (EN PRIMEUR): yes
EST 2000. Area under vine: 0,4ha (100% red). Annual production: 3,5 tons, 2 050 litres

LISTED WINES

RED: Pinot (W); Shiraz (W)

TOP-OF-THE-HILL ▷ UITERWYK
TORTOISE HILL ▷ GLEN CARLOU
TORTUE ▷ STOUMANN'S
TOWERKOP ▷ LADISMITH
TRADEAUX ▷ BARRYDALE
TRADITION ▷ VILLIERA
TRADOUW ▷ BARRYDALE
TRANQUILLE ▷ CABRIÈRE

TRAVINO

OLIFANTS RIVER MAP

CO-OP: Klei Pan Rd, off N7
PO Box 2, Klawer 8145
TEL: 027 216-1616
FAX: 027 216-1425
travino@kingsley.co.za
CHAIRMAN: Deon du Toit
GM: Alkie van der Merwe
C'MASTER: Alkie van der Merwe
V'CULTURIST: Jeff Joubert
SALES: Mon–Sat 07:30–13:00 & 14:00–17:00
TASTINGS: no fee
TOURS: by appointment
EST 1969. Vines 80% white. Annual production: 10 000 tons
FLAGSHIPS: Robert Alexander wines

LISTED WINES

Carlton Hill

RED: Pinotage (W)

Robert Alexander

D WHITE: Chardonnay (W)
RED: Merlot (W); Shiraz (W)

Travino

D WHITE: Chardonnay (U); S Blanc (U)
SS WHITE: Special Nectar (U, Chenin)
PINK: Rosé (U, Shiraz)
RED: Grand Rouge (W, Pinotage & Ruby Cab); Merlot (W); Shiraz (W)
DESSERT: Knight (W, White Muscadel); Musketeer (W, Red Muscadel)

TRESUVA ▷ JANÉZA
TRIA CORDA ▷ OVERGAAUW

TUKULU

Grapes grown at Papkuilsfontein farm in Darling, wine made at Distell's Nederburg cellar in Paarl

PO Box 46, Stellenbosch 7599
TEL: 021 809-7569 (sales); 021 877-5145 (Nederburg cellar); see also Distell
FAX: 021 887-7318, 021 883-2603
info@distell.co.za
www.tukulu.co.za
OWNER: Papkuilsfontein Vineyards (Pty) Ltd, Distell
DIRECTORS: EG Le Roux (MD), AG Steyn, ME Moleka, AM Manyika
W'MAKER: Wellington Metshane
V'CULTURIST: Hannes van Rensburg
SALES: at Oude Libertas, Distell (Stellenbosch); Mon–Sat 08:30–17:00
TASTINGS: fee of R8,50
EST 1999. Annual production: 31 000 litres
FLAGSHIP: Pinotage

LISTED WINES

D WHITE: Chenin (U)
RED: Pinotage (W)

HIGHLY RATED

Tukulu Chenin Blanc
2001 – W4 02
Tukulu Pinotage
2000 – VD 01, PT 00
1999 – PT 00

TULBAGH

TULBAGH MAP

CO-OP: Drostdy, Tulbagh
PO Box 85, Tulbagh 6820
TEL: 023 230-1001
FAX: 023 230-1358
tkw@tulbaghwine.co.za
DIRECTORS: William Theron (MD), Aalwyn Theron, Danie Kotze, Danie Viljoen, Pierre Grobler, Pieter Theron
GM: Marius Burger
C'MASTER: Stephan Smit
SALES: Mon–Fri 08:00–17:00, Sat 09:00–13:00
TASTINGS: no fee
EST 1900 (producing wine since 1906). Area under vine: 550ha (75% white). Annual production: 7 500 tons, 7m litres

LISTED WINES

Tulbagh

D SPARKLING: Brut (U, S Blanc)
S SPARKLING: Vin Doux (U, Colombar & Pinotage)

D WHITE: Chardonnay (W); Chenin (U)
RED: Cab (U); Merlot (U); Pinotage (U); Shiraz (U)
PORT: Port (U, Pinotage)

Village Collection

D WHITE: Blanc de Blanc (U, S Blanc & Chenin & Colombard)
OD WHITE: Extra Light (U, Fernão Pires & Clairette Blanche)
SS WHITE: Late Harvest & Stein (U, S Blanc & Chenin & Colombad)
S WHITE: Natural Sweet (U)
RED: Classic Red (U, Ruby Cab & Pinotage)

TUMARA ▷ BELLEVUE
TURTLE CREEK ▷ MAKRO

TWEE JONGE GEZELLEN

TULBAGH MAP

ESTATE: Roodezand Rd, Tulbagh
PO Box 16, Tulbagh 6820
TEL: 023 230-0680
FAX: 023 230-0686
tjg@mweb.co.za
www.tjwines.co.za
OWNER: Nicky Krone (Nicolas Charles)
C'MASTER: Nicky Krone
SALES: Mon–Fri 09:00–16:00, Sat 10:00–14:00
VISITS: by appointment for large groups, formal tastings
TASTINGS: no fee for casual tastings, fee on request for formal tastings
TOURS: fees as above; Mon–Fri 11:00, 15:00, Sat 11:00
EST 1710 (producing wine since 1937). Area under vine: 120ha (75% white). Annual production: 1 100 tons, 660 000 litres
FLAGSHIP: Krone Borealis

LISTED WINES

MCC SPARKLING: Krone Borealis Brut (U, Chardonnay & Pinot)
PINK SPARKLING: Rosé (U, Chardonnay & Pinot & Chenin)
D WHITE: Light (U, French & German varieties); Thirty-Nine (U, Riesling & Chenin & S Blanc & Chardonnay)
OD WHITE: Schanderl (U, Muscat de F)
S WHITE: Night Nectar (U, Chenin & Gewürz, Natural Sweet)
DESSERT: Engeltjipipi (U, Chenin & Sémillon, Noble Late Harvest)
FORTIFIED: Balm of the Night (U, Muscat de F)

HIGHLY RATED

Twee Jonge Gezellen Krone Borealis Brut
1997 – PB 01
1994 – W4 99
Twee Jonge Gezellen

Rosé Brut
NV – W3½ 00

TWO OCEANS ▷ BERGKELDER
TWO ROADS ▷ FLAGSTONE

UITERWYK

STELLENBOSCH MAP

ESTATE: Stellenboschkloof Rd, Vlottenburg
PO Box 15, Vlottenburg 7604
TEL: 021 881-3711
FAX: 021 881-3776
info@uiterwyk.co.za
www.uiterwyk.co.za
OWNERS: De Waal brothers
DIRECTORS: Pieter De Waal (MD), Chris De Waal, Daniël De Waal
C'MASTERS: Chris De Waal (whites), Daniël De Waal (reds)
SALES: Oct–Apr Mon–Sat 10:00–16:30; May–Sept 10:00–12:30 & 14:00–16:30
TASTINGS: no fee
TOURS: by appointment
N MONUMENT: manor house (1791)
EST 1765. Area under vine: 120ha (50% red). Annual production: 1 000 tons, 760 000 litres
FLAGSHIPS: Uiterwyk Estate, Top-of-the-Hill Pinotage
SINGLE VYD WINES: S Blanc, Top-of-the-Hill Pinotage, Viognier

LISTED WINES

D WHITE: Res Chardonnay (W); S Blanc (U); Viognier (U)
RED: Cab (W); Cab Franc (W); Merlot (W); Pinotage (W); Shiraz (W); Top-of-the-Hill Pinotage (W); Uiterwyk Estate (W, Pinotage & Merlot & Cab Franc)

HIGHLY RATED

Uiterwyk Estate
1997 – W4 01
1996 – W4 99
Uiterwyk Merlot
1998 – W3½ 01
1996 – W4 99
Uiterwyk Pinotage
1998 – W3½ 00
1997 – W3½ 99
1995 – VD 97
Uiterwyk Top-of-the-Hill Pinotage
2000 – W4 02, VD 01, PT 01
1997 – VD 01, PT 99, W4 99
1996 – JP 99, W4½ PC 98, PT 98, VD 98

UITKYK

STELLENBOSCH MAP

ESTATE: R44 between Stellenbosch & Paarl
PO Box 3, Elsenburg 7607
TEL: 021 884-4416 (tastings, sales); 021 884-4710 (cellar)
FAX: 021 884-4717
info@uitkyk.co.za
www.uitkyk.co.za

OWNER: Lusan Holdings (Pty) Ltd
DIRECTORS: R Jones (MD); A Diedericks, HA Hëyl, BCM Tan, Prof CJ van Wyk
GM: Rudi Buys
C'MASTER: Estelle Swart
V'CULTURISTS: Rudi Buys, Jaco van den Berg
SALES: Mon–Fri 09:00–17:00, Sat–Sun 10:00–16:00
TASTINGS: fee of R10
TOURS: by appointment
FUTURES (EN PRIMEUR): yes
N MONUMENTS: murals (late 18th Century)
PICNICS: available for purchase
EST 1929. Area under vine: 170ha (50% red). Annual production: 700 tons, 420 000 litres

LISTED WINES

D WHITE: Chardonnay (W); S Blanc (U)
RED: Cab (W); Cab Shiraz (W)
BRANDY: 10 Year Old (Cinsaut & Clairette Blanche & Cab)

HIGHLY RATED

Uitkyk Carlonet
1996 – W4 99
1992 – VD 97

UITVLUCHT

ROBERTSON MAP
CO-OP: Bath St, Montagu
PO Box 332, Montagu 6720
TEL: 023 614-1340
FAX: 023 614-2113
uitvluchtwine@lando.co.za
DIRECTORS: EE Meyer (chairman), HJ Joubert, J Knel, D Nel, S Joubert
GM: JA Laubscher
C'MASTER: JA Laubscher
SALES: Mon–Fri 08:30–17:30, Sat 09:00–13:30
TASTINGS: no fee
EST 1942. Area under vine: 330ha (90% white). Annual production: 6 000 tons, 4m litres
FLAGSHIPS: Merlot, Muscat de F

LISTED WINES

OD SPARKLING: Vin Sec (U, S Blanc)
SS SPARKLING: Vin Doux (U, S Blanc & Chenin & Colombard)
D WHITE: Blanc de Blanc (U, S Blanc & Colombard)
OD WHITE: Colombard (U)
SS WHITE: Chenin (U); Late Harvest (U, Chenin)
RED: Derdeheuwel Rood (W, Cab & Ruby Cab); Merlot (W)
DESSERT: Red Muskadel (U); White Muskadel (U)
PORT: Port (U, Ruby Cab)

UITZICHT ▷ REYNEKE
UMFIKI ▷ GOUDINI
UN DEUX TROIS ▷ SEIDELBERG

UPLAND

WELLINGTON MAP

ESTATE: Fisantekuil, Blouvlei Rd, Wellington
PO Box 152, Wellington 7654
TEL: 082 731-4774
FAX: 021 873-5724
edmund@oettle.com
OWNER: E Oettlé
C'MASTER: Edmund Oettlé
VISITS: by appointment only
FUTURES (EN PRIMEUR): yes
ACCOM: Fisantekuil guest cottage
EST 1990 (producing wine since 1996). Area under vine: 12ha (84% white). Annual production: 40 tons, 30 000 litres
FLAGSHIP: Cabernet
SINGLE VYD WINE: Cab
ORGANIC WINE: Cab

LISTED WINES

RED: Cab (W)
SPIRITS: Brandy, Grappa

HIGHLY RATED

Upland Cabernet Sauvignon 1998 – W3½ 02

URBANE ▷ B&E NÉGOCIANTS

UVA MIRA

STELLENBOSCH MAP

P PRODUCER: Annandale Rd, off R44, Stellenbosch
PO Box 1511, Stellenbosch 7599
TEL: 021 880-1682
FAX: 021 880-1682
uva.mira@softswitch.co.za
OWNER: DW Weedon
GM: Ian Sutherland
W'MAKING: at Distell & Tokara
V'CULTURIST: Ian Sutherland
VISITS: by appointment only
EST 1994. Area under vine: 30ha (60% red)
FLAGSHIP: S Blanc
ORGANIC WINES: all

LISTED WINES

D WHITE: S Blanc (U)
RED: Cab (W); Merlot (W)

VALENTINE ▷ HUGUENOT

VAN LOVEREN

ROBERTSON MAP

P CELLAR: R317 between Robertson & Bonnievale
PO Box 19, Klaasvoogds 6707
TEL: 023 615-1505
FAX: 023 615-1336
vanloveren@lando.co.za
OWNER: Wynand & Nico Retief
DIRECTORS: Phillip Retief (MD), Wynand Retief, Nico Retief, Hennie Retief, Bussell Retief, Neil Retief
C'MASTER: Bussell Retief
V'CULTURIST: Neil Retief
SALES: Mon–Fri 08:30–17:00, Sat 09:30–13:00
TASTINGS: no fee
PICNICS: welcome to bring
EST 1937 (producing wine since 1972). Area under

vine: 250ha (60% red). Annual production: 2 570 tons, 1,8m litres
FLAGSHIPS: Ltd Cab, Ltd Shiraz
SINGLE VYD WINE: Spes Bona Chardonnay

LISTED WINES

Papillon

D SPARKLING: Brut (U, Colombar & S Blanc)
SS SPARKLING: Demi-Sec (U, Colombar & Riesling & White Muscadel)
S SPARKLING: Vin Doux (U, Muscat de F)

Van Loveren

D WHITE: Blanc de Blanc (U, Colombar & S Blanc); Cape Riesling (U, Crouchen); Colombar Chardonnay (W); Four Cousins Dry White (U, Colombar & S Blanc & Muscadel); Pinot Gris (U); Res Chardonnay (W); S Blanc (U); Sémillon (U); Spes Bona Chardonnay (U)
OD WHITE: Colombar (U); Fernão Pires (U); Vino Blanc (U, Colombar & S Blanc)
S WHITE: Special Late Harvest (U, Gewürz)
S PINK: Four Cousins Rosé (U, Natural Sweet)
SS PINK: Blanc de Noir Muscat (U, Red Muscadel)
D PINK: Blanc de Noir (U, Shiraz)
RED: Cab Shiraz (W); Four Cousins Dry Red (U, Shiraz & Ruby Cab & Merlot); Ltd Cab (W); Ltd Shiraz (W); Merlot (W); Pinotage (W); River Red (U, Shiraz & Ruby Cab & Merlot)
FORTIFIED: Rooi Muskadel (U)

HIGHLY RATED

Van Loveren Reserve Chardonnay
2001 – W3½ 02

VAN SCHOOR

ROBERTSON MAP

P CELLAR: Klipdrif, Robertson, off Bonnievale Rd
PO Box 65, Robertson 6705
TEL: 023 626-6455 (cellar, sales); 023 626-3316 (tastings)
FAX: 023 626-5073
mike@vanschoorwines.co.za
www.vanschoorwines.co.za
OWNER: Van Schoor Trust
DIRECTORS: Melt van Schoor (MD), Mike van Schoor
GM: Mike van Schoor
W'MAKERS: Mike van Schoor, Hilko Hegewisch
VITICULTURIST: Francois van Schoor
SALES: Mon–Fri 09:00–17:00, Sat 09:30–13:00
TASTINGS: no fee
TOURS: yes
EST 1984 (producing wine since 2001). Area under vine:

120ha (60% red). Annual production: 250 tons, 170 000 litres

LISTED WINES

D WHITE: Chardonnay (W)

RED: Pinotage (W)

VANSHA ▷ RIDGEBACK

VAN ZYLSHOF

ROBERTSON MAP

ESTATE: Main St, Bonnievale
PO Box 64, Bonnievale 6730
TEL: 023 616-2401
FAX: 023 616-3503
vanzylshof@lando.co.za
OWNERS: Chris & Andri van Zyl
C'MASTER: Andri van Zyl
SALES: Mon–Fri 09:00–17:00, Sat 09:00–13:00
TASTINGS: no fee
TOURS: on request
EST 1930 (producing wine since 1940). Area under vine: 30ha (84% white). Annual production: 450 tons

LISTED WINES

D WHITE: Chardonnay (W); Chardonnay Riverain (U); Chenin (U); S Blanc (W)

RED: Cab Merlot (W)

VAUGHAN JOHNSON'S

NÉGOCIANT: wine shop on Dock Rd, Waterfront, Cape Town
PO Box 50012, Waterfront 8002
TEL: 021 419-2121
FAX: 021 419-0040
vjohnson@mweb.co.za
www.vaughanjohnson.com
OWNER: Vaughan Johnson
SALES: Mon–Fri 09:00–18:00, Sat 09:00–17:00, Sun 10:00–15:00

OWN LABELS

D WHITE: Good Everyday Cape White (U, Chenin & Colombar & Sémillon & S Blanc); Seriously Good Plonk (U, Chenin & S Blanc); Sunday Best (U, Chenin & Chardonnay); Waterfront Collection Great White (U, Chenin)

RED: Good Everyday Cape Red (Cinsaut & Ruby Cab & Tinta B); Seriously Good Plonk (Cinsaut & Pinotage); Sunday Best (Cab & Merlot); Waterfront Captain's Claret (Tinta B)

VEENWOUDEN

PAARL MAP

P CELLAR: R45 N Paarl
PO Box 7086, N Paarl 7623
TEL: 021 872-6806
FAX: 021 872-1384
veenwouden@intekom.co.za
OWNER: Deon van der Walt
GM: Charles van der Walt
C'MASTER: Marcel van der Walt
V'CULTURISTS: Charles &

Marcel van der Walt
VISITS: by appointment only
EST 1989 (producing wine since 1993).
Area under vine: 14,5ha (100% red). Annual production: 80 tons, 60 000 litres
FLAGSHIPS: Classic, Merlot
SINGLE VYD WINE: Thornhill

LISTED WINES

Thornhill

RED: Shiraz (W)

Veenwouden

RED: Classic (W, Cab & Merlot & Cab Franc & Malbec); Merlot (W, Merlot & Cab); Vivat-Bacchus (W, Merlot & Malbec & Cab Franc)

HIGHLY RATED

Veenwouden Classic
1999 – W4½ 02
1998 – JP 01, VD 00, W4 00
1996 – W4½ 99, PB 98
1995 – JP 99
Veenwouden Merlot
1999 – W3½ 02
1998 – W4½ 01, PB Trophy 00, VD 00
1997 – W4 00
1995 – W4 98
Veenwouden Vivat Bacchus
1999 – VD 01, W4 01

VENDÔME

PAARL MAP
P CELLAR: S Paarl, off N1 (exit 59)
PO Box 36, Huguenot 7645
TEL: 021 863-3905
FAX: 021 863-0094
lerouxjg@icon.co.za
www.vendome-wines.com
OWNER: JG Le Roux
C'MASTER: Jannie Le Roux Jnr
SALES: Mon–Fri 09:30–16:30, Sat 09:30–12:30
TASTINGS: R5pp
TOURS: by appointment
PICNICS: welcome to bring
EST 1897. Area under vine: 35ha (60% white).
Annual production: 50 tons, 32 500 litres
FLAGSHIP: Cabernet

LISTED WINES

D WHITE: Chardonnay (U)
RED: Cab (W); Le Roux Merlot Cab (W)

VERA CRUZ ▷ DELHEIM
VERDI ▷ EIKENDAL
VERDUN ▷ ASARA

VERGELEGEN

SOMERSET WEST MAP
P CELLAR: Lourensford Rd, Somerset West
PO Box 17, Somerset West 7129
TEL: 021 847-1334 (switchboard); 021 847-1337 (tastings)
FAX: 021 847-1608
vergelegen@amfarms.co.za
www.vergelegen.co.za

OWNER: Anglo American
DIRECTORS: Don Tooth (MD), MW Spicer, PM Baum, EK Diack, JS Knight, Hon JH Steyn, AJ Trahar, VP Urew
C'MASTER: André van Rensburg
V'CULTURIST: Neil Rossouw
SALES: Mon–Sun 09:30–16:00
TASTINGS: fee of R5
TOURS: by appointment, at a fee of R10 – 10:30, 11:30, 15:00 (May–Oct Mon–Fri & Sat mornings only)
N MONUMENTS: Camphor trees (300 years old)
RESTAURANT: Lady Phillips 09:30–16:00, Rose Terrace – lunches Nov–Apr; Tel 021 847-1346
PICNICS: available for purchase Nov–Apr
EST 1989 (producing wine since 1992). Area under vine: 103ha (60% red). Annual production: 1 000 tons, 580 000 litres
FLAGSHIP: Vergelegen (red blend)
SINGLE VYD WINES: Res Chardonnay, Res S Blanc

LISTED WINES

D WHITE: Chardonnay (W); Res Chardonnay (W); Res S Blanc (U); S Blanc (U); Vin de Florence (U, Chenin & Muscat d A & S Blanc)
RED: Cab (W, Cab & Merlot & Cab Franc); Cab Franc Merlot (W); Merlot (W, Merlot & Cab Franc); Mill Race (W, Cab & Merlot & Cab Franc); Shiraz (W); Vergelegen (W, Cab & Merlot & Cab Franc)

HIGHLY RATED

Vergelegen Chardonnay
2001 – W4 02
2000 – W3½ 01
1999 – W4 00
1997 – W3½ 98
Vergelegen Reserve Chardonnay
2001 – W4 02
2000 – W4 02
1999 – VD 01, MG 01, W4 01
1998 – PB 99, MG 99
1997 – W4 98
1996 – SA Trophy 98
Vergelegen Sauvignon Blanc
2001 – SA Certificate 02
2000 – JC 00
1999 – W4 99
1998 – W3½ 98
Vergelegen Reserve Sauvignon Blanc
2001 – TT 02, W5 02, JP 02
2000 – W3½ 01
1999 – W3½ 99
1998 – JP 99 W4 99, MG 98
1996 – VD 97
Vergelegen Reserve Sémillon
1999 – W4½ 02, PB 00
Vergelegen
1999 – W4½ 02, SA Trophy 02, JP 02, PB 01
1998 – W4 01, JP 01

Vergelegen Cabernet Franc Merlot
1999 – W4 01
1998 – PB 99
Vergelegen Cabernet Sauvignon
1999 – W3½ 02
1998 – TG 02, W5 02, JC 00
Vergelegen Mill Race Red
1999 – W4 01
Vergelegen Merlot
1998 – W4½ 01, JC 00
Vergelegen Shiraz
1999 – W4 01
Vergelegen Sémillon Noble Late Harvest
2000 – W4½ 02
1998 – W4 01, MG Trophy 99

VERGENOEGD

HELDERBERG MAP

ESTATE: off R310 (Bayden Powell Drive), Faure
PO Box 1, Faure 7131
TEL: 021 843-3248
FAX: 021 843-3118
enquiries@vergenoegd.co.za
OWNERS: JA Faure Trust / ISF Trust Property – Vergenoegd Trust
DIRECTORS: J Strauss, JA Faure
GM: JA Faure
C'MASTER: John Faure
SALES: Mon–Fri 08:30–16:00, Sat 09:30–12:30
TASTINGS: no fee
TOURS: by appointment
N MONUMENT: homestead (1773)
EST 1800 (producing wine under own label since 1972). Area under vine: 100ha (95% red). Annual production: 500 tons
FLAGSHIP: Reserve

LISTED WINES

RED: Cab (W); "New Label" (W, Cab & Cab Franc & Merlot & Shiraz); Merlot (W); Reserve (W, Cab & Merlot & Cab Franc); Shiraz (W)
PORT: Vintage (W, Tinta B)

HIGHLY RATED

Vergenoegd Cabernet Sauvignon
1999 – W3½ 02
1995 – W4 99
1992 – W4 99
Vergenoegd Merlot
1999 – W4 02
1998 – W4 02, JC 01
1997 – W3½ 99
1996 – W3½ 99
Vergenoegd Reserve
1999 – W3½ 02
1996 – W3½ 99
1995 – W4 98
Vergenoegd Shiraz
1998 – W4 01
1997 – W4 00
Vergenoegd Vintage
1995 – W4 99
1994 – W4 98

VERSUS ▷ STELLENBOSCH VYDS
V GENERATIONS ▷ CEDERBERG
VICTORY ▷ WAMAKERSVALLEI

VILJOENSDRIFT

ROBERTSON MAP

P CELLAR: 12km out of Robertson on Bonnievale Rd (R317)
PO Box 653, Robertson 6705
TEL: 023 615-1901
FAX: 023 615-3417
viljoensdrift@lando.co.za
www.viljoensdrift.co.za
OWNERS: Fred & Manie Viljoen
C'MASTER: Fred Viljoen
V'CULTURIST: Manie Viljoen
SALES: Mon–Fri 09:00–16:00, Sat 10:00–14:00 at river tasting area
TASTINGS: no fee
TOURS: by appointment
PICNICS: available for purchase; welcome to bring
EST 1863 (producing wine since 1998). Area under vine: 80ha (50% red). Annual production: 500 tons, 300 000 litres
FLAGSHIPS: River Grandeur wines

LISTED WINES

D WHITE: Colombar Chenin (U); River Grandeur Chardonnay (W); River Grandeur Chenin (U); River Grandeur S Blanc (U); Sémillon (U)
RED: Merlot Shiraz (W); River Grandeur Cab (W); River Grandeur Pinotage (W); River Grandeur Shiraz (W)

HIGHLY RATED

Viljoensdrift River Grandeur Chenin Blanc
2001 – W3½ 02
Viljoensdrift River Grandeur Shiraz
2001 – W3½ 02
2000 – W4 02

VILLAGE RED ▷ MCGREGOR

VILLIERA

STELLENBOSCH MAP

P CELLAR: cnr R304 & Old Paarl Rd, Koelenhof (±2km from N1, Stellenbosch exit)
PO Box 66, Koelenhof 7605
TEL: 021 865-2002/3
FAX: 021 865-2314
wine@villiera.com
www.villiera.com
OWNERS: Grier family
DIRECTORS: Jeff Grier (MD), Simon Grier, Cathy Brewer, Xander Grier
C'MASTER: Jeff Grier
V'CULTURIST: Simon Grier
SALES: Mon–Fri 08:30–17:00, Sat 08:30–13:00
TASTINGS: no fee
TOURS: self-guided
PICNICS: welcome to bring
EST 1983. Area under vine: 300ha (60% white). Annual production: 2 000 tons, 1,3m litres
FLAGSHIPS: Bush Vine S Blanc, Cru Monro, Monro Brut

SINGLE VYD WINES: Bush Vine S Blanc

WINE WITHOUT SULPHUR: Brut Natural

LISTED WINES

Blue Ridge

D WHITE: Blanc (U, Chenin & S Blanc)

RED: Rouge (W, Shiraz & Carignan)

Tradition

MCC SPARKLING: Brut (W, Pinot & Chardonnay & Pinotage & Chenin); Rosé Brut (U, Pinot & Chardonnay & Pinotage)

Villiera

MCC SPARKLING: Brut Natural (U, Chardonnay); Monro Brut (W, Pinot & Chardonnay)

D WHITE: Blanc Fumé (W, S Blanc); Bush Vine S Blanc (U); Cellar Door Chenin (W); Chardonnay (W); Chenin (U); S Blanc (U); R Riesling (U)

OD WHITE: Gewürz (U); Sonnet (U, Muscat Ottonel & Chenin)

RED: Cab (W); Cru Monro (W, Cab & Merlot); Down to Earth (W, Shiraz & Carignan & Gamay & Merlot & Pinotage); Merlot (W); Merlot Pinotage (W); Pinotage (W); Res Merlot (W); Shiraz (W)

DESSERT: Inspiration (W, Chenin, Noble Late Harvest)

PORT: Port (W, Shiraz & Pinotage & Gamay)

HIGHLY RATED

Villiera Brut
1995 – TT 02, W5 02

Villiera Brut Natural Chardonnay
1999 – W4 02
1998 – W4½ 01

Villiera Tradition Brut
NV – W3½ 01

Villiera Tradition Brut Rosé
NV – W4 01, W4 00, W3½ 98

Villiera Tradition Carte d'Or Cuvée Brut
NV – SA Trophy 99

Villiera Chenin Blanc
2001 – DT 01, VD 01
2000 – JC 01
1999 – W3½ 00
1996 – W4 CC 99

Villiera Cellar Door Chenin Blanc
2001 – W4½ 02

Villiera Rhine Riesling
2000 – JC 01

Villiera Sauvignon Blanc
1997 – W4 97, VD 97
1991 – W3½ 98

Villiera Traditional Bush Vine Sauvignon Blanc
1997 – SA Certificate 98, DW 97, W4 97, VD 97

Villiera Blanc Fumé
2001 – W4 02
1999 – W4 00

1997 – W4 97
Villiera Cru Monro
1996 – VD 98
1995 – VD 97
Villiera Merlot Pinotage
2000 – W3½ 02
1999 – W4 01
1998 – W3½ 00
Villiera Reserve Merlot
1998 – W4 01
Villiera Pinotage
2000 – W3½ 02
1997 – W4 99, VD 98
1996 – W3½ 98
Villiera Inspiration
2000 – W4 02
Villiera LBV
1997 – W3½ 01
Villiera Cape Vintage
1995 – VD 98, W3½ 98
Woolworths Brut Reserve (Villiera)
NV – W3½ 01
Woolworths Brut Rosé (Villiera)
NV – W4 01, W3½ 00, W3½ 98
Woolworths 2000 Collection Brut Reserve (Villiera)
NV – W3½ 00
Woolworths Rhine Riesling (Villiera)
2000 – W3½ 01
Woolworths Reserve Cabernet Sauvignon (Villiera)
1997 – W4 99

VILLIERSDORP

WORCESTER MAP

CO-OP: Main Rd, Villiersdorp
PO Box 14, Villiersdorp 6848
TEL: 028 840-1151
FAX: 028 840-1833
kelkiewyn@freemail.absa.co.za
GM: Hannes Cilliers
C'MASTER: Danie Conradie
SALES: Mon–Fri 08:00–17:00, Sat 08:00–13:00
TASTINGS: no fee
TOURS: by appointment
RESTAURANT: Kelkiewyn farm-stall & coffee shop
EST 1922 (producing wine since 1980s). Area under vine: 470ha (65% white). Annual production: 7 500 tons, 1,8m litres

LISTED WINES

SS SPARKLING: Demi Sec (U)
D WHITE: Colombar chardonnay (W); S Blanc (U)
OD WHITE: Chenin (U); Colombar (U)
SS WHITE: Late Vintage (U)
SS PINK: Rosé (U, Ruby Cab)
RED: Cab (W); Cab Merlot (U); Merlot (U)
FORTIFIED: Hanepoot Jerepiko (U)
PORT: Port (U)

VINAY ▷ SLANGHOEK
VIN DE CAREME ▷ ASHANTI
VIN DE CONSTANCE ▷ K CONSTANTIA
VIN DE ERSTELLE ▷ L'AVENIR
VIN DE FLORENCE ▷ VERGELEGEN
VIN DE GLACIERE ▷ SIGNAL HILL

VIN DE LA TORTUE ▷ STOUMANN'S
VIN DE L'EMPEREUR ▷ SIGNAL HILL
VIN DE LIZA ▷ SIMONSIG
VIN DE MEURVEUR ▷ L'AVENIR

VINDIGO

NÉGOCIANT: 108 Sarel Cilliers St, Napier
TEL: 028 423-3069
FAX: 028 423-3069
vindigo@isat.co.za
OWNERS: Nick & Terie Yell
W'MAKER: Niels Verburg (Beaumont)
V'CULTURIST: Sebastian Beaumont
SALES: Mon–Fri 09:30–17:30, Sat–Sun 09:30–16:30
TASTINGS: no fee
EST 1999. Annual production: 150 cases

LISTED WINES

De Hoopen
RED: Merlot (W); Pinotage (W)

VINFRUCO

STELLENBOSCH MAP

COMPANY: Oude Molen, Distillery Rd, Stellenbosch
PO Box 12730, Die Boord 7613
TEL: 021 886-6458
FAX: 021 886-6589
info@vinfruco.co.za
www.arnistonbay.co.za
DIRECTORS: LB Kriel (exec chairman), JHH Coetzee, Dr PF de V Clüver, AJ de Haast, AP de Wet, A du Toit, TG Keuzenkamp, AW Mostert, NW Oosthuizen, AJ van Velden, NC Versfeld
C'MASTER: Nicky Versfeld
VISITS: not open to the public
EST 1992
FLAGSHIP: Arniston Bay

LISTED WINES

Arniston Bay
D WHITE: Chenin Chardonnay (U)
PINK: Rosé (U, Pinotage)
RED: Ruby Cab Merlot (U)

VINTNER'S CHOICE ▷ MAKRO
VIRGINIA ▷ DISTELL
VISION ▷ KAAPZICHT
VIVAT BACCHUS ▷ VEENWOUDEN
VLAEBERG ▷ DOMINION
VLAKTEPLAAS ▷ KANGO

VLOTTENBURG

STELLENBOSCH MAP

CO-OP: cnr R310 & Old Vlottenburg Rd
PO Box 40, Vlottenburg 7604
TEL: 021 881-3828/9
FAX: 021 881-3357
vlottenb@netactive.co.za
www.vlottenburg.co.za
GM: Kowie Du Toit
C'MASTER: Kowie du Toit
V'CULTURIST: PG Slabbert
SALES: Mon–Fri 09:00–17:00, Sat 09:00–12:30

TASTINGS: fee of R10
EST 1945. Area under vine: 1 000ha (60% red). Annual production: 7 500 tons, 5,6m litres
FLAGSHIP: Ltd Release Cab

LISTED WINES

D WHITE: Chardonnay (W); S Blanc (U)
OD WHITE: Blanc de Blanc (U, Chenin & S Blanc); Chenin (U)
SS WHITE: Late Harvest (U, S Blanc)
RED: Cab (W); Ltd Rel Cab (W); Merlot (W); Pinotage (W); Reserve (W, Shiraz & Cab); Rouge (U, Cinsaut & Pinotage); Shiraz (W)
DESSERT: Muscat de Hambourg (U)

HIGHLY RATED

Vlottenburg Merlot
1999 – W4 01
Vlottenburg Pinotage
1998 – W3½ 00
1997 – PT 98

VON ORTLOFF

FRANSCHHOEK MAP

P CELLAR: Dassenberg farm, Franschhoek
PO Box 341, Franschhoek
7690
TEL: 021 876-3432
FAX: 021 876-4313
vortloff@mweb.co.za
OWNERS: Evi & Georg Schlichtmann
C'MASTERS: Evi & Georg Schlichtmann
VISITS: by appointment only
EST 1992. Area under vine: 15ha (50% red). Annual production: 80 tons
FLAGSHIP: Cabernet
SINGLE VYD WINES: all

LISTED WINES

D WHITE: Chardonnay (W); No 5 S Blanc (U)
RED: Cab (W, Cab & Merlot); No 7 Merlot (W)

VREDENDAL

OLIFANTS RIVER MAP

CO-OP: 17 Sirkel Rd, Vredendal
PO Box 75, Vredendal
8160
TEL: 027 213-1080
FAX: 027 213-3476
vredwine@namaqua.co.za
OWNER: West Corp – see separate listing
GM: Izak Visagie
C'MASTER: Pieter Verwey
V'CULTURIST: PLD Koegelenberg
SALES: Mon–Fri 08:00–17:30, Sat 08:30–12:30
TASTINGS: no fee
TOURS: Feb–Apr, 10:00 & 15:00
PICNICS: welcome to bring
EST 1946 (producing wine since 1952)

LISTED WINES
See separate listing under West Corp – reds made at Spruitdrift winery, whites made at Vredendal winery

VREDENHEIM

STELLENBOSCH MAP

P CELLAR: R310, Vlottenburg
PO Box 369, Stellenbosch 7600
TEL: 021 881-3637
FAX: 021 881-3296
trendsetter@vredenheim.co.za
www.vredenheim.co.za
OWNERS: Bezuidenhout family
DIRECTORS: AD Le Hanie (MD), JP le Hanie, CJ Bezuidenhout, JE Bezuidenhout
C'MASTER: JP le Hanie
SALES: Mon–Fri 09:00–17:00, Sat 09:30–14:00
TASTINGS: fee of R5 for three tastings, R7 for five tastings
N MONUMENTS: manor house and grounds (1975)
RESTAURANT: Barrique, lunch & dinner Tues–Sat, 12:00–15:00 & 18:00–22:00, Tel 021 881-3001
PICNICS: available for purchase
EST 1987. Area under vine: 90ha (60% red)
FLAGSHIP: Vredenheim Res

LISTED WINES

Angels
S WHITE: Angels (U, Chenin)

Vredenheim
SS SPARKLING: Vredenvonkel (U, S Blanc & Chenin)
D WHITE: Chenin (W)
RED: Cab Merlot (W); Res (W, Shiraz & Cab & Merlot)

VRIESENHOF ▷ PARADYSKLOOF

WABOOMSRIVIER

WORCESTER MAP

CO-OP: R43 Rd, Wolseley
PO Box 24, Breede River 6858
TEL: 023 355-1730
FAX: 023 355-1731
DIRECTORS: SF du Toit (MD), P van der Merwe, J Pieterse, M Basson, D Zeeman, A Roos, W Marais
GM: CH van der Merwe
W'MAKERS: CH van der Merwe, W Viljoen
V'CULTURISTS: S du Toit, J van Rooyen
SALES: Mon–Fri 08:00–17:00, Sat 08:00–10:00
TASTINGS: no fee
TOURS: yes
FUTURES (EN PRIMEUR): yes
EST 1949. Area under vine: 851ha (42% red). Annual production: 11 396 tons, 8,7m litres

LISTED WINES

Wagenboom
OD SPARKLING: Rubellite (U, Pinotage &

Chenin, Demi Sec)
D WHITE: Blanc de Blanc (U, Chenin & S Blanc)
SS WHITE: Chenin Laat Oes (U)
RED: Cab (W); Pinotage (W); Rodewagen (U, Pinotage & Cinsaut & Cab); Ruby Cab (U)
DESSERT: Soet Hanepoot (W)
PORT: Port (W, Ruby Cab)

WAGENBOOM ▷ WABOOMSRIVIER

WAMAKERSVALLEI

WELLINGTON MAP

CO-OP: Distillery Rd, Wellington
PO Box 509, Wellington 7654
TEL: 021 873-1592 (tastings, sales); 021 873-1582 (cellar)
FAX: 021 873-3194
sales@wamakers.co.za
www.wamakersvallei.co.za
MD: Ernest Brink
GM: Chris Roux
C'MASTER: Bennie Wannenburg
V'CULTURIST: De Wet Viljoen
SALES: Mon–Fri 08:00–13:00 & 14:00–17:00, Sat 08:30–11:30
TASTINGS: no fee
TOURS: by appointment
PICNICS: welcome to bring
EST 1941. Area under vine: 950ha (55% red). Annual production: 10 000 tons, 50m *l*
FLAGSHIPS: La Cave wines
SINGLE VYD WINES: La Cave

LISTED WINES

SPARKLING: 33° South Sparkling Wine (U, S Blanc)
D WHITE: Bain's Way Chardonnay (U); Bain's Way Chenin (U); Bain's Way S Blanc (U); 33° South Dry White (U, S Blanc)
RED: Bain's Way Cab (W); Bain's Way Pinotage (W); Cinsaut (U); Duke Victory Red (W, Pinotage & Cab); La Cave Cab (W); La Cave Merlot (W); La Cave Pinotage (W); La Cave Shiraz (W); Merlot (W); 33° South Dry Red (U, Cinsaut & Pinotage)
DESSERT: Duke Muscat (U, Natural Sweet)

HIGHLY RATED

Wamakersvallei La Cave Cabernet Sauvignon
2001 – W4 02
Wamakersvallei La Cave Pinotage
2001 – W4 02

WANDSBECK ▷ AGTERKLIPHOOGTE
WARRANT ▷ WILDEKRANS

WARWICK

STELLENBOSCH MAP

ESTATE: R44, between Klapmuts & Stellenbosch
PO Box 2, Muldersvlei 7607
TEL: 021 884-3146 (tastings); 021 884-4410 (sales)

FAX: 021 884-4025
info@warwickwine.co.za
www.warwickwine.co.za
OWNERS: ES Ratcliffe family
MD: Michael Ratcliffe
C'MASTER: Louis Nel
V'CULTURIST: Ronald Spies
SALES: Mon–Fri 10:00–16:00, Sat Nov–Apr 09:00–13:00
TASTINGS: no fee
TOURS: by appointment
PICNICS: available for purchase by arrangement; welcome to bring
EST 1964 (producing wine since 1982). Area under vine: 55ha (90% red). Annual production: 250 tons
FLAGSHIP: Trilogy
SINGLE VYD WINES: Cab Franc, Pinotage, Merlot

LISTED WINES

D WHITE: Chardonnay (W); S Blanc (U)
RED: Cab (W); Cab Franc (W); Merlot (W); Old Bush Vines Pinotage (W); Three Cape Ladies (W, Pinotage & Cab & Merlot); Trilogy (W, Cab & Merlot & Cab Franc)

HIGHLY RATED

Warwick Chardonnay
1999 – W3½ 00
1998 – W4 99
Warwick Cabernet Franc
1998 – W3½ 01
1995 – W4 98
1994 – W5 98
Warwick Cabernet Sauvignon
2000 – W3½ 02
1999 – W3½ 02
Warwick Merlot
1999 – W3½ 02
Warwick Old Vines Pinotage
2000 – JC 01
Warwick Trilogy
1998 – W3½ 00
1996 – VD 98
1995 – W3½ 98

WATERFORD

STELLENBOSCH MAP

P CELLAR: Blaauwklippen Rd, Stellenbosch
PO Box 635, Stellenbosch 7600
TEL: 021 880-0496
FAX: 021 880-1007
waterfordhil@icon.co.za
www.waterfordwines.com
OWNERS: Jeremy & Leigh Ord
PARTNER: Kevin Arnold
C'MASTER: Kevin Arnold
V'CULTURIST: Bob Hobson
SALES: Mon–Fri 08:00–17:00
TASTINGS: Mon–Fri 09:00–16:30, no fee
TOURS: yes
EST 1998. Area under vine: 40ha (70% red). Annual production: 60 tons, 45 000 litres

LISTED WINES

Kevin Arnold
RED: Shiraz (W)

Peacan Stream
D WHITE: Chenin (U, Chenin & Chardonnay & S Blanc)
RED: Cab (W)

Waterford
D WHITE: Chardonnay (W); S Blanc (U)
RED: Cab (W)

HIGHLY RATED
Kevin Arnold Shiraz
1999 – W3½ 01, JC 00
1998 – W4 99
Waterford Chardonnay
2001 – W4 02
Waterford Cabernet Sauvignon
2000 – W4 02
1999 – W4½ 02
1998 – W4 00

WATERFRONT ▷ V JOHNSON'S
WATERSIDE ▷ GRAHAM BECK

WEBERSBURG

STELLENBOSCH MAP

CELLAR: Annandale Rd, Stellenbosch
PO Box 3428, Somerset West 7129
TEL: 021 851-7417
FAX: 021 852-5280
weber@iafrica.com
OWNER: Fred Weber
W'MAKER: Giorgio Dalla Cia (consulting)
V'CULTURIST: Braam Steyn
VISITS: by appointment only
EST 1996. Area under vine: 38ha (100% red). Annual production: 60 tons, 42 000 *l*

LISTED WINES
RED: Webersburg (W, Cab)

WELGEGUND

PRODUCER: apex of Perdeskoen Rd, Wellington
PO Box 683, Wellington 7654
TEL: 021 864-1185 (manager); 021 873-2123 (manor house)
FAX: 021 873-2683
rac@icon.co.za
OWNER: Alex Camerer
MANAGER: Bernardt Kruger
W'MAKER: Johan Joubert (Bovlei)
V'CULTURIST: Johan Pienaar (consultant)
VISITS: by appointment only
MONUMENT: cottages (1792)
ACCOM: B&B
EST 1930s (producing wine since 1998). Area under vine: 28ha (63% red). Annual production: wine made at Bovlei Co-op

LISTED WINES
RED: Carignan (W)

WELGEMEEND

PAARL MAP

ESTATE: Old Paarl Rd, Klapmuts
PO Box 1408, S Paarl 7624
TEL: 021 875-5210
FAX: 021 875-5239
welgemeend@worldonline.co.za
OWNER: Welgemeend Wine CC

DIRECTORS: Ursula Hofmeyr (MD), Jan Hofmeyr, Ilna Hofmeyr, Louise Hofmeyr
C'MASTER: Louise Hofmeyr
V'CULTURIST: Ursula Hofmeyr
SALES: Wed 14:00–16:00, Sat 09:00–12:30
TASTINGS: no fee
TOURS: yes
EST 1974 (producing wine since 1979). Area under vine: 12,5ha (100% red). Annual production: 54 tons
FLAGSHIP: Estate Reserve

LISTED WINES

RED: Amadé (W, Shiraz & Grenache & Pinotage); Douelle (W, Cab & Merlot & Cab Franc & Malbec); Estate Reserve (W, Cab & Merlot & Cab Franc); Soopjeshoogte (W, Cab & Merlot & Cab Franc)

WELLINGTON

WELLINGTON MAP

CO-OP: Distillery Rd, Wellington
PO Box 520, Wellington 7654
TEL: 021 873-1163
FAX: 021 873-2423
wellwyn@iafrica.com
DIRECTORS: MW Malan (chairman), P du T Malan, H le Roux, JF le Roux, PF Malan
GM: Gert Boerssen
W'MAKERS: Gert Boerssen, Koos Carstens
V'CULTURIST: Bertus Albertyn
SALES: Mon–Fri 08:00–13:00 & 14:00–17:00
TASTINGS: no fee
EST 1934. Area under vine: 1 600ha (60% white). Annual production: 10 000 tons, 7,3m litres
FLAGSHIPS: Cab, Pinotage
SINGLE VYD WINES: Cab, Merlot, Pinotage, Port, S Blanc, Shiraz

LISTED WINES

D WHITE: Chardonnay (W); Chenin (U); S Blanc (U)
SS WHITE: Stein (U, Chenin)
RED: Cab (W); Cinsaut Ruby Cab (W); Merlot (W); Pinotage (W); Shiraz (W)
DESSERT: Hanepoot Jerepiko (U)
PORT: Port (U, Tinta B)

HIGHLY RATED

Wellington Cabernet Sauvignon
1998 – VD 00

WELMOED

STELLENBOSCH MAP

COMPANY: R310, Lynedoch
PO Box 465, Stellenbosch 7599
TEL: 021 881-3870
FAX: 021 881-3434
E-MAIL, WEBSITE: see Stellenbosch Vineyards
OWNER: Stellenbosch Vineyards
C'MASTER: Chris Kelly

V'CULTURIST: F de Villiers
SALES: Mon–Fri 09:00–17:30, Sat 09:00–17:00, Sun 10:00–16:00
TASTINGS: R1 per glass
RESTAURANT: Duck Pond, lunches Mon–Sun 12:00–16:00 (closed on Tues Jun–Jul), Tel/Fax 021 881-3310

LISTED WINES

D WHITE: Blanc de Blanc (U, Chenin & S Blanc); Res S Blanc (U); S Blanc Selection (U); Sémillon (W)
OD WHITE: Chenin Selection (U)
RED: Cab (W); Merlot (W); Pinotage (W); Shiraz (W)

HIGHLY RATED

Welmoed Chardonnay
1998 – W3½ 99
Welmoed Reserve Sauvignon Blanc
1998 – W4 98
1997 – W4 97
Welmoed Merlot
1999 – W3½ 01
1998 – W3½ 99
Welmoed Shiraz
2000 – W3½ 01

WELTEVREDE

ROBERTSON MAP

ESTATE: on R317 from Robertson, 5km outside Bonnievale
PO Box 6, Bonnievale 6730
TEL: 023 616-2141 (tastings, sales, cellar); 023 616-2895 (manor house)
FAX: 023 616-2460
info@weltevrede.com
www.weltevrede.com
OWNER: LN Jonker
DIRECTORS: Lourens Nicolas Jonker (MD), Philip Jonker
GM: P Jonker
C'MASTER: P Jonker
V'CULTURISTS: P Jonker, Francois Viljoen
SALES: Mon–Fri 08:00–17:00, Sat 09:0–15:00
TASTINGS: no fee
TOURS: by appointment
N MONUMENT: Oupa se Wingerd (1926)
ACCOM: Ons Huisie, Belvedere
RESTAURANT: Die Parsmandjie, Mon–Fri 08:00–17:00, Sat 09:00–15:30, Tel 023 616-2141
EST 1926 (producing wine since 1975). Area under vine: 100ha (70% white). Annual production: 1 400 tons, 980 000 litres
FLAGSHIP: Oude Weltevreden Chardonnay
SINGLE VYD WINES: Oude Weltevreden Chardonnay, Oude Weltevreden Merlot Cab, Weltevrede S Blanc

LISTED WINES

Oude Weltevreden

D WHITE: Chardonnay (W)
RED: Merlot Cab (W)

Weltevrede

MCC SPARKLING: Philip Jonker Brut (U, Chardonnay)
D WHITE: Chardonnay (W); Prive du Bois (W, S Blanc & Chardonnay); River's Edge Chardonnay (U); River's Edge Blanc de Blanc (U, Colombard & Chenin); River's Edge Cape Riesling (U, Crouchen); River's Edge S Blanc (U); S Blanc (U)
OD WHITE: R Riesling (U); River's Edge Colombard (U); River's Edge Muscat Colombard (U, Muscat de F & Colombard)
SS WHITE: Gewürz (U)
S WHITE: Ovation R Riesling (U, Natural Sweet); Therona SLH (U)
PINK: River's Edge Blanc de Noir (U, Muscat de F)
RED: River's Edge Tricolore (W, Cab & Merlot & Syrah)
DESSERT: Cape Muscat (U, Muscat de Hambourg)
FORTIFIED: Ouma se Wyn (U, Muscat de F, White Muscadel); Oupa se Wyn (U, Muscat de F, Red Muscadel)

HIGHLY RATED

Oude Weltevreden Chardonnay
1998 – W4 99
Weltevrede Philip Jonker Brut
1998 – W3½ 01
1995 – W4 98
Weltevrede Cape Muscat
2000 – W3½ 02
Weltevrede Ouma se Wyn
2000 – W3½ 02
1999 – W3½ 00
1997 – VD 98
Weltevrede Oupa se Wyn
2000 – W3½ 02

WELVANPAS

WELLINGTON MAP

P CELLAR: Bovlei, Wellington
PO Box 75, Wellington 7655
TEL: 021 864-1238
FAX: 021 864-1239
welvanpas@xinet.co.za
OWNER: Dan Retief Snr
C'MASTER: Dan Retief Jnr
SALES: Mon–Fri 09:00–12:00 & 13:30–16:30, Sat 09:00–13:00
TASTINGS: no fee
FLAGSHIP: Cabernet

LISTED WINES

D WHITE: Amity (U, Chardonnay & S Blanc)
SS WHITE: Bukettraube (U)
RED: Cab

WESKUS ▷ WINKELSHOEK

WEST CORP

COMPANY: amalgamation of Vredendal & Spruitdrift co-ops
PO Box 75, Vredendal 8160
TEL: 027 213-1080
FAX: 027 213-3476
GM: Izak Visagie
TASTINGS & SALES: at

Vredendal & Spruitdrift Co-ops – see separate listings
EST 2002 (producing wine at Vredendal Co-op since 1952, at Spruitdrift Co-op since 1972). Area under vine: 4 780ha. Annual production: ±90 000 tons
FLAGSHIPS: Goiya wines, Mount Maskam wines

LISTED WINES

Whites made at Vredendal Co-op, reds made at Spruitdrift Co-op

Cape Rock

OD WHITE: White (U, Chardonnay & S Blanc)
SS WHITE: Natural Sweet (U, Colombard & Chardonnay & Chenin & S Blanc)
SS PINK: Rosé (U, Chenin & Pinotage)

Gôiya

OD WHITE: Kgeisje (U, Chardonnay & S Blanc)
RED: G!aan (U, Shiraz & Pinotage)

Mount Maskam

D WHITE: Chardonnay (U); S Blanc (U)
RED: Cabaret (W, Cab & Ruby Cab); Merlot (W); Pinotage (W); Shiraz (W)
FORTIFIED: Port (W, Ruby Cab)

Namaqua

D WHITE: Grand Cru (U, Chenin & Colombard); Lite (U, various cultivars)
SS WHITE: Johannisberger (U, Chenin & Chardonnay & S Blanc); Late Harvest (U, Chenin); Stein (U, Chenin & Colombard)
SS PINK: Rosé (U, Colombard & Pinotage)
RED: Dry Red (U, Cab & Cab Franc & Shiraz & Merlot)
FORTIFIED: Sherry (Chenin & Colombard & Chardonnay & S Blanc)

Spruitdrift

S SPARKLING: Vin Doux (u, Chenin)
D WHITE: Grand Cru (U, Chenin & Colombard)
OD WHITE: Colombard Chardonnay (U)
SS WHITE: Chenin (U); Natural Sweet (U, Chenin & Chardonnay & Colombard & S Blanc)
PINK: Blanc de Noir (U, Red Muscadel)
RED: Cab (U); Merlot (U); Pinotage (W); Shiraz (U)
FORTIFIED: Hanepoot (U); Red Jerepigo (U, Red Muscadel); White Jerepigo (U, White Muscadel)

Vredendal

S SPARKLING: Spumanté (U, Hanepoot)

WESTERN WINES

NÉGOCIANT: based at Doornbosch Centre, Stellenbosch
PO Box 769, Stellenbosch 7599
TEL: 021 882-8177
FAX: 021 882-8176
manager@westerwines.co.za
www.western-wines.com
OWNER: Roger Gabb
DIRECTORS: Mike Paul (MD), Ed Adams, James Reid, Steve Smith
GM: James Reid
W'MAKERS: Rhyan Wardman, Ben Jordaan
VISITS: not open to the public
EST 1998. Annual production: 9m litres
FLAGSHIP: Kumala Reserve wines
SINGLE VYD WINES: Kumala Single Varietal Cab, Kumala Single Varietal S Blanc

LISTED WINES

Kumala

D WHITE: Chardonnay Colombard (W); Chenin Chardonnay (W); S Blanc Colombard (U); Sémillon Chardonnay (W); Single Varietal S Blanc (U)
RED: Res Cab (W); Res Merlot (W); Res Shiraz (W); Single Varietal Cab (W)

WHALEHAVEN

WALKER BAY MAP

P CELLAR: Hemel-en-Aarde Valley, Hermanus
Private Bag X14, Suite 10, Hermanus 7200
TEL: 028 316-1633
FAX: 028 316-1640
whwines@hermanus.co.za
DIRECTORS: S Kreusch (MD), R Middelmann, M Middelmann, N Middelmann, Dr A Bottega
C'MASTER: Storm Kreusch
SALES: Mon–Fri 09:30–17:00, Sat 10:30–14:00
TASTINGS: fee of R10pp for groups of five or more
EST 1995. Annual production: 80 tons, 45 000 litres
FLAGSHIP: Pinot

LISTED WINES

D WHITE: Chardonnay (U); Chardonnay (W)
RED: Baleine Noire (W, Cab & Cab Franc); Cab (W); Merlot (W); Pinot (W)

HIGHLY RATED

WhaleHaven Pinot Noir
1998 – W3½ 00
1997 – W3½ 01
1996 – W4 98

WHOLE BERRY ▷ SPRINGFIELD
WIDE RIVER ▷ ROBERTSON

WILDEKRANS

ELGIN MAP

ESTATE: off Houw Hoek Mountain Rd
PO Box 200, Elgin 7180
TEL: 021 859-5587 (tastings, sales); 028 284-9829 (cellar)
FAX: 021 859-5587
houw@mweb.co.za
OWNER: Dr Bruce Elkin
C'MASTER: Dr B Elkin
V'CULTURIST: Barry Anderson
SALES: at estate and Wildekrans Wine Shop & Art Gallery at Orchard Farmstall on N2, Grabouw – Mon–Fri 07:30–17:30 (estate), Mon–Sun 08:30–17:30 (farmstall)
TASTINGS: no fee
TOURS: on request
PICNICS: welcome to bring to the estate
EST 1985 (producing wine since 1993). Area under vine: 50ha (50% red). Annual production: 100 tons, 65 000 litres
FLAGSHIPS: Barrel Selection Pinotage, S Blanc, Warrant
SINGLE VYD WINES: Barrel Selection Pinotage, Cab, Chardonnay, Merlot, S Blanc, Sémillon

LISTED WINES

D WHITE: Caresse Marine (W, S Blanc & Chenin); Chardonnay (W); Res Chenin (W); S Blanc (U); Sémillon (W)
RED: Barrel Sel Pinotage (W); Cab (W); Cab Franc Merlot (W); Merlot (W); Pinotage (W); Warrant (W, Cab & Merlot)

HIGHLY RATED

Wildekrans Pinotage
1996 – W3½ 98
Woolworths Terroir Walker Bay Wildekrans Merlot
1997 – W4 99

WILD KWAGGA ▷ MEERHOF
WILD YEAST ▷ SPRINGFIELD

WILHELMSHOF

P PRODUCER: wines made at Saxenburg
PO Box 12200, Stellenbosch 7613
TEL: 021 903-9507 (house)
FAX: 021 903-9507
OWNERS: Nico & Dr Petra van der Merwe
GM: Dr Petra van der Merwe
W'MAKER: Nico van der Merwe
SALES: by arrangement, and at retail stores
EST 1999. Annual production: 70 tons, 50 000 litres
FLAGSHIP: Mas Nicolas Cape

LISTED WINES

D WHITE: Robert Alexander (W, Chardonnay – different varieties selected

each vintage)
RED: Mas Nicolas Cape (W, Shiraz & Cab)

WILLIAM EVERSON ▷ JA CLIFT
WIND ▷ HARTSWATER

WINDFALL

ROBERTSON MAP

P CELLAR: Agterkliphoogte, Robertson
PO Box 802, Robertson 6705
TEL: 023 626-4498 (cellar), 033 394-6377 (sales)
FAX: 033 342-2531
windfallwinefarm@webmail.co.za
OWNER: Robert Alexander
C'MASTER: Helmut Hanekom
V'CULTURIST: Nico van der Westhuizen
VISITS: not open to the public
EST 1992 (producing wine since 2000).
Area under vine : 14ha (60% red). Annual production: 35 tons, 19 000 litres
FLAGSHIP: Pinotage

LISTED WINES

D WHITE: Chenin (U)
RED: Merlot (U); Pinotage (U) ; Ruby Cab (U)

WINDMEUL

SEE PAARL MAP

CO-OP: R44 (Agter-Paarl Rd)
PO Box 2013, 7630
TEL: 021 869-8100/8043
FAX: 021 869-8614
windmeul@iafrica.com
DIRECTORS: Wilhelm Dreyer (MD), A Kotze, WP Dreyer, L Vorster, D Morkel, D Stoffberg, N Stofberg
GM: Danie Marais
C'MASTER: Danie Marais
V'CULTURISTS: Danie Marais, Paul Wallace
SALES: Mon–Fri 08:00–17:00, Sat 09:00–13:00
TASTINGS: no fee
TOURS: by appointment
EST 1944.
Area under vine: 1 295ha (58% white). Annual production: 9 000 tons, 16,6m litres
FLAGSHIP: Cabernet Merlot

LISTED WINES

D WHITE: Chenin (U); S Blanc (U)
RED: Cab (W); Cab Merlot (W); Merlot (W); Pinotage (W); Shiraz (W)

HIGHLY RATED

Windmeul Cabernet Sauvignon
2000 – W3½ 02
Windmeul Cabernet Sauvignon Merlot
2000 – W3½ 02
Windmeul Merlot
1999 – W3½ 01

WINDS OF CHANGE ▷ AFRICAN TERROIR

WINECORP

PO Box 99, Lynedoch 7603
TEL: 021 881-3690
FAX: 021 881-3699
winecorp@iafrica.com
www.winecorp.co.za
OWNERS: Spier Group 62%, balance by public
CHAIRMAN: Raisal Rahmatallah
CEO: Johan van Reenen
CHIEF W'MAKER: Frans Smit
V'CULTURIST: Johan Smit

LISTED WINES

See separate listings for Spier, Longridge and Savanha – the Ashwood cellar in Franschhoek, closed to the public, also falls within the Winecorp group

WINKELSHOEK

SWARTLAND MAP

P CELLAR: N7, Piketberg
PO Box 395, Piketberg 7320
TEL: 022 913-1092
FAX: 022 913-1095
winhoek@intekom.co.za
OWNER: HAC Hanekom, JJ Brand
C'MASTER: HAC Hanekom
SALES: Mon–Fri 09:00–17:00, Sat 09:00–13:00
TASTINGS: fee of R1
RESTAURANT: Mon–Sun 09:00–18:00

LISTED WINES

Weskus

D WHITE: Grand Cru (U, Chenin)
OD WHITE: Blanc de Blanc (U)
SS WHITE: Late Harvest (U, Chenin)
RED: Dry Red (U, Pinotage)

WITHOEK

LITTLE KAROO MAP

P CELLAR: Wesoewer Rd, off Voortrekker St, Calitzdorp
PO Box 181, Calitzdorp 6660
TEL: 044 213-3639
FAX: 044 213-3690
OWNER: Koos Geyser
C'MASTER: Koos Geyser
V'CULTURIST: Francois Steenkamp
VISITS: by appointment only
EST 1996. Area under vine: 25ha (82% white). Annual production: 8 tons, 4 800 litres

LISTED WINES

RED: Droë Rooi (W, Tinta B & Pontac & Ruby Cab)
FORTIFIED: Hanepoot (U); Muskadel (U)
PORT: Ruby (W, Tinta B & Pontac & Ruby Cab)

WOODLANDS

NÉGOCIANT: 10 Harewood Ave, Somerset West
TEL: 021 855-2249
FAX: 021 855-4924

rbuchner@worldonline.co.za
OWNER: Raimund Buchner
DIRECTORS: R Buchner, J Ressi
C'MASTER: Kosie Möller
VISITS: by appointment only
ACCOM: Assegai guest house
EST 2000.
Annual production: 32 tons, 22 500 litres

LISTED WINES

Assegai

RED: Cab (W); Merlot (W); Pinotage (W); Shiraz (W)

WOOLWORTHS

NÉGOCIANT: national retail chain (public company)
Head office at 93 Longmarket St, Cape Town
PO Box 680, Cape Town 8000
TEL: head office 021 407-2777
FAX: head office 021 407-3846
allanmullins@woolworths.co.za
www.woolworths.co.za
CEO: Simon Susman
DIRECTOR FOOD: Richard Butt
WINE SEL MGR: Allan Mullins
WINE BUYING MGR: Ivan Oertle
SALES: from branches country-wide
FLAGSHIPS: Limestone Hill Chardonnay (De Wetshof), Merlot Cab (Morgenhof), Res Cab Merlot (Neil Ellis), Res Chardonnay (Neil Ellis), Signature Series
ORGANIC WINES: Organic range (Sonop)

LISTED WINES

Limited Release

OD WHITE: Gewürz (Bergsig)
RED: Merlot (Villiera), Pinot (Jordan)

Organic

D WHITE: S Blanc
RED: Pinotage; Shiraz

Premier

D WHITE: S Blanc (Darling Cellars)

Reserve

D WHITE: Barrel Res Chenin (Ken Forrester); Chardonnay (Neil Ellis);

Constantia

S Blanc (Buitenverwachting)
RED: Cab (Villiera); Cab Merlot (Neil Ellis); Shiraz Cab (La Motte); Grenache Syrah (Ken Forrester)
DESSERT: Noble Late Harvest Chenin (Ken Forrester)

Selected

D WHITE: Cape White (Rooiberg)
SS SWEET: Late Harvest (Simonsvlei)
RED: Cape Red (Simonsvlei)

Bubbly

MCC SPARKLING: Brut (Villiera); Rosé (Villiera); Vintage (Villiera)
SPARKLING: Spumanté (Rooiberg); Spumanté Brut (Rooiberg); Spumanté Rosé (Rooiberg)

Signature Series
D WHITE: Chardonnay (Danie De Wet)
RED: Merlot Cab (Morgenhof)
Longmarket Range
D WHITE: Blanc de Blanc (Villiera); Cape Riesling (Weltevrede); Chardonnay (Robertson); Chenin (Rooiberg); Chardonnay Sémillon (Delheim); S Blanc (Robertson)
PINK: Blanc de Noir (Swartland); Rosé (Villiera)
RED: Bush Vine Shiraz (Swartland); Cab (Bergsig); Merlot Cab Franc (Simonsvlei); Pinotage (Rooiberg); Maison Rouge (Bergsig)
DESSERT: Muscat Colombard (Weltevrede)
Other:
D WHITE: Bianca Light (Delheim); Chardonnay Pinot (Cabrière); Premier Grand Cru; R Riesling (Villiera)
OD WHITE: Bouquet Blanc; Zesty White
SS WHITE: Stein
PINK: Perky Pink
RED: Dry Red; Grand Rouge (La Motte); Juicy Red; Merlot (La Motte); Shiraz (Rooiberg)
GLUHWEIN: Gluwein (Winecorp)

HIGHLY RATED

Brut Reserve (Villiera)
NV – W3½ 01
Brut Rosé (Villiera)
NV – W4 01, W3½ 00, W3½ 98
Brut Reserve 2000 Celebration
NV – W3½ 00
2000 Collection Brut Reserve (Villiera)
NV – W3½ 00
Ken Forrester Barrel Chenin Blanc
2001 – W3½ 02, JC 02
Rhine Riesling (Villiera)
2000 – W3½ 01
Weisser Riesling (Lievland)
1996 – VD 97
Exclusive Reserve Sauvignon Blanc
2001 – W3½ 01
Vin de Cuvée Sauvignon Blanc (Robertson)
2000 – W4½ 00
Reserve Cabernet Sauvignon (Villiera)
1997 – W4 99
Signature Series Cabernet Sauvignon (Flagstone)
1999 – W3½ 01
Cabernet Sauvignon Merlot (Neil Ellis)
1999 – W4 01
Terroir Stellenbosch Kaapzicht Cabernet Sauvignon Merlot
1997 – W3½ 99
Terroir Walker Bay

Wildekrans Merlot
1997 – W4 99
Reserve Pinotage (Kaapzicht)
1998 – W3½ 00
Reserve Groenekloof Shiraz
1999 – JC 02
Fields of Gold Natural Sweet
NV – W4 01
Barrel Reserve Chenin Blanc Noble Late Harvest (Ken Forrester)
1999 – W3½ 00

WORKERS VINTAGE ▷ HELDERKRUIN
WRITER'S BLOCK ▷ FLAGSTONE

YELLOW CELLAR ▷ HELDERKRUIN

YONDER HILL

HELDERBERG MAP

P CELLAR: R44 between Stellenbosch & Somerset West
PO Box 914, Stellenbosch 7599
TEL: 021 855-1008 (tastings, sales, cellar); 021 855-0863 (manor house)
FAX: 021 855-1006
wines@yonderhill.co.za
www.yonderhill.co.za
OWNER: FS & DM Naudé
GM: Frikkie Naudé
C'MASTER: David Lockley
VISITS: by appointment only
EST 1989 (producing wine since 1993). Under vine: 10ha (97% red). Annual production: 105 tons, 73 500 litres
FLAGSHIP: Merlot

LISTED WINES

D WHITE: Chardonnay (U)
RED: Cab (W); Inanda (W, Cab & Merlot & Cab Franc); Merlot (W)

HIGHLY RATED

Yonder Hill Inanda
2000 – W3½ 02
1998 – W4½ 01
Yonder Hill Merlot
1999 – W3½ 01

ZANDDRIFT

PAARL MAP

P CELLAR: R301, S Paarl – closed for replanting
PO Box 1302, S Paarl 7624
FAX: 021 863-2081
zanddrift@yahoo.com
www.zanddrift.co.za
CHAIRMAN: Koh Seow Chuan (Singapore)
MD: Koos de Waal
SALES: Cellar currently not in production – replanting commenced 2002, red only

ZANDVLIET

ROBERTSON MAP

ESTATE: Ashton, via N1 to Worcester, through Robertson
PO Box 36, Ashton 6715
TEL: 023 615-1146
FAX: 023 615-1327
shiraz@lando.co.za
www.zandvliet.co.za
OWNERS: Paul & Dan de Wet

GM: Wouter Theron
W'MAKERS: Paul de Wet, Johan van Wyk
V'CULTURISTS: Dan de Wet, Wouter Theron
SALES: Mon–Fri 09:00–17:00, Sat 09:00–13:00
TASTINGS: no fee
PICNICS: welcome to bring
EST 1867 (first bottled vintage in 1975). Area under vine: 155ha (50% red). Annual production: 1 300 tons
FLAGSHIP: Kalkveld Syrah
SINGLE VYD WINE: Kalkveld Syrah

LISTED WINES

Astonvale

D WHITE: Chardonnay (U); Crème (U, Colombard & S Blanc); S Blanc (U)
OD WHITE: Ferent (U, Colombard & Chardonnay)
RED: Old Paddock Cab Shiraz (W); Ruby Cab (U); Shiraz (U)

Equus

D WHITE: Chardonnay (W)
RED: Cab Merlot Shiraz (W)

Zandvliet

D WHITE: Chardonnay (W)
RED: Kalkveld Shiraz American Oak; Kalkveld Syrah French Oak; Shiraz (W)

HIGHLY RATED

Zandvliet Chardonnay
1998 – W3½ 99
1997 – W4 98
Zandvliet Equus Cabernet Sauvignon Merlot Shiraz
1999 – W3½ 01

ZANDWIJK

PAARL MAP

P CELLAR: Pieter Hugo St, Courtrai, S Paarl
PO Box 2674, Paarl 7620
TEL: 021 863-2368
FAX: 021 863-1884
zandwijk@capegate.co.za
www.zandwijk.co.za
OWNER: Cape Gate
GM: Neil Schnoor
C'MASTER: Neil Schnoor
V'CULTURIST: Frank Pietersen
SALES: Mon–Fri 08:00–12:30 & 13:30–17:00
TASTINGS: no fee
TOURS: in off-season, by appointment
PICNICS: welcome to bring
EST 1983. Area under vine: 9ha (50% red). Annual production: 150 tons, 90 000 litres
FLAGSHIP: Kleine Draken Cab

LISTED WINES

Kleine Draken

D WHITE: Chardonnay (W); S Blanc (U)
S WHITE: Kiddush (U, Cinsaut, Natural Sweet)
RED: Cab (W); Dry Red (W, Merlot & Malbec & Cab Franc)

ZANTSI ▷ DARLING CELLARS
ZAXIMUS ▷ CITRUSDAL

ZESTY ▷ WOOLWORTHS

ZEVENWACHT

STELLENBOSCH MAP

P CELLAR: Langverwacht Rd, Kuils River
PO Box 387, Kuils River 7579
TEL: 021 903-5123
FAX: 021 903-3373
info@zewenwacht.co.za
www.zewenwacht.co.za
OWNERS: Johnson family
PROPRIETOR: Harold Johnson
C'MASTER: Karl Lambour
V'CULTURISTS: Kevin Watt, Jacques Viljoen
SALES: Mon–Fri 08:30–17:00, Sat–Sun 09:30–17:00
TASTINGS: no fee
TOURS: by appointment
N MONUMENT: manor house (1800)
ACCOM: country inn, Tel 021 906-1560, Fax 021 906-1570
RESTAURANT: Mon–Sun 07:00–12:00, Tel 021 903-5257
PICNICS: available for purchase
EST 1800. Area under vine: 200ha (50% red). Annual production: 1 000 tons, 675 000 litres
EST Merlot, S Blanc, Shiraz
SINGLE VYD WINE: Tin Mine Chenin

LISTED WINES

D WHITE: Blanc de Blanc (U, Chenin & S Blanc); Chardonnay (W); Chenin (U); S Blanc (U); Tin Mine Chenin (U)
OD WHITE: Bouquet Blanc (U, Gewürz & Riesling & Hanepoot)
RED: Cab (W); Chevalier Cab Merlot (W); Merlot (W); Shiraz (W); Zevenrood (W, Cab & Merlot & Shiraz)

HIGHLY RATED

Zevenwacht Chardonnay
1996 – VD 97
Zevenwacht Cabernet Sauvignon Merlot
1996 – W4 99, SA Certificate 98
Zevenwacht Merlot
1999 – W3½ 00
Zevenwacht Reserve Merlot
1996 – W4 99
Zevenwacht Pinotage
1997 – W4 99
Zevenwacht Shiraz
1997 – W4 98

ZONDERNAAM ▷ TOKARA
ZONNEBLOEM ▷ DISTELL

Regional Wine Route Offices

BREEDE VALLEY WINES
Tel: 023 342-8710
Fax: 023 342-2294
manager@winelands.co.za
www.worcesterwinelands.co.za

CALITZDORP WINE ROUTE
Tel: 044 213-3312
Fax: 044 213-3295
kannalandc@telkom.sa.net
www.calitzdorp.org.za

CONSTANTIA WINE ROUTE
Tel: 021 794-1810
Fax: 021 794-1812
wine@icon.co.za,
www.constantiauitsig.co.za

DURBANVILLE WINE ROUTE
Tel: 083 310-1228
Fax: 021 976-5631
nitida@mweb.co.za

HELDERBERG WINE ROUTE
Tel: 021 852-6166
Fax: 021 851-1497
hwr@mweb.co.za
www.helderbergwineroute.co.za

KLEIN KAROO WINE ROUTE
Tel/Fax: 028 572-1284
info@kleinkarroowines.co.za
www.kleinkaroowyne.co.za

OLIFANTS RIVER WINE TRUST
Tel/Fax: 027 213-3126
olifantsrivwineroute@kingsley.co.za

ORANGE RIVER WINE TRUST
Tel: 054 337-8800
Fax: 054 332-4408
marketing@owk.co.za
www.owk.co.za

PAARL VINTNERS
Tel: 021 872-3605
Fax: 021 872-3841
paarl@wine.co.za
www.paarlwine.co.za

ROBERTSON VALLEY WINE ROUTE
Tel: 023 626-3167
Fax: 023 626-1054
info@robertsonwinevalley.co.za
www.robertsonwinevalley.co.za

STELLENBOSCH WINE ROUTE
Tel: 021 886-4310
Fax: 021 886-4330
info@wineroute.co.za
www.wineroute.co.za

SWARTLAND WINE ROUTE
Tel: 022 487-1133
Fax: 022 487-2063
swartlandinfo@westc.co.za
www.swartlandwineroute.co.za

TULBAGH WINE TRUST
Tel/Fax: 023 230-1348
tulbaghinfo@lando.co.za
www.tulbagh.com

VIGNERONS DE FRANSCHHOEK
Tel: 021 876-3062
Fax: 021 876-2964
franschhoek@wine.co.za
www.franschhoekwines.co.za

WELLINGTON WINE ROUTE
Tel: 021 873-4604
Fax: 021 873-4607
welltour@cis.co.za
www.visitwellington.com

STELLENBOSCH

1 African Terrior 8G
2 Alto 11G
3 Amani 9B
4 Annandale 11G
5 Asara 9F
6 Audacia 10G
7 Bellevue 5C
8 Bergkelder 7H
9 Beyerskloof 5F
10 Bilton 11H
11 Blaauwklippen 9H
12 Blue Creek 10H
13 Bonfoi 8D
14 Boschkloof 9D
15 Bouwland 5E
16 Bowe-Joubert 9C
17 Camberley 7L
18 Carisbrooke 9E
19 Clos du Ciel 12H
20 Clos Malverne 7F
21 Clovelly 5E
22 Coppoolse Finlayson 4A
23 Delaire 7K
24 Delheim 4J
25 De Meye 1F
26 De Toren 9C
27 De Trafford 12J
28 Devonair 6E
29 Devon Hill 6E
30 Dornier 11I
31 Du Preez 8H
32 EagleVlei 1E
33 Eersterivier 9E
34 Fort Simon 7D
35 Goede Hoop 6C
36 Graceland 10G
37 Hartenberg 6E
38 Hazendal 6B
39 Helderkruin 12H
40 Hoopenburg 1F
41 Jacobsdal 9B
42 JC le Roux 6E
43 Jordan 8C
44 Kaapzicht 6C
45 Kanonkop 3H
46 Kanu 9D
47 Klawervlei 3F
48 Klein Gustrouw 8J
49 Kleine Zalze 9G
50 Knorhoek 4I
51 Koelenhof 2F
52 Laibach 3H
53 Lanzerac 8J

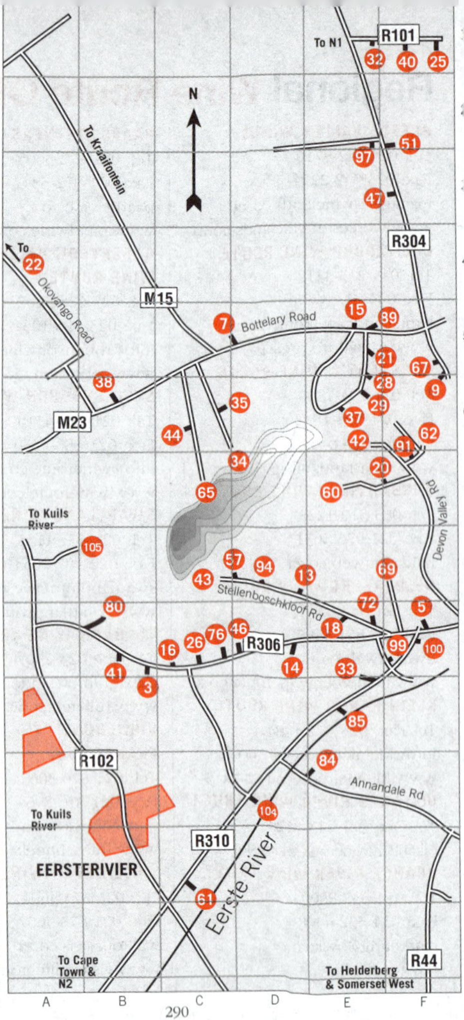

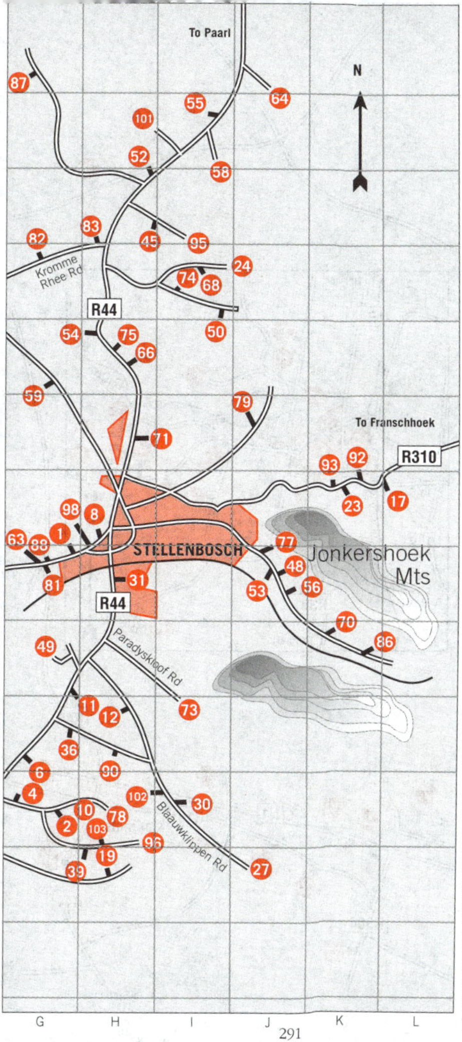

54	L'Avenir	5H
55	Le Bonheur	2I
56	Le Riche	8J
57	L'Emigré	8D
58	Lievland	2I
59	Louiesenhof	5G
60	Louisvale	7E
61	Meerlust	12C
62	Meinert	6F
63	Middelvlei	8G
64	Mont Destin	1J
65	Mooiplaas	7C
66	Morgenhof	5H
67	Mulderbosch	5F
68	Muratie	4I
69	Neethlingshof	8F
70	Neil Ellis	9K
71	Nietvoorbij	6H
72	Overgaauw	9E
73	Paradyskloof	10I
74	Quoin Rock	4I
75	Remhoogte	5H
76	Reyneke	9C
77	Rozendal	8J
78	Rust en Vrede	11H
79	Rustenberg	6J
80	Saxenburg	9B
81	SFW/Distillers -now Distell	8G
82	Simonsig	4G
83	Slaley	3H
84	Spier Cellars	11E
85	Spier (sales)	10E
86	Stark-Condé	K9
87	Starke	G1
88	Stellekaya	8G
89	Stellendrift	5E
90	Stellenzicht	10H
91	SylvanVale/Devon Valley Hotel	7F
92	Thelema	7K
93	Tokara	7K
94	Uiterwyk	8D
95	Uitkyk	3I
96	Uva Mira	11H
97	Villiera	2E
98	Vin Fruco	7G
99	Vlottenburg	9F
100	Vredenheim	9F
101	Warwick	2H
102	Waterford	11H
103	Webersburg	11H
104	Welmoed	11D
105	Zevenwacht	8B

PAARL & WELLINGTON

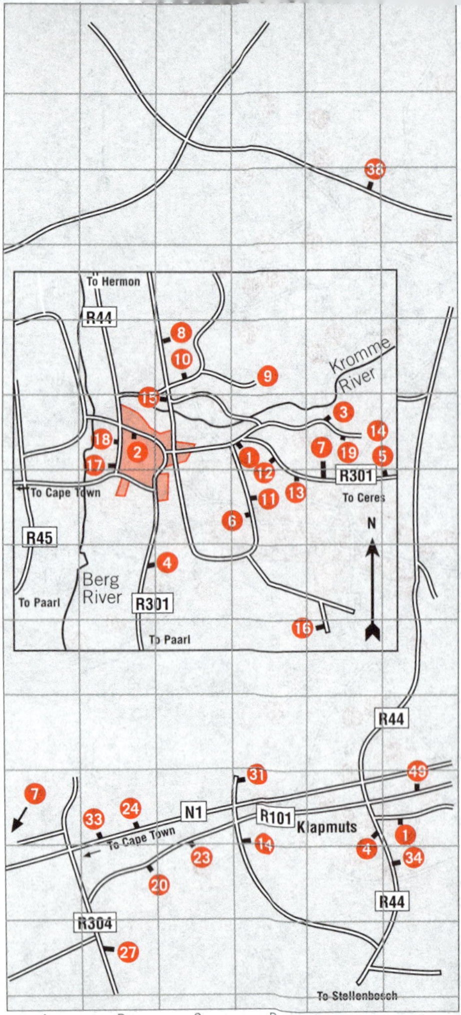

WELLINGTON MAP (INSET):

1	Bovlei	6D
2	Cape Wine	6B
3	De Compagnie	6E
4	Diemersfontein	7D
5	Doolhoof	7D
6	Hildenbrand	7D
7	Huguenot	6E
8	Jacaranda	5C
9	Linton Park	5D
10	Mischa	5C
11	Mont Du Toit	7D
12	Napiêr	6D
13	Oude Wellington	7D
14	Siyabonga	6E
15	Stormberg	5B
16	Upland Estate	9E
17	Wamakersvallei	6B
18	Wellington	6B
19	Welvanpas	6E

PAARL MAP:

1	Anura	11F
2	Ashanti	7L
3	Avondale	9K
4	Avondvrede	11E
5	Backsberg	13G
6	Bernheim	4J
7	Bodega	11A
8	Boland (NPaarl)	5I

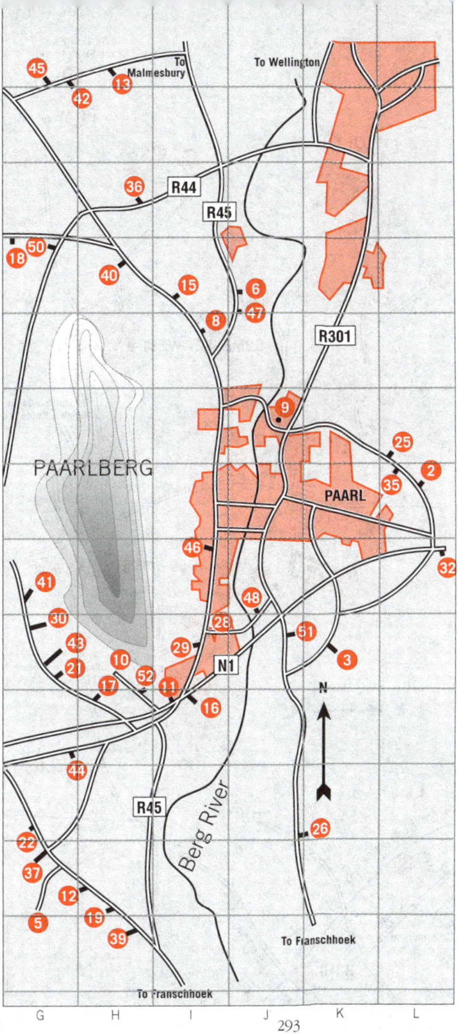

9 Boland (Dal Josafat) 6J
10 Brenthurst 9H
11 Coleraine 10I
12 Cowlin 12H
13 David Frost 1H
14 De Meye 11D
15 De Villiers 4I
16 De Zoete Inval 10I
17 Diamant 10H
18 Domaine Brahms 4G
19 Drakensig 12H
20 EagleVlei 12B
21 Fairview 9G
22 Glen Carlou 11G
23 Hoopenberg 11C
24 Joostenberg 11B
25 Kleinvallei 6L
26 Klompzicht 11K
27 Koelenhof 13B
28 KWV 9I
29 Laborie 9I
30 Landskroon 9G
31 Leidersburg 11D
32 Mellasat 8L
33 Mijn-Burg 11B
34 Mont Destin 12F
35 Nederburg 7L
36 Nelson's Creek 3H
37 Niel Joubert 12G
38 Perdeberg 3E
39 Rupert & Rothschild 13H
40 Rhebokskloof 4H
41 Ruitersvlei 8G
42 Scali 1G
43 Seidelberg 9G
44 Simonsvlei 10G
45 Sonop 1G
46 Tommorrow 8I
47 Veenwouden 4J
48 Vendome 8J
49 Welgemeend 11F
50 Windmeul 4G
51 Zanddrift 9J
52 Zandwijk 9H

HELDERBERG & SOMERSET WEST

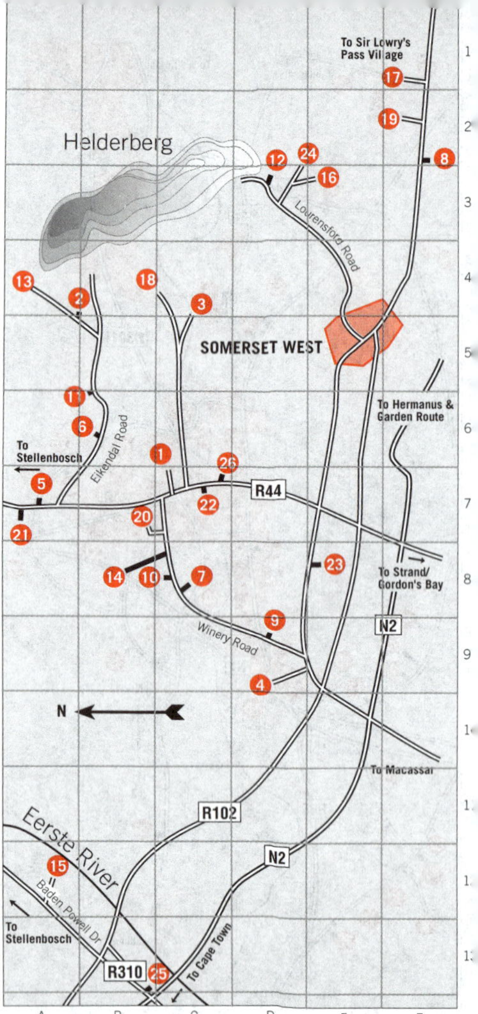

1	Avontuur	6C
2	Clos du Ciel	4A
3	Cordoba	4C
4	Dellrust	9D
5	Eikendal	7A
6	Grangehurst	6B
7	Helderberg	8C
8	Ingwe	2F
9	JP Bredell	9D
10	Ken Forrester (96 Winery Rd)	8C
11	Longridge	5B
12	Lourensford	3D
13	Lushof	4A
14	Lyngrove	8C
15	Meerlust	12A
16	Morgenster	3E
17	Mount Rozier	1F
18	Nooitgedacht	4B
19	Onderkloof	2F
20	Post House	7B
21	Somerbosch	7A
22	Stonewall	7C
23	Topaz	8E
24	Vergelegen	2D
25	Vergenoegd	13B
26	Yonder Hill	7C

FRANSCHHOEK & SIMONDIUM

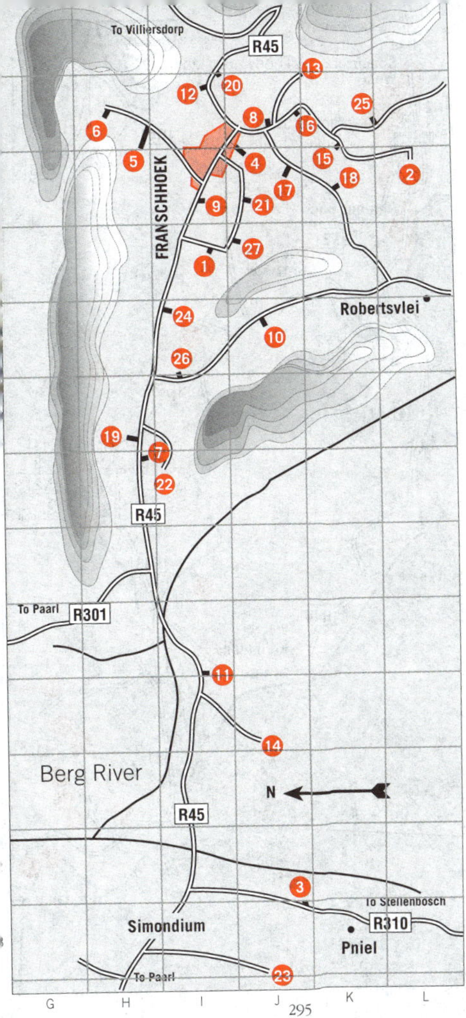

1	Agusta	4I
2	Boekenhoutskloof	3L
3	Boschendal	13J
4	Cabrière	3J
5	Chamonix	3H
6	Dieu Donne	2H
7	Eikehof	7H
8	Elephant Pass	2J
9	Franschhoek V	3I
10	Glenwood	5J
11	Graham Beck	9I
12	Haute Cabrière	2I
13	Jean Daneel	2J
14	L'Ormarins	10J
15	La Bourgogne	3K
16	La Bri	2J
17	La Couronne	3J
18	Landau du Val	3K
19	La Motte	6H
20	La Petite Ferme	2I
21	Mont Rochelle	3J
22	Môreson	7I
23	Plaisir de Merle	13J
24	Rickety Bridge	5I
25	Stony Brook	2K
26	TenFiftySix	5I
27	Von Ortloff	4J

WALKER BAY & ELGIN

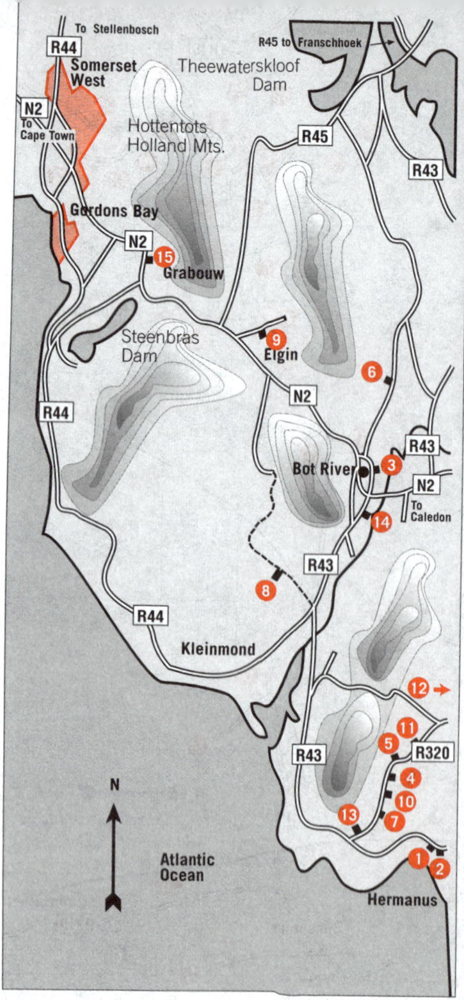

1 Bartho Eksteen
2 B&E Negotiants
3 Beaumont
4 Bouchard Finlayson
5 Cape Bay/Newton Johnson
6 Goedvertrouw
7 Hamilton Russell
8 Iona
9 Paul Cluver
10 Southern Right
11 Sumaridge
12 Teslaarsdal
13 WhaleHaven
14 Wildekrans (farm)
15 Wildekrans (tastings)

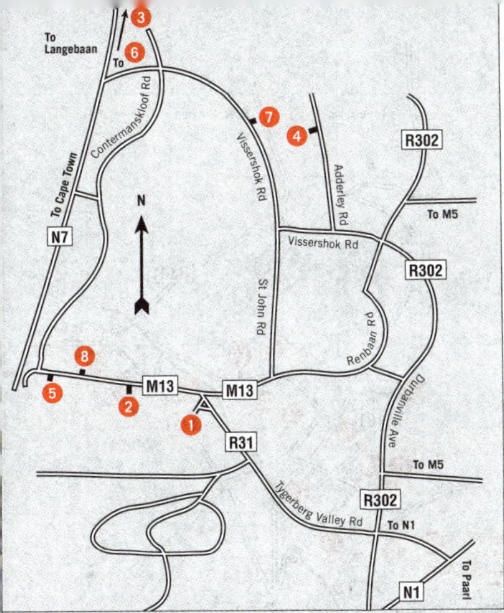

DURBANVILLE

1 Altydgedacht
2 Bloemendal
3 Capaia
4 Diemersdal
5 Durbanville Hills
6 Havana Hills
7 Meerendal
8 Nitida

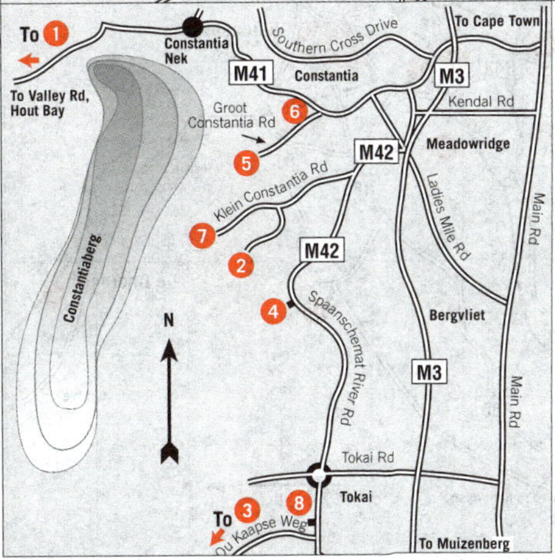

CONSTANTIA

1 Ambeloui
2 Buitenverwachting
3 Cape Point
4 Constantia Uitsig
5 Groot Constantia
6 High Constantia
7 Klein Constantia
8 Steenberg

WORCESTER

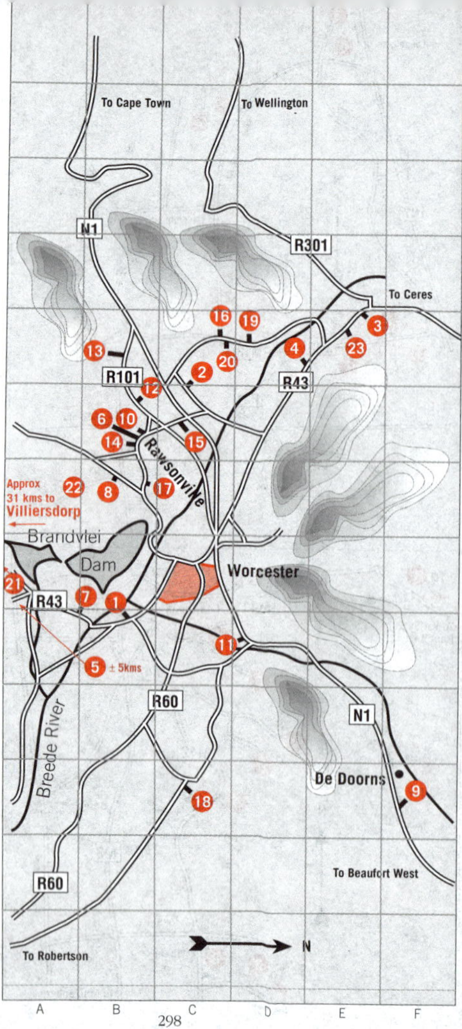

1 Aan de Doorns 8B
2 Badsberg 6C
3 Bergsig 5E
4 Botha 5D
5 Brandvlei 9B
6 Cape Vineyards 6B
7 Cilmor 8B
8 Daschbosch 7B
9 De Doorns 11F
10 Deetlefs 6B
11 De Wet 9D
12 Du Preez 6B
13 Du Toitskloof 5B
14 Goudini 6B
15 Groot Eiland 6C
16 Lee & Jones 5C
17 Merwida 7B
18 Nuy 11C
19 Opstal 5D
20 Slanghoek 5C
21 Stettyn 8A
22 Villiersdorp 6A
23 Waboomsrivier 5E

ROBERTSON

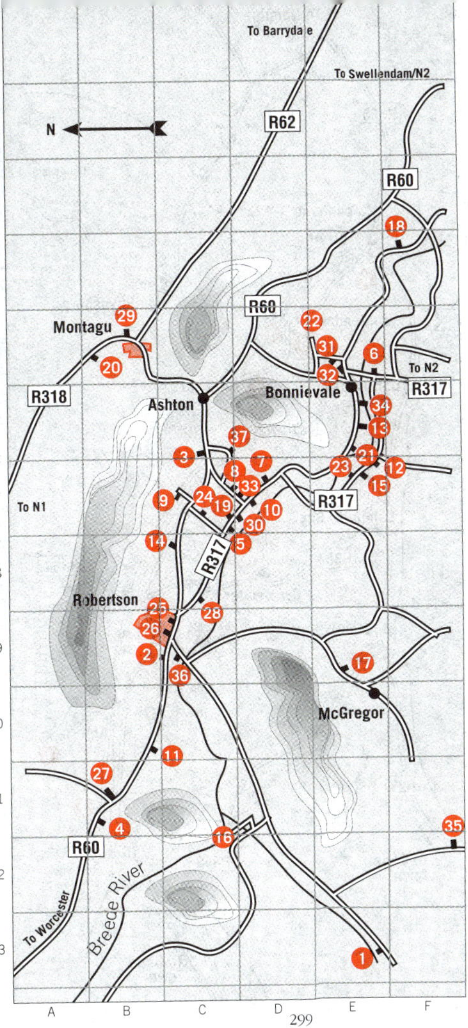

1	Agterkliphoogte	13E
2	Appelsdrift	9B
3	Ashton	6C
4	Bon Cap	11B
5	Bon Courage	7D
6	Bonnievale	5E
7	De Wetshof	7D
8	Excelsior	7D
9	Fraai Uitzicht	7C
10	Goedverwacht	7D
11	Graham Beck	10B
12	Janéza	7E
13	Jonkheer	6E
14	Kranskop	7C
15	Langverwacht	7E
16	Le Grand Chasseur	11C
17	McGregor	9E
18	Merwespont	4F
19	Mon Don	7D
20	Montagu Co-op	5B
21	Mooiuitsig	6E
22	Nordale	5E
23	Quando	7E
24	Rietvallei	7C
25	Robertson	9C
26	Roodezandt	9C
27	Rooiberg	11B
28	Springfield	8C
29	Uitvlucht	5B
30	Van Loveren	7D
31	Van Schoor	5E
32	Van Zylshof	5E
33	Viljoensdrift	7D
34	Weltevrede	6E
35	Windfall	12F
36	Wonderfontein	9C
37	Zandvliet	6D

OLIFANTS RIVER

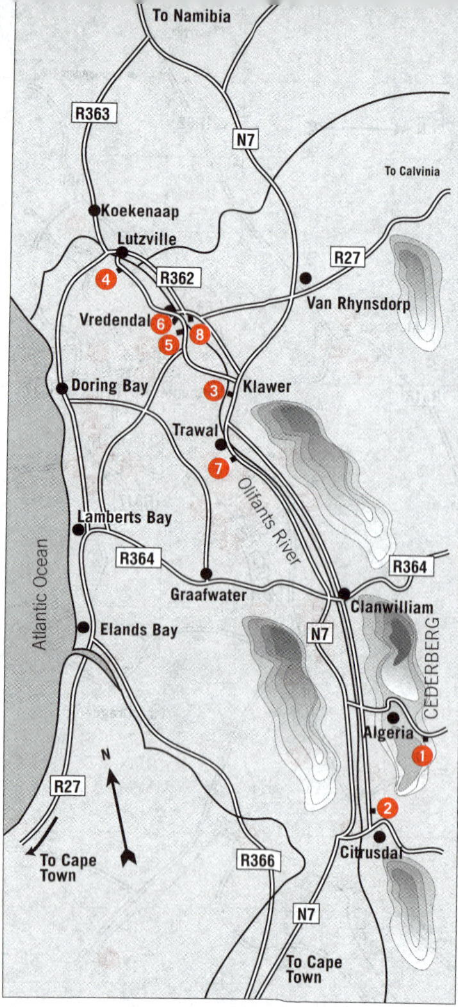

1 Cederberg
2 Citrusdal
3 Klawer
4 Lutzville
5 Spruitdrift
6 Stoumann's
7 Travino
8 Vredendal

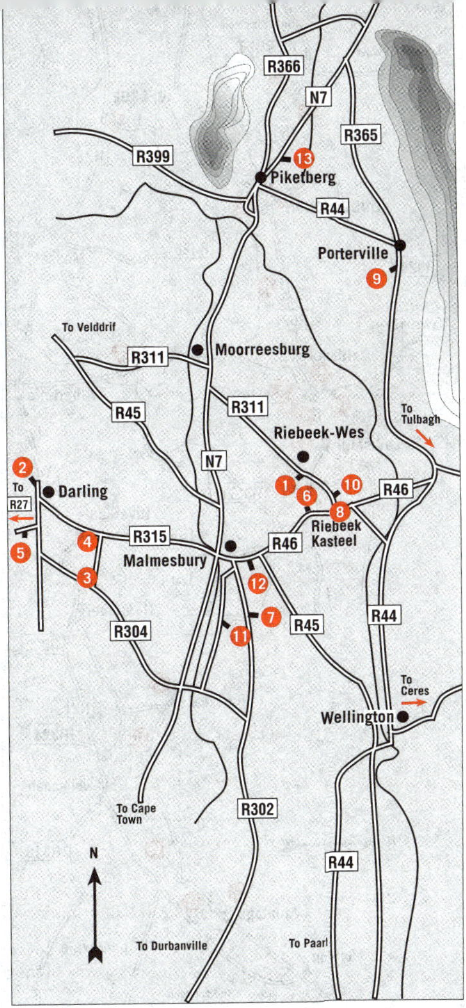

1 Allesverloren
2 Alexanderfontein
3 Cloof
4 Darling Cellars
5 Groote Post
6 Kloovenburg
7 Lammershoek
8 Meerhof
9 Porterville
10 Riebeek
11 Spice Route
12 Swartland
13 Winkelshoek

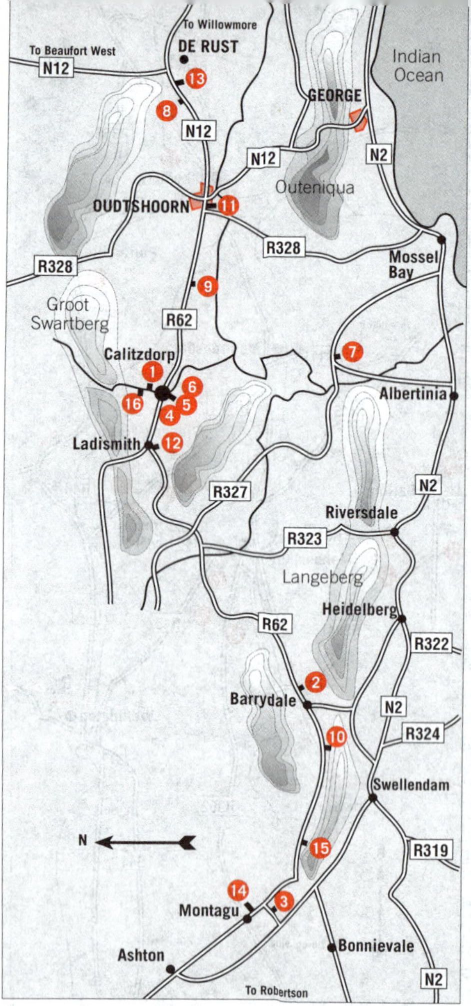

1 Axe Hill
2 Barrydale
3 Bloupunt
4 Boplaas
5 Calitzdorp
6 Die Krans
7 Die Poort
8 Domein Doornkraal
9 Grundheim
10 Joubert-Tradauw
11 Kango
12 Ladismith
13 Mons Ruber
14 Montagu
15 Rietrivier
16 Withoek

TULBAGH

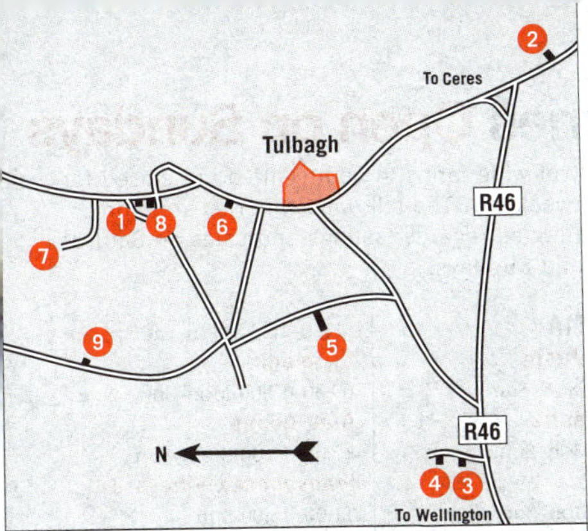

1 Drostdy
2 De Heuvel
3 Kloofzicht
4 Lemberg
5 Mont Pellier
6 Rijk's
7 Theuniskraal
8 Tulbagh
9 Twee Jonge Gezellen

KEY TO MAPS

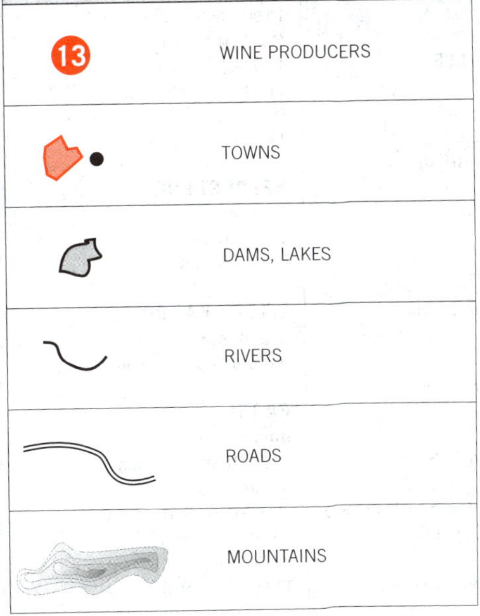

Wineries Open on Sundays

The majority of wine farms in South Africa are closed for most of the weekend. The following wineries are more accommodating as regards tastings and sales on Saturday afternoons and Sundays:

CONSTANTIA

Constantia Uitsig
10:00–16:00 Sat & Sun

Groot Constantia
09:00–18:00 Mon–Sun (Dec–Apr)
10:00–17:00 Mon–Sun (May–Nov)

DURBANVILLE

Bloemendal
11:00–15:00 Sun
09:00–16:00 Sat

Durbanville Hills
11:00–15:00 Sun
09:30–14:30 Sat

ELGIN

Iona
"Any reasonable hour" Mon–Sun

Wildekrans
08:30–17:30 Mon–Sun (farm stall)

FRANSCHHOEK

Agusta
10:00–18:00 Mon–Sun (summer)
09:00–17:00 (winter)

Boschendal
09:00–12:30 Sun (Nov–Apr)
08:30–16:30 Mon–Sat

Chamonix
09:30–17:00 Mon–Sun

Dieu Donné
10:00–15:00 Sat & Sun

Franschhoek Vineyards
11:00–15:00 Sun
10:00–16:00 Sat

Mont Rochelle
11:00–13:00 Sun
11:00–17:00 Mon–Sat

Môreson
11:00–17:00 Mon–Sun

HELDERBERG

Eikendal
10:30–16:00 Sun
09:00–16:00 Sat

LITTLE KAROO

Grundheim
08:00–17:00 Mon–Sun

PAARL

Anura
09:30–16:00 Mon–Sun

Ashanti
09:30–16:00 Sun (in season)
10:00–16:00 Sat

Backsberg
11:00–14:00 Sun

08:30–14:00 Sat

Glen Carlou

09:00–16:45 Mon–Sun

Laborie

09:00–17:00 Mon–Sun (summer)

09:00–17:00 Mon–Sat (winter)

Nederburg

11:00–16:00 Sun (Nov–Mar)

10:00–16:00 Sat (Nov–Mar)

Rheboksklоof

09:00–17:00 Mon–Sun

Ruitersvlei

09:00–17:00 Mon–Sun

Seidelberg

10:00–16:00 Sat & Sun

Simonsvlei

11:00–15:00 Sun

08:30–16:30 Sat

ROBERTSON

Bon Cap Organic

08:00–18:00 Mon–Sun

SOMERSET WEST

Vergelegen

09:30–16:00 Mon–Sun

STELLENBOSCH

Annandale

10:00–17:00 Mon–Sun

Audacia

12:00–16:00 Sun (Oct–Apr)

12:00–17:00 Sat

Blaauwklippen

08:30–17:00 Mon–Sun

Delaire

10:00–16:00 Sun

10:00–17:00 Sat

Delheim

10:30–15:30 Sun (Oct–Apr)

09:00–15:30 Sat

Hazendal

10:00–15:00 Sat & Sun

JC le Roux

10:00–15:00 Sun (Sept–Apr)

10:00–16:00 Sat

Kleine Zalze

11:00–16:00 Sun

09:00–17:00 Mon–Sat

Louiesenhof

10:00–18:00 Sun (summer)

10:00–18:00 Sat

Morgenhof

10:00–15:00 Sat & Sun (May–Oct)

10:00–17:00 Sat & Sun (Nov–Apr)

Muratie

10:00–16:00 Sat & Sun

Neethlingshof

10:00–16:00 Sat & Sun

Saxenburg

10:00–16:00 Sun & Sat

Spier

09:00–17:00 Mon–Sun

Stellenzicht

10:00–16:00 Sat & Sun

SylvanVale

07:00–21:30 Mon–Sun (Devon Valley Hotel)

Uitkyk

10:00–16:00 Sat & Sun

Welmoed

10:00–16:00 Sun

09:00–17:00 Sat

Zevenwacht
09:30–17:00 Sat & Sun

SWARTLAND
Lammershoek
09:00–17:00 Mon–Sun

TULBAGH
Drostdy
14:30–17:00 Sun (museum)
09:00–17:00 Sat

Kloofzicht
09:00–16:00 Mon–Sun
Paddagang
09:00–16:00 Mon–Sun

WELLINGTON
Diemersfontein
10:00–16:30 Mon–Sun

Saturday Afternoons

The following cellars aren't open on Sundays as are those above, but do offer wine tastings and sales well into the afternoon (after 3pm) on Saturdays:

CEDERBERG
Cederberg
08:00–12:30 & 14:00–17:00

DURBANVILLE
Diemersdal
09:00–16:00

FRANSCHHOEK
Plaisir de Merle
10:00–16:00 (Oct–May)

FREE STATE
Goudveld
08:00–17:00

LITTLE KAROO
Die Krans
09:00–16:00

MONTAGU
Bloupunt
09:00–16:00

OLIFANTSRIVIER
Travino
07:30–13:00 & 14:00–17:00

PAARL
De Zoete Inval
09:00–17:00
KWV
09:00–16:30 (Wine Emporium)
Ridgeback
08:00–17:00

ROBERTSON
Springfield
09:00–16:00

STELLENBOSCH

Asara
10:00–17:00 Nov–Apr
Bilton
10:00–16:00
Distell
08:30–17:00 (Oude Libertas)
Helderkruin
10:00–17:00
Knorhoek
10:00–17:00
L'Avenir
10:00–16:00
L'Émigré
09:30–16:30
Le Bonheur
10:00–16:00
Middelvlei
10:00–16:00
Simonsig
09:30–16:00
Uiterwyk
10:00–16:30 (Oct–Apr)
14:00–16:30 (May–Sept)

TULBAGH

Rijk's
10:00–16:00

VREDENDAL

Stoumann's
14:00–17:00

Lunching & Dining on the Wine Farms

Want to dine, lunch or snack in the winelands, at the wineries, among the vines? Look no further...

CONSTANTIA

Buitenverwachting
Tue–Fri lunch & dinner, Sat dinner
Tel: 021 794-3522
Constantia Uitsig
Mon–Sun 10:00–21:00
Tel: 021 794-4480
Jonkershuis, Groot Constantia
Mon 10:00–16:00, Tue–Sat 09:00–22:00, Sun 09:00–17:00
Tel: 021 794-6255
La Colombe, Constantia Uitsig
Mon–Sun 10:00–21:00
Tel: 021 794-2390
Spaanschemat River Café, Constantia Uitsig
Mon–Sun 08:30–17:00
Tel: 021 794-3010
Steenberg
Mon–Sun, breakfast, lunch & dinner
Tel: 021 713-2222
Tavern, Groot Constantia
Mon–Sun 10:00–17:00

(Apr–Oct), 10:00–18:00
(Nov–Mar)
Tel: 021 794-1144

DURBANVILLE

@ the Hills, Durbanville Hills
Tue–Sun 12:00–15:00
Tel: 021 558-1300

Bloemendal
Mon–Sat, buffet breakfast, lunch & dinner, Sun lunch
Tel: 021 975-7575

Die Wijnhuis, Bloemendal
Mon–Fri 09:00–17:00, Sat 09:00–16:00, Sun 11:00–15:00 (cheese platters)
Tel: 021 976-2682

FRANSCHHOEK

Boschendal
Mon–Sun buffet lunch
Tel: 021 870-4273

Bread & Wine, Môreson
Mon–Sun 12:00–16:00
Tel: 021 876-3692

Haute Cabrière
Mon–Sun lunch, dinner seasonal
Tel: 021 876-3688

La Brasserie, Agusta
Tue–Sun 11:30–until late
Tel: 021 876-4143

La Cotte, Franschhoek Vyds
Tue–Sun lunch
Tel: 021 876-2521

La Couronne
Mon–Sun – breakfast, lunch & dinner
Tel: 021 876-2770

La Petite Ferme
Mon–Sun 12:00–16:00
Tel: 021 876-3016

Le Café, Boschendal
Mon–Sun light lunches (10:00–17:00)
Tel: 021 870-4273

La Maison de Chamonix
Tue–Sun 12:30–17:00, Fri 18:00 till late
Tel: 021 876-2393

HELDERBERG

Avontuur
Mon–Fri lunch
Tel: 021 855-3450

Helderberg, Stellenbosch Vyds
Mon–Sat 12:00–14:00
Tel: 021 842-2012

96 Winery Road, Ken Forrester
Mon–Sun lunch, dinner Mon–Sat
Tel: 021 842-2020

LITTLE KAROO

Die Poort, Herbertsdale
Mon–Sun lunch
Tel: 028 735-2406

Jemima, Doornkraal
Tue–Sat 12:00–14:00 & 18:30–22:00, Sun 18:30–22:00
Tel: 044 272-0808

R62 Deli & Restaurant, Joubert-Tradauw
Mon–Sun 10:00–16:00
Tel: 028 572-1619

NORTHERN CAPE

Die Wynhuis, Hartswater
Mon–Fri 09:00–17:00
Tel: 053 474-0907

Die Geelhuis, Douglas Winery
Mon–Sun 10:00–22:00
Tel: 053 298-2579

PAARL

Anura
Mon–Sun 09:30–16:00
Tel: 021 875-5360

Backsberg
Tue–Fri 10:00–16:00, Sat & Sun 11:00–15:00 (light meals)
Tel: 021 875-5141

Icampanelli, Laborie
Tue–Sun lunch 12:00–16:00, Tues-Sat dinner 18:30 till late
Tel: 021 807-3095

Il Casale, Ashanti
Tue–Sat 11:00–23:00, Sun lunch in season
Tel: 021 862-6288

Olive & Vine, Seidelberg
Mon–Fri 09:00–17:00, Sat 10:00–16:00
Tel: 021 863-3495

Pot Luck, Bernheim
Opens in 2003
Tel: 021 872-5618

R101, Simonsvlei
Tue–Sun lunch, dinner Tue–Sun
Tel: 021 863-2486

Rhebokskloof
Mon–Sun lunch, teas 09:30 & 14:30, Thurs–Mon dinner
Tel: 021 869-8606

Rosemarie's, Mont Destin
Mon–Fri 08:30–17:00, Fri–Sat 19:30–21:30
Tel: 021 875-5443

Ruitersvlei Bistro
Tue–Sat lunch & dinner
Tel: 021 863-3959

ROBERTSON

Café Maude, Bon Courage
Mon–Fri 09:00–17:00, Sat 09:30–16:00
Tel: 023 626-4178

Die Parsmandjie, Weltevrede
Mon–Sat 12:00–14:00
Tel: 023 616-2141

Fraai Uitzicht 1798
Wed–Sun 11:00–22:00
Tel: 023 626-6156

SOMERSET WEST

L'Auberge du Paysan
Tue–Sat lunch, dinner Mon–Sat
Tel: 021 842-2008

Lady Phillips, Vergelegen
Mon–Sun 09:30–16:00, teas & lunches
Tel: 021 847-1346

Rose Terrace, Vergelegen
Mon–Sun lunch (Nov–Apr)
Tel: 021 847-1346

STELLENBOSCH

Auberge Rozendal
Mon–Sun dinner, lunch by appointment
Tel: 021 883-8737

Barrique, Vredenheim
Tue–Sun 12:00–15:00 &
18:00–22:00
Tel: 021 881-3001

Bergkelder Bistro
Mon–Sat 12:00–14:00
Tel: 021 809-4892

Blaauwklippen
Mon–Sat 09:00–17:00,
Sun 09:00–16:00
Tel: 021 880-0133

Bottelary Kos & Wyn, Stellendrift
Mon–Sun 09:00–17:00
Tel: 021 865-2204

Dame Jeanne Coffee Shop, Morgenhof
Mon–Sun breakfast
(09:00–11:30)
Tel: 021 889-5510

Delheim Garden Restaurant
Mon–Sat 12:00–15:00, Sun
12:00–15:00 (Oct–Apr)
Tel: 021 888-4607

Duck Pond, Welmoed
Mon–Sun 12:00–14:00 (closed
Tues Jun & Jul)
Tel: 021 881-3310

Eikendal
Mon–Fri 12:00–14:00
(Nov–Apr), Sun 12:00–14:00
(Nov–Apr)
Tel: 021 855-1422

Figaro's, Spier
Mon–Sun breakfast
(06:00–10:30), dinner
(18:00–22:00)
Tel: 021 809-1100

Governer's Hall & The Terrace, Lanzerac
Mon–Sun breakfast, lunch &
dinner
Tel: 021 887-1132

Green Door, Delaire
Mon–Sun 12:00–15:00
Tel: 021 885-1149

Guinea Fowl, Saxenburg
Wed–Mon lunch,
dinner Wed–Sat
Tel: 021 906-5232

Hartenberg
Mon–Sat 12:00–14:00 (light
lunches)
Tel: 021 865-2541

Hermitage, Hazendal
Mon–Sun 12:00–14:30
Tel: 021 903-5112

JC le Roux
Tue–Sat 10:00–15:00, Sun
10:00–13:00 (Sep–Apr)
Tel: 021 882-2590

Jonkershuis, Spier
Mon–Sun 12:30–15:30 &
18:00–22:00
Tel: 021 809-1100

Kleine Zalze
Mon–Sun lunch
Tel: 021 880-0717

Lord Neethling
Sep–Apr: Mon–Sun 09:00–23:00.
May–Aug: lunch & dinner
Fri & Sat
Tel: 021 883-8966

Louiesenhof
Mon–Sun 10:00–15:00
(light meals)

Tel: 021 865-2632

Morgenhof Gazebo

Mon–Sun – 12:00–14:30 (May–Oct), 12:00–15:00 (Nov–Apr)

Tel: 021 889-5510

Oude Libertas, Distell

Mon–Fri lunch, dinner Mon–Sat

Tel: 021 809-7495

Riverside Pub, Spier

Mon–Sun 11:00–18:00

Tel: 021 809-1100

Spier Farm Stall

Mon–Sun – 09:00–17:00 (Apr–Sep), 09:00–20:30 (Oct–Mar)

Tel: 021 809-1100

Taphuis, Spier

Mon–Sun lunch (12:30–15:30)

Tel: 021 809-1100

Tokara

Tue–Sat lunch & dinner

Tel: 021 808-5959

Toto's, Spier

Mon–Sun 12:30–15:30

Tel: 021 809-1100

Vineleaf, SylvanVale

Devon Valley Hotel

Mon–Sun 07:00–21:30

Tel: 021 865-2012

Zevenwacht

Mon–Sun 07:00–22:00

Tel: 021 903-5123

SWARTLAND

Meerhof

Mon–Fri 09:00–17:00, Sat 09:00–13:00

Tel: 082 786-1843

Winkelshoek

Mon–Sun 09:00–18:00

Tel: 022 913-1092

TULBAGH

Paddagang

Mon–Sun 09:00–16:00

Tel: 023 230-0394

Rijk's

Mon–Sun 07:30–10:00, 12:00–14:00 & 19:00–21:00

Tel: 023 230-1006

WELLINGTON

Hildenbrand's Table

Mon–Sun 12:00–15:00

Tel: 021 873-4115

The River Café, Oude Wellington

Sun lunch, Tue–Sat dinner (Sun dinner in season)

Tel: 021 873-2009

The Picnic alternative

Picnic hampers are available for purchase from the following farms and cellars – mostly by prior arrangement:

Agusta, Franschhoek

Tel: 021 876-3195

Blaauwklippen, Stellenbosch

Tel: 021 880-0133

Boland, Paarl

Tel: 021 862-6190 (Dec–Jan)
Boschendal, Franschhoek
Tel: 021 870-4273 (Nov–Apr)
Buitenverwachting, Constantia
Tel: 021 794-3522 (in season)
Delaire, Stellenbosch
Tel: 021 885-1149 (Sep–Apr)
Diemersfontein, Wellington
Tel: 021 864-5050
Distell, Stellenbosch
Tel: 021 809-7495 (Oude Libertas)
Domein Doornkraal, Little Karoo
Tel: 044 251-6715
Eikendal, Helderberg
Tel: 021 855-1422
Groot Constantia, Constantia
Tel: 021 794-1144
Groote Post, Swartland
Tel: 022 492-2825
Hazendal, Stellenbosch
Tel: 021 903-5112
Joubert-Tradauw, Little Karoo
Tel: 028 572-1619
Klawer, Olifants River
Tel: 027 216-1530
Kleine Zalze, Stellenbosch
Tel: 021 880-0717
Laborie, Paarl
Tel: 021 807-3390
L'Ormarins, Franschhoek
Tel: 021 874-1026
Louiesenhof, Stellenbosch
Tel: 021 882-2896
Lutzville, Olifants River
Tel: 027 217-1516
Meerhof, Swartland
Tel: 082 786-1843
Mont Destin, Paarl
Tel: 021 875-1008
Mont Rochelle, Franschhoek
Tel: 021 876-3000
Nederberg, Paarl
Tel: 021 862-3104 (Nov–Mar)
Nelson's Creek, Paarl
Tel: 021 869-8453
Oude Wellington, Wellington
Tel: 021 873-1008
Paul Cluver, Elgin
Tel: 021 844-0605
Seidelberg, Paarl
Tel: 021 863-3495
Simonsig, Stellenbosch
Tel: 021 888-4900
Spier Farm Stall, Stellenbosch
Tel: 021 809-1100
Springfield, Robertson
Tel: 023 626-3661
Stellendrift, Stellenbosch
Tel: 021 887-6561
Sumaridge, Walker Bay
Tel: 028 312-1097
Uitkyk, Stellenbosch
Tel: 021 884-4416
Vergelegen, Somerset West
Tel: 021 847-1334 (Nov–Apr)
Viljoensdrift, Robertson
Tel: 023 615-1901
Vredenheim, Stellenbosch
Tel: 021 881-3637
Warwick, Stellenbosch
Tel: 021 884-3146
Zevenwacht, Stellenbosch
Tel: 021 903-5123

Accommodation in the Winelands

Want to spend a night or more on a wine farm? The following producers offer accommodation of various kinds:

CONSTANTIA

Constantia Uitsig
Country lodge
Tel: 021 794-6500
Fax: 021 794-7605
reservations@uitsig.co.za

Steenberg
Country hotel
Tel: 021 713-2222
Fax: 021 713-2221
reservations@steenberghotel.com

FRANSCHHOEK

Blueberry Hill
Self-catering cottages
Tel: 021 876-3362
Fax: 021 876-2114
blwine@iafrica.com

Cape Chamonix
Guest houses
Tel: 021 876-2494
Fax: 021 876-3237
marketing@chamonix.co.za

Grand Provence, Agusta
Guest house
Tel: 021 876-3195
Fax: 021 876-2652
guesthouse@agustawines.co.za

La Couronne Hotel
Tel: 021 876-2770
Fax: 021 876-4168
reservations@lacouronnehotel.co.za

La Petite Ferme
Guest house
Tel: 021 876-3016
Fax: 021 876-3624
lapetite@iafrica.com

Owl Cottage, Elephant Pass
Self-catering
Tel: 021 876-3280
Fax: 021 876-2219

HELDERBERG

Assegaai, Woodlands
Guest house
Tel: 021 855-2249
Fax: 021 855-4924
rbuchner@worldonline.co.za

Eikendal Lodge
Tel: 021 855-3617
Fax: 021 855-3862
lodge@eikendal.com

Goose Cottage, Ken Forrester
Tel: 021 855-2374
Fax: 021 855-2373
marketing@kenforresterwines.com

LITTLE KAROO

Bloupunt
Self-catering cottages
Tel/Fax: 023 614-2385
bloupunt@lando.co.za

Lentelus, Joubert-Tradauw
B & B

Tel: 028 572-1619
Fax: 028 572-1315
helenaj@lando.co.za

Mons Ruber
Self-catering
Tel: 044 251-6550
Fax: 044 251-6550
monsr@lantic.net

Sanddrif, Cederberg
Self-catering
Tel: 027 482-2827
Fax: 027 482-1188
cederwyn@iafrica.com

N CAPE/ FREE STATE

Rene's Guesthouse, Douglas
Tel: 082 266-9038
wynkelder@gwk.co.za

Veenwouden, Landzicht
Guest house
Tel: 053 591-0237
landzicht@inext.co.za

PAARL

Die Wildegans, Brahms
Tel: 021 863-8555
Fax: 021 863-8555
brahms@iafrica.com

Landskroon
Self-catering cottage
Tel: 021 863-1039
Fax: 021 863-2810
landskroon@mweb.co.za

Mont Destin
Guest apartments
Tel/Fax: 021 875-5040
wine@montdestin.co.za

Ruitersvlei
Guest rooms
Tel: 021 863-1517
Fax: 021 863-1443
ruitersv@iafrica.co.za

Sonop
Guest house
Tel: 021 869-8723
Fax: 021 886-4838
office@africanterroir.co.za

ROBERTSON

Fraai Uitzicht 1798
Tel/Fax: 023 616-6156
info@fraaiuitzicht.co.za

Merwenstein, Merwespont
B & B
Tel: 023 616-2800
Fax: 023 616-2734
merwespont@lando.co.za

Ons Huisie & Belvedere, Weltevrede
Guest cottages
Tel: 023 616-2895
Fax: 023 616-2460
info@weltevrede.com

Weltevrede Guest Farm, Bon Cap Organic
Tel: 023 616-1629
Fax: 023 626-1895
boncap@intekom.co.za

STELLENBOSCH

Auberge Rozendal
Tel: 021 887-8737
Fax: 021 883-8738
rozendal@mweb.co.za

Camberley Cottage
Tel/Fax: 021 885-1176
camberleywine@hotmail.com

The Cellars

Mulderbosch, Starke
B & B

Tel: 021 884-4433
Fax: 021 883-4324
info@muldersvlei.co.za

De Kleine Hoop, Le Riche
Guest house
Tel: 021 887-8958
lerichewines@adept.co.za

Delaire Mountain Chalets
Self-catering
Tel: 021 885-1756
Fax: 021 885-1270
delaire@iafrica.com

Devon Valley Hotel, Sylvanvale
Tel: 021 882-2012
Fax: 021 882-2610
devon@iafrica.com

Graceland
Tel: 021 881-3121
Fax: 021 881-3341
pmcn@iafica.com

Kleine Zalze
Guest cottages
Tel: 021 880-0717
Fax: 021 880-0716
quality@kleinezalze.co.za

Knorhoek
Guest house
Tel: 021 865-2114
Fax: 021 865-2627
guesthouse@knorhoek.co.za

L'Avenir
Guest house
Tel: 021 889-5001
Fax: 021 889-5258
lavenir@adept.co.za

L'Émigré
Self-catering
Tel: 021 881-3702
Fax: 021 881-3773

La Provence, Bowe Joubert
Guest cottages
Tel: 021 881-3716
Fax: 021 881-3859
info@bowejoubert.com

Lanzerac Hotel
Tel: 021 887-1132
Fax: 021 887-6998
info@lanzerac.co.za

Lyngrove Country House
Tel: 021 842-2116
Fax: 021 880-0851
www.lyngrove.co.za

Overgaauw
Guest cottage
Tel: 021 881-3815
Fax: 021 881-3436
info@overgaauw.co.za

Palm Cottage, Simonsig
Tel: 021 883-2881
Fax: 021 888-4944
wine@simonsig.co.za

Uitzicht, Reyneke
B & B
Tel: 021 881-3261
wine@reynekewines.co.za

Village at Spier
Hotel
Tel: 021 809-1100
Fax: 021 809-1144
info@spier.co.za

Vineyard Cottage, Kaapzicht
Tel: 082 737-8329
Fax: 021 906-1622
kaapzicht@mweb.co.za

Zevenwacht Country Inn
Tel: 021 906-1560
Fax: 021 906-1570
reservations@zevenwacht.co.za

TULBAGH

Lemberg
Guest house
Tel: 023 230-0659
Fax: 023 230-0661
schindler@lando.co.za

Rijk's Country Hotel
Tel: 023 230-1006
Fax: 023 230-1125
bookings@rijks.co.za

WALKER BAY/ ELGIN

The Cottage, Teslaarsdal
Tel: 021 312-3591
Fax: 028 312-3591

Iona
Guest house
Tel: 028 284-9678
Fax: 028 284-9078
gunn@corpdial.co.za

Paul Cluver
Guest house
Tel: 021 844-0605
Fax: 021 844-0150
info@cluver.co.za

Mill & Peppertree, Beaumont
Self-catering cottages
Tel: 028 284-9194
Fax: 028 284-9733
beauwine@netactive.co.za

Sumaridge Guest House
Self-catering
Tel: 028 312-1097
Fax: 028 312-2824
www.sumaridge.co.za

WELLINGTON

Diemersfontein
Country house
Tel: 021 873-2671
Fax: 021 873-4526
wine@diemersfontein.co.za

Fisantekuil, Upland
Guest farm
Tel: 021 864-1184
Fax: 021 873 5724
edmund@oettle.com

Jacaranda Cottage
Self-catering
Tel: 021 864-1235
jacaranda@iafrica.com

Klein Rheboksskloof,
Hildenbrand
Tel/Fax: 021 873-4115
info@wine-estate-
hildenbrand.co.za

Oude Wellington
Guest house
Tel: 021 873-2262
Fax: 021-873 4639
info@kapwein.com

Welgegund
B & B
Tel: 021 864-1185
Fax: 021 873-2683
rac@icon.co.za

WORCESTER

Breede Valley
Self-catering chalets
Tel: 023 342-8710
Fax: 023 342-2294
manager@worcesterwinelands.co.za

Merwida
Country lodge
Tel: 023 349-1144
Fax: 023 349-1953
wines@merwida.co.za

Cape Winelands Tour Operators

Those that specialise in both day-trips and overnight visits to the Cape winelands – inclusive of wine tastings and meals:

ADAMASTOR & BACCHUS

3 Highworth Rd, Sea Point 8005
Tel: 021 439-3169
Fax: 021 511-4667
Cell: 083 229-1172
johnford@iafrica.com

HYLTON ROSS TOURS

31 Neptune Street, Paarden Eiland
PO Box 287, Paarden Eiland 7420
Tel: 021 511-1784
Fax: 021 511-2401
info@hyltonross.co.za
www.hyltonross.co.za

JUDY'S TOURS

13 Van Zyl Street, Somerset West 7130
Tel/Fax: 021 851-4205
Cell: 084 500-1941
judyk@zsd.co.za

MOTHER CITY TOURS

1 Drake Street, Observatory
PO Box 13601, Mowbray 7705
Tel: 021 448-3817
Fax: 021 448-3844
fatima@mctours.co.za
www.mctours.co.za

SPECIALIZED TOURS

99 High Strand Complex, Boundary Road, Green Point
PO Box 49, Green Point 8051
Tel: 021 425-3259
Fax: 021 425-3329
specialt@iafrica.com
www.specialtours.co.za

VINEYARD VENTURES

5 Hanover Road, Fresnaye 8005
PO Box 554, Sea Point 8060
Tel: 021 434-8888
Fax: 021 434-9999
Cell 082 893-5387/082 920 2825
vinven@iafrica.com
www.vineyardventures.co.za

VINTAGE CAPE TOURS

94 Mill Street, Paarl 7646
PO Box 162, Main Street, Paarl 7622
Tel: 021 872-9252
Fax: 021 872-0174 or 862-1484
vctours@adept.co.za
www.vintagecape.co.za

Wine-routing differently

Want a bird's-eye view of the cape winelands? Choose from a variety of helicopter flips, short trips to stop-overs for tastings and/or lunch. What about hot-air-ballooning? Or perhaps you'd like to do it on horseback, or on a motor-bike, Harley style...

CIVAIR HELICOPTERS

V&A Waterfront:
East Pier Rd, after
Table Bay Hotel
Tel: 021 419-5182
Fax: 021 419-5183
Cape Town
International Airport:
Tel: 021 934-4488
Fax: 021 934-4462
civair@mweb.co.za
www.civair.co.za

WINELAND BALLOONING

Seasonal – mid-summer only
Tel/Fax: 021 863-3192

SPIER HORSE TRAILS

Spier Farm, Route R310,
Stellenbosch:
Cell 083-627-2282
Tel: 021 881-3683
Fax: 021 881-3683
horserides@spier.co.za

WINE VALLEY HORSE TRAILS

Rhebokskloof Wine Estate, off
R44, Paarl
Tel: 021 869-8687
Cell: 083-226-8735
gavmic@worldonline.co.za

MONT ROCHELLE EQUESTRIAN CENTRE

Mont Rochelle Wine Estate,
Franschhoek
Cell: 083-300-4368
Fax: 021 876-2362

PARADISE STABLES

Robertsvallei Farm (next to
Forestry Dept), Franschhoek
Tel/Fax: 021 876-2160

CLASSIC TWIN TOURS

Harley rental, operating from
the Antique Tractor Guest
House
R304 Koelenhof, Stellenbosch
Tel: 021 865 2558
Fax: 021 887 9678
Cell: 083 3393196
enquiries@classictwintours.com
www.classictwintours.com

Futures, *En Primeur*

The following producers offer to sell their wines (or some of them) prior to release – sometimes at prices lower than what they will retail for when they come onto the market:

Agusta
Allesverloren
Alto
Altydgedacht
Annandale
Anthill
Asara
Bartho Eksteen
Bergkelder
Blaauwklippen
Bloemendal
Bloupunt
Bonfoi
Brahms
Camberley
Chamonix
Cloof
De Toren
Deetleefs
Eikehof
Flagstone
Glen Carlou
Graceland
Havana Hills
High Constantia
Katbakkies
Kloofzicht
Lammershoek
Lushof
Maiden Wines
Meerhof
Neethlingshof
Nordale
Oude Wellington
Reyneke
Rhebokskloof
Ridgeback
Seidelberg
Signal Hill
Slaley
Springfield
Stellenzicht
Sumaridge
Topaz
Uitkyk
Upland
Waboomsrivier

Specialist Wine Shops

EASTERN CAPE

DAVINO'S

Beach Hotel, Marine Drive,
Summerstrand,
Port Elizabeth, 6001
PO Box 319, Century Hill, 6000
Tel: 041 583-2166
Fax: 041 583-6220

PRESTON'S

9th Avenue Shopping Centre
Walmer, Port Elizabeth, 6070
PO Box 5070, Walmer, 6065
Tel/Fax: 041 581-1993
www.pinmac.co.za

GAUTENG

BOOTLEGGERS

Shop L1, Fourways Crossing
Retail Centre, c/o William Nicol
& Witkoppen Roads, Fourways
PO Box 2060, Witkoppen 2068
Tel: 011 465-9777
Fax: 011 465-9528
bootleggers@mweb.co.za

CENTRAL LIQUORS

47 4th Street, Springs, 1559
PO Box 152, Springs 1560
Tel: 011 812-1428
Fax: 011 812-1910
val@centralliquors.co.za
www.centralliquors.co.za

CHEERS WINE & SPIRIT

48 Grosvenor Ave, Bryanston
PO Box 79266,
Senderwood, 2145
Tel: 011 706 4652

GLENFAIR HYPER LIQUOR

Shop 53 Glenfair Centre,
Lynnwood Road, Lynnwood
Manor, Pretoria
Posnet Suite#122, Private Bag
X025, Lynnwood Ridge 0040
Tel/Fax: 012 361-4563
gfl.mkuyper@lantic.net
www.glenfairliquor.co.za

JOHNNY'S

Atterbury Value Mart, Atterbury
Road, Faerie Glen, Pretoria
PO Box 110, Faerie Glen,
Pretoria 0043
Tel: 012 991-4999
Fax: 012 991-4996
mandy@johnnysliquor.co.za

MAKRO CENTURION

1 Bloukranz Street, Highveld
Extension 2, Centurion
PO Box 7496, Centurion 0046
Tel: 012 673-3700
Fax: 012 665 1798
mprinsl@makro.co.za
www.makro.co.za

MAKRO WOODMEAD

Woodmead Value Mart,

Waterfall Crescent South,
Woodmead Ext 5, Sandton
PO Box 1771, Kelvin 2054
Tel: 011 208-9000
Fax: 011 804-7330
mhardy@makro.co.za
www.makro.co.za

NORMAN GOODFELLOW'S

192 Oxford Road, Illovo,
Johannesburg
PO Box 55162, Northlands 2116
Tel: 011 788-4814
Fax: 011 442-8868
goodfellow@icon.co.za

PICARDI REBEL SANDTON

Village Walk Centre, 125
Rivonia Road, Sandton 2196
PO Box 784030, Sandton, 2146
Tel: 011 884-2151
Fax: 011 884-1067

RIVONIA CELLARS

349 Rivonia Boulevard,
Rivonia 2128
PO Box 1968, Rivonia 2128
Tel: 011 803-6121/2
Fax: 011 803-7600
rcellars@global.co.za

THE WINE ROOM

Bryan Wedge Shopping Centre,
Main Road, Bryanston
Tel: 011 482-8274
wineroom@iafrica.com
www.wineroom.co.za

KWAZULU-NATAL

THE CELLAR

Shop 26, The Workshop,
Commercial Road,
Durban 4001
PO Box 3050, Durban, 4000
Tel/Fax: 031 304-6713

LA CAVE

Umhlanga Centre, Ridge Road,
Umhlanga Rocks
PO Box 50289, Musgrave, 4062
Tel: 031 561-6792
Fax: 031 561-4487

MAKRO SPRINGFIELD

90 Electron Road, Durban, 4001
PO Box 172, Durban 4000
Tel: 031 203-2896
Fax: 031 263-1305
ljvrens@makro.co.za
www.makro.co.za

THE VILLAGE VINEYARD

Kloof Village Mall, 33 Village
Road, Kloof
PO Box 977, Kloof, 3640
Tel: 031 764-6679
Fax: 031 764-7196

WINE DOMAIN

Shop 1, 22 Salisbury Ave,
Westville, 3630
Tel: 031 266-4394
Fax: 031 266-6124
sales@winedomain.co.za

WINES OF RIDGEVIEW

314 –317 Umhlanga Road,
Umhlanga Ridge, 4320
PO Box 22151,
Glenashley 4022
Tel: 031 566-5969
Fax: 031 566-3486
info@ridgeviewwines.co.za

WESTERN CAPE

ALPHEN KWIKSPAR

Alphen Village Centre,
Constantia
Tel: 021 794-5986
Fax: 021 794-4705

AROMA CANAL WALK

Shop 277, Canal Walk,
Century City, 7441
Tel: 021 551-7511

AROMA ALPHEN CELLARS

Shop 16, Old Village Centre,
Constantia
Tel: 021 794-3143
Fax: 021 794-8694
Alphen@aroma.co.za

AROMA SOMERSET WEST

Next to Somerset Mall,
Somerset West
Tel: 021 852-7551
Fax: 021 852-7552
aromasom@aroma.co.za

BREEDE VALLEY WINES

Kleinplasie, off Robertson Rd,
Worcester
PO Box 59, Worcester 6849
Tel: 023 342-8710
Fax: 023 342-2294
manager@worcesterwinelands.co.za
www.worcesterwinelands.co.za

CAROLINE'S, CAPE TOWN

15 Long Street, Cape
Town 8001
Tel: 021 419-8984
Fax: 021 419-8985
carowine@mweb.co.za

CAROLINE'S, SOMERSET WEST

Shop 236A, Entrance 3,
Somerset Mall, Somerset
West 7130
Post Net Suite 104, Private Bag
X15, Somerset West, 7129
Tel: 021 851-3690
Fax: 021 851-3684
carowine@mweb.co.za

CAROLINE'S, V&A WATERFRONT

Tel: 021 425-5701
Fax: 021 425-5702
carowine@mweb.co.za

CHAPMAN'S PEAK LIQUOR STORE

Main Road, Hout Bay 7800
Tel: 021 790-1088
Fax: 021 790-1089
lidian@iafrica.com

CONSTANTIA UITSIG WINE SHOP

Spaanschemat River Road,
Constantia Uitsig Farm
PO Box 402, Constantia 7848
Tel: 021 794-1810
Fax: 021 794-1812
wine@icon.co.za
www.constantiauitsig.co.za

EASTCLIFF KWIKSPAR

Cnr Main & Fairways Roads,
Hermanus
Tel: 028 313-1095
Fax: 028 313-2502

LA COTTE

Main Road, Franschhoek
PO Box 137, Franschhoek 7690
Tel: 021 876-3775
Fax: 021 876-3036
info@lacotte.co.za
www.lacotte.co.za

THE LIQUOR STORE, CANAL WALK

Shop 131, Canal Walk,
Century City, 7441
Tel: 021 551 7563

PALMS FINE WINE SHOP

Palms Centre, 145 Sir Lowry
Road, Cape Town, 8001
PO Box 488, Constantia, 7848
Tel: 021 461-7266
Fax: 021 461-9547
anouschka@cvi.co.za
www.palmsfinewineshop.co.za

PICARDI REBEL, CLAREMONT

218 Main Road, Claremont 7700
Tel: 021 671-9918
Fax: 021 683-9025

PICARDI REBEL TYGERVALLEY

Shop 496 Tygervalley Centre,
Willie van Schoor Avenue,
Durbanville
Tel: 021 914-1649/50
Fax: 021 914-2420

PLATTEKLOOF KWIKSPAR

Hendrik Verwoerd Drive,
Panorama
Tel: 021 939-1472
Fax: 021 939-2273

SOMERSET WEST KWIKSPAR

150 Main Road, Somerset
West, 7130
PO Box 2835, Somerset
West, 7129
Tel: 021 851-5795
Fax: 021 851-5799

SPIER WINE CENTRE

Baden Powell Drive, Lynedoch,
Stellenbosch
PO Box 1078,
Stellenbosch 7599
Tel: 021 809-1100
Fax: 021 881-3634
info@spier.co.za
www.spier.co.za

STEVEN ROM, SEA POINT

Checkers Galleria Centre,
76 Regent Road, Sea Point
PO Box 613, Sea Point 8060
Tel: 021 439-6043
Fax: 021 434-0401
rom@iafrica.com
www.wineseller.co.za

STEVEN ROM, THREE ANCHOR BAY

Bay Point, Stanley Place,
Three Anchor Bay
PO Box 613, Sea Point 8060
Tel: 021 439-1112
E-mail/Website as per Sea Point

VAUGHAN JOHNSON'S

V&A Waterfront, Cape Town
PO Box 50012, Waterfront 8002
Tel: 021 419-2121
Fax: 021 419-0040
vjohnson@mweb.co.za
www.vaughanjohnson.co.za

WINE & COMPANY

7 High Street, Hermanus
Suite 47, Private Bag X15,
Hermanus 7200
Tel: 028 313-2047
Fax: 028 312-4029
winenco@telkomsa.net

WINE CONCEPTS, NEWLANDS

Cnr Kildare Rd & Main St,
Newlands, Newlands 7700
Tel: 021 671-9030
Fax: 021 671-9031
sales@wineconcepts.co.za
www.wineconcepts.co.za

WINE CONCEPTS ON KLOOF STREET

50 Kloof Street, Cape Town
Tel: 021 426 4401
Fax: 021 426 4402
E-mail/Website as per Newlands

WINE VILLAGE

Hemel-en-Aarde Road,
Hermanus
PO Box 465, Hermanus 7200
Tel: 028 316-3988
Fax: 028 316-3989
wine@hermanus.co.za
www.wine-village.co.za

DELIVERIES, DOOR-TO-DOOR

STELLENBOSCH WINES DIRECT

Vlottenburg Road, Vlottenburg
PO Box 12641, Die Boord, 7613
Tel: 021 881-3791
Fax: 021 881-3740
info@33.co.za
www.33.co.za

VINEYARD CONNECTION

Delvera, cnr R44 and
Muldersvlei Road
Tel: 021 884-4360
Fax: 021 884-4361
info@vineyardconnection.co.za
www.vineyardconnection.co.za

WINEmail

PO Box 1401, Stellenbosch, 7599
Tel: 021 883-9398
Fax: 021 886-8833
winemail@rsp.co.za
www.winemail.co.za

WORLD WINE TRADE

Wine Gift Express
Goedehoop Road,
Franschhoek
PO Box 251, Franschhoek,
7690
Tel: 021 876-3465
Fax: 021 876-3501
wwt.iafrica.com
www.wwt.co.za

Cellaring Space To Rent

BERGKELDER VINOTEQUE

Only Distell and Lusan wines
accommodated
R18,47 per case per annum
Adam Tas Road R44),
Stellenbosch
PO Box 184, Stellenbosch 7599
Tel: 021 809-7000
Fax: 021 887-9533
info@vinoteq.co.za
www.vinoteq.co.za

EIKENDAL VINEYARDS

R300 per 10-case locker
per annum
R250 per 8-case locker per annum
Strand Rd (R44), between
Stellenbosch and Somerset West
PO Box 2261, Dennesig 7601
Tel: 021 855-1422
Fax: 021 855-1027
eikendal@netactive.co.za
www.eikendal.com

GLENFAIR HYPER LIQUOR

R15 per 6-case bin
per month
R20 per 9-case bin
per month
Sanlam Centre, Lynnwood
Road, Pretoria
Postnet Suite #122, Private bag
X025, Lynnwood Ridge 0040
Tel: 012 361-4563
Fax: 012 361-4509
Gfl.mkuyper@lantic.net

WINE CELLAR CLUB

R2,50 per bottle per annum
Unit 4, Prices Industrial Park,
Nelson Road, Observatory,
Cape Town
PO Box 13439,
Observatory 7935
Tel/Fax: 028 312-1663
info@winecellar.co.za
www.winecellar.co.za

Top Restaurant Winelists

BLOEMFONTEIN

De Oude Kraal ◆
Tel: 051 564-0636

CAPE TOWN

Buitenverwachting, ◆★
(Constantia)
Tel: 021 794-3522
Caroline's (City Centre) ✧
Tel: 021 419-8984
Cellars Hotel (Constantia) ◆
Tel: 021 794-2137
De Oude Welgemoed ◆
(Bellville)
Tel: 021 913-4962
Emily's (V&A Waterfront) ◆
Tel: 021 421–1133
La Villa (Gardens) ◆★
Tel: 021 462-1999
Tides, Bay Hotel ◆ ★
(Camps Bay)
Tel: 021 439-4444

DURBAN

Rainbow Terrace,
Hilton Hotel ◆
Tel: 031 336-8239

JOHANNESBURG

Browns (Rivonia) ✧
Tel: 011 803-7533
Cento's (Kensington) ◆
Tel: 011 622-7272
Fairlawns (Morningside) ◆
Tel: 011 804-2540
Shiekh's Palace (Rivonia) ◆
Tel: 011 807–4119

LANGEBAAN

Farmhouse Hotel ◆
Tel: 022 772-2062

PAARL

Bosman's, Grande Roche ◆★
Tel: 021 863-2727

PRETORIA

Il Restorante Ritrovo ◆
Tel: 012 460–4367

SOMERSET WEST

Golden Tulip, Lord Charles ◆
Tel: 021 855-1040

STELLENBOSCH

Vineleaf, Devon Valley Hotel ◆
Tel: 021 865-2012

◆ Recipient of a **Diamond Award** at the 2002 Diners Club Winelist of the Year Awards
✧ Recipient of a **Cellar Diamond Award** at the 2002 Diners Club Winelist of the Year Awards
★ **5 Star** rating in WINE magazine's 2003 guide to the Top 100 Restaurants in South Africa

Diners Club Winelist of the Year Diamond Awards are presented to restaurants on the basis of various criteria, including the quality and variety of wines on offer, whether the restaurant has a qualified sommelier or trained wine advisor to assist diners, and whether the cellar is air-conditioned – Cellar Diamond Awards differ inasmuch as the recipient restaurants don't have printed winelists.

Wine Courses, Tutored Tastings

for both novices and aficionados

Whether you just want to know what's in the glass, or are intent on becoming an expert on the subject, there are a number of wine education options in South Africa

Cape Wine Academy Courses, Diplomas

www.capewineacademy.co.za

Centres where CWA courses are offered:

STELLENBOSCH

Juan Lombaard
Private Bag X 5054,
Stellenbosch 7600
Tel: 021 809-7597
Fax: 021 883-9179
jlombaard@cwa.org.za

GAUTENG

Marilyn Cooper
PO Box 785084, Sandton 2146
Tel: 011 783-4585/6
Fax: 011 883-2356
cwa@iafrica.com

PRETORIA

Glyn de Jager
PO Box 11428,
Queenswood 0121
Tel/Fax: 012 333-1978
Cell 083 263-4384
cwapta@iafrica.com

DURBAN

Geoff Willis
81 Kensington Drive, Durban
North 4051
Tel/Fax: 031 564-5067
cwadur@iafrica.com

BLOEMFONTEIN

Sandra Lotz
PO Box 13516, Noordstad,
Bloemfontein 9302
Tel: 051 403-8500
Fax: 051 448-6076
Cell 082 924-7254
wineworx@bigfoot.com

PORT ELIZABETH

Jaco Schoeman
PO Box 13780,
Humewood, 6013
Tel: 041 504-3872
Fax: 041 504-3744
Cell 082 575-7578
jaco@petech.ac.za

Accolade Wine Appreciation

Introductory courses offered by Claude Felbert
Tel: 021 712-4245
Fax: 021 712-4245
accolade@ibox.co.za

Basic Wine Appreciation

Offered by Charles Hardie
Tel: 011 447-9699
outtolunch@wol.co.za

Michael Fridjhon's Wine Master Classes

Contact Alex Mason-Gordon at Out Sorceress Marketing
Tel: 011 482-5936
Fax: 011 482-2272
alex@outsorceress.co.za